Textbooks in Telecommunication Engineering

Series Editor
Tarek S. El-Bawab,
Professor and Dean,
School of Engineering,
American University of Nigeria

Telecommunications have evolved to embrace almost all aspects of our everyday life, including education, research, health care, business, banking, entertainment, space, remote sensing, meteorology, defense, homeland security, and social media, among others. With such progress in Telecom, it became evident that specialized telecommunication engineering education programs are necessary to accelerate the pace of advancement in this field. These programs will focus on network science and engineering; have curricula, labs, and textbooks of their own; and should prepare future engineers and researchers for several emerging challenges. The IEEE Communications Society's Telecommunication Engineering Education (TEE) movement, led by Tarek S. El-Bawab, resulted in recognition of this field by the Accreditation Board for Engineering and Technology (ABET), November 1, 2014. The Springer's Series Textbooks in Telecommunication Engineering capitalizes on this milestone, and aims at designing, developing, and promoting high-quality textbooks to fulfill the teaching and research needs of this discipline, and those of related university curricula. The goal is to do so at both the undergraduate and graduate levels, and globally. The new series will supplement today's literature with modern and innovative telecommunication engineering textbooks and will make inroads in areas of network science and engineering where textbooks have been largely missing. The series aims at producing high-quality volumes featuring interactive content; innovative presentation media; classroom materials for students and professors; and dedicated websites. Book proposals are solicited in all topics of telecommunication engineering including, but not limited to: network architecture and protocols; traffic engineering; telecommunication signaling and control; network availability, reliability, protection, and restoration; network management; network security; network design, measurements, and modeling; broadband access; MSO/cable networks; VoIP and IPTV; transmission media and systems; switching and routing (from legacy to next-generation paradigms); telecommunication software; wireless communication systems; wireless, cellular and personal networks; satellite and space communications and networks; optical communications and networks; free-space optical communications; cognitive communications and networks; green communications and networks; heterogeneous networks; dynamic networks; storage networks; ad hoc and sensor networks; social networks; software defined networks; interactive and multimedia communications and networks; network applications and services; e-health; e-business; big data; Internet of things; telecom economics and business; telecom regulation and standardization; and telecommunication labs of all kinds. Proposals of interest should suggest textbooks that can be used to design university courses, either in full or in part. They should focus on recent advances in the field while capturing legacy principles that are necessary for students to understand the bases of the discipline and appreciate its evolution trends. Books in this series will provide high-quality illustrations, examples, problems and case studies. For further information, please contact: Dr. Tarek S. El-Bawab, Series Editor, Professor and Dean, School of Engineering, American University of Nigeria, telbawab@ieee.org; or Mary James, Senior Editor, Springer, mary.james@springer.com.

More information about this series at http://www.springer.com/series/13835

Rolando Herrero

Fundamentals of IoT Communication Technologies

Rolando Herrero
Northeastern University
Boston, MA, USA

The solutions and slides for this book can be found at https://www.springer.com/us/book/9783030700805

ISSN 2524-4345 ISSN 2524-4353 (electronic)
Textbooks in Telecommunication Engineering
ISBN 978-3-030-70082-9 ISBN 978-3-030-70080-5 (eBook)
https://doi.org/10.1007/978-3-030-70080-5

This Springer imprint is published by the registered company Springer Nature Switzerland AG
The registered company address is: Gewerbestrasse 11, 6330 Cham, Switzerland

Preface

What Is This Book About?

The *Internet of Things* (IoT) is often described as a group of standardized technologies, systems, protocols, and design principles associated with Internet-connected entities and things that interact with the physical environment. Although IoT relies on many moving parts, a basic classification can serve to identify three main components: (1) devices that interact with the physical environment, (2) applications that perform analytics and extract knowledge from devices in order to make decisions, and (3) Internet-based communication that provides the "glue" that ties these two other components together. This book is about this complex latter component.

Topologies of most IoT architectures are affected by power and computational complexity constraints that render most traditional communication mechanisms ineffective. IoT communication technologies attempt to overcome these problems by relying on several overlapping standard and non-standard protocols designed and developed by multiple organizations ranging from industry consortia to standardization bodies. This book explores many of these state-of-the-art protocols and maps them to functionality associated with the different layers of the well-known layered architectures with special focus on their interaction. Moreover, the book introduces these mechanisms from a perspective of protocol stacks that support diverse IoT solutions and use cases like home automation or *Industrial IoT* (IIoT). Additionally, other aspects of IoT communication technologies like routing, security, and resource management are also explored.

To summarize, understanding IoT communication technologies is key to understanding how IoT works. There is no IoT without dedicated protocols that overcome the limitations of IoT access networks. This book presents, describes, and investigates the relationship between the many communication technologies that are essential to the successful deployment of most IoT solutions.

Why Did I Write This Book?

The motivation for this book is the need of finding a single book that addresses state-of-the-art IoT networking and communication technologies. Specifically, while teaching "Fundamentals of IoT" at Northeastern University in Boston, I noticed the lack of "student-ready" material that can be used to present the most relevant IoT communication technologies. Moreover, because of the dynamic nature of these protocols, most of the material I use in the class comes from industry standards. Unfortunately, these standards are not "student-friendly" as they are usually intended for developers and architects. It is impossible to find a single book, or many for that matter, that address all the topics relevant to the class. Yes, there are many IoT books out there; however, there are very few books that focus on IoT communication and networking. Those that do focus on these topics, however, typically miss the big picture, and their content is fairly compartmentalized.

The main motivation is to introduce a book that:

- provides exclusive focus on IoT communication and networking technologies from a layered architecture perspective. IoT protocols are presented and classified based on physical, link, network, transport, and application layer functionality.
- presents and discusses the main families of networking architectures that rely on the IoT protocols (i.e. LWPAN vs. WPAN).
- introduces use cases and examples that focus on protocol interaction to build network stacks.
- analyzes the impact of the IoT mechanisms on network and device performance with special emphasis on power consumption and computational complexity.

Of course, these goals comply with the information presented in the previous section.

Intended Audience

The book is intended for advanced undergraduate-level or introductory graduate-level students as well as for practicing engineers, technologists, and system architects who wish to learn and understand the principles of IoT communication and networking technologies. The assumption is that the reader is familiar with mainstream communication and networking fundamentals including some basic knowledge of signals and systems.

Book Organization

This book follows a progression that starts with a description of IoT and how IoT fits in the context of conventional communication systems; it continues by presenting IoT protocols from a perspective of the layered architecture and concludes by focusing on advanced topics associated with IoT. This content is delivered in three parts:

- Part I—Understanding IoT Communications: It deals with the bases of IoT communications that are the building blocks needed to understand architectures and topologies. It includes two chapters. Chapter 1 provides an introduction that details the evolution of IoT, discussing components and the main differences when compared to traditional networking. Chapter 2 focuses on the topologies and the interaction between the different players involved in IoT networking.
- Part II—IoT Network Layers: It explores the different communication layers of standard IoT solutions. It includes three chapters. Chapter 3 introduces the physical and link layers of wireless and wireline IoT technologies. Chapter 4 explores the network and transport layers including adaptation mechanisms to enable IoT physical and link layer protocols to support IPv6. Chapter 5 deals with the application layers that support the establishment and maintenance of IoT sessions.
- Part III—Advanced IoT Networking Topics: It reviews the advanced IoT networking topics that complement the protocols presented in Part II. It includes four chapters. Chapter 6 explores the different mechanisms used for resource identification and management. Chapter 7 focuses on the technologies that provide traffic routing and message forwarding. Chapter 8 presents a wide range of full and hybrid IoT LPWAN industry standards. Chapter 9 introduces Thread, a popular home automation WPAN architecture.

It is recommended that these parts and chapters are read in sequential order as there are many dependencies among them.

Boston, MA, USA
October 2020

Rolando Herrero

Contents

Acronyms

3GPP	3rd Generation Partnership Project
5G NSA	5G Non-Standalone
5G SA	5G Standalone
6LoBAC	IPv6 over BACnet
6LoBTLE	IPv6 over Low power Bluetooth Low Energy
6LoPLC	IPv6 over Power Line Communication
6LoWPAN	IPv6 over Low power Wireless Personal Area Networks
6P	6top Protocol
6TiSCH	IPv6 over TSCH
6top	6TiSCH Operational Sublayer
AA	Authoritative Answer
AC	Alternating Current
ACE	Authentication and Authorization for Constrained Environments
ACL	Access Control List
ADC	Analog to Digital Converter
AES	Advanced Encryption Standard
AGC	Automatic Gain Control
AH	Authentication Header
AI	Artificial Intelligence
AIFS	Arbitrary Interframe Spacing
AMI	Alternate Mark Inversion
AMQP	Advanced Message Queuing Protocol
ANCOUNT	Answer Count
AoA	Angle of Arrival
AoD	Angle of Departure
API	Application Program Interfaces
APS	Application Support Layer
ARCOUNT	Additional Count
ARM	Advanced RISC Machine
ARP	Address Resolution Protocol
ARQ	Automatic Repeat reQuest
AS	Access Stratum
ASCII	American Standard Code for Information Interchange
ASK	Amplitude Shift Keying
ASN	Absolute Slot Number
ATM	Adaptive Tone Mapping
ATP	Association Timeout Period
AWGN	Additive White Gaussian Noise
B5G	Beyond 5G
BAN	Body Area Network
BER	Bit Error Rate
BFD	Bidirectional Forwarding Detection

BLE	Bluetooth Low Energy
bppb	bits per pixel per band
bps	bits per second
BPSK	Binary PSK
BSS	Basic Service Set
BSSID	BSS Identifier
CA	Certificate Authority
CAP	Contention Access Period
CBC	Cipher Block Chaining
CBOR	Concise Binary Object Representation
CDMA	Code Division Multiple Access
CFS	Contention Free Slot
CIoT	Celullar IoT
CISF	Contention Interface Spacing
CNAME	Canonical Name
CoAP	Constrained Application Protocol
COBS	Consistent Overhead Byte Stuffing
codec	Coder/Decoder
CONNACK	Connection Acknowledgment
CoSIP	Constrained SIP
CPI	Cyclic Prefix Insertion
CPS	Cyber Physical System
CPU	Central Processing Unit
CRC	Cyclical Redundancy Checking
CS	Circuit Switching
CSM	Central Management System
CSMA/CA	Carrier Sense Multiple Access with Collision Avoidance
CSMA/CD	Carrier Sense Multiple Access with Collision Detection
CSRC	Contributing Source
CSS	Chirp Spread Spectrum
ct	content type
CTA	Channel Time Allocation
CTAP	Channel Time Allocation Period
CTS	Clear to Send
D7AAvP	D7A Advertising Protocol
D7ANP	D7A Network Protocol
D7AP	DASH7 Alliance Protocol
D7ASP	D7A Session Protocol
D7ATP	D7A Transport Protocol
DA	Destination Address
DAC	Digital to Analog Converter
DAD	Duplicate Address Detection
DAE	Destination Address Encoding
DAM	Destination Address Mode
DAO	DODAG Advertisement Object
DBPSK	Differential BPSK
DC	Direct Current
DCF	Distributed Coordination Function
DCI	Downlink Control Information
DECT ULE	DECT Ultra Low Energy
DECT	Digital Enhanced Cordless Telecommunications
DIFS	Distributed Inter Frame Spacing
DIO	DODAG Information Object

DIS	DODAG Information Solicitation
DLC	Data Link Control
DMRS	Demodulation Reference Signal
DNS	Domain Name System
DODAG	Destination Oriented Directed Acyclic Graph
DOFDM	Distributed OFDM
DoS	Denial of Service
DPA	Direct Peripheral Access
DPSK	Differential PSK
DRS	Dynamic Rate Shifting
DSAP	Destination Service Access Point
DSCP	Differentiated Services Code Point
DSP	Digital Signal Processing
DSSS	Direct Sequence SS
DS-UWB	Direct Sequence Ultra Wideband
DTLS	Datagram Transport Layer Security
DTX	Discontinuous Transmissions
DV	Distance Vector
ECC	Elliptic-Curve Cryptography
EC-GSM-IoT	Extended Coverage GSM IoT
ECN	Explicit Congestion Notification
EDCF	Enhanced DCF
EDR	Enhanced Data Rate
eDRX	Extended Discontinuous Reception
EID	Extension Identifier
EMI	Electromagnetic Interference
eMTC	Enhanced Machine Type Communication
eNB	eNode Base Station
EoF	End of Frame
EPC	Evolved Packet Core
EPS	Evolved Packet System
ESP	Encapsulating Security Payload
ESS	Extended Service Set
ETag	Entity Tag
ETSI	European Telecommunications Standards Institute
ETX	Expected Transmission
EUI	Extended Unique Identifier
E-UTRAN	Evolved UMTS Terrestrial Radio Access Network
FCS	Frame Checksum
FDD	Frequency Division Duplex
FDMA	Frequency Division Multiple Access
FEC	Forward Error Correction
FED	Full End Device
FFD	Full Function Devices
FHC	Frame Control Header
FHSS	Frequency Hopping SS
FIB	Forwarding Information Base
FOV	Field of View
FQDN	Fully Qualified Domain Name
FSF	Frame Control Field
FSK	Frequency Shift Keying
Gbps	Gigabits per second
GFSK	Gaussian FSK

GMSK	Gaussian Minimum Shift Keying
GPIO	General Purpose Input and Output
GPS	Global Positioning System
GPU	Graphical Processing Unit
GSM	Global System for Mobile Communications
GUA	Global Unicast Addresses
H2M	Human-to-Machine
HAL	Hardware Abstraction Layer
HAN	Home Area Network
HARQ	Hybrid Automatic Repeat reQuest
HC1	Header Compression 1
HC2	Header Compression 2
HEC	Hybrid Error Correction
HIP BEX	HIP Base Exchange
HIP DEX	HIP Diet Exchange
HIP	Host Identity Protocol
HPCW	High Priority Contention Window
HS	High Speed
HSS	Home Subscriber Server
HTML	Hypertext Markup Language
HTTP	HyperText Transfer Protocol
HTTPU	HTTP over UDP
HVAC	Heating, Ventilation and Air conditioning
I^2C	Inter Integrated Circuit
I/O	Input/Output
IANA	Internet Assigned Numbers Authority
IBSS	Independent BSS
ICMP	Internet Control Message Protocol
IE	Information Element
IETF	Internet Engineering Task Force
if	interface description
IID	Interface Identifier
IIoT	Industrial Internet of Things
IKE	Internet Key Exchange
IoT	Internet of Things
IP TTL	IP Time to Live
IP	Internet Protocol
IPHC	IP Header Compression
IPSec	IP Security
IPv4	Internet Protocol version 4
IPv6	Internet Protocol version 6
ISI	Inter Symbol Interference
ISM	Instrument, Scientific and Medical
ISO	International Organization for Standardization
IV	Initialization Vector
JID	Jabber Identifier
JTAG	Joint Test Action Group
Kbps	Kilobits per second
KC	Key Control
KEK	Key Encryption Key
KI	Key Index
KIM	Key Identifier Mode

KS	Key Source
L2CAP	Logical Link Control and Adaptation Protocol
LAN	Local Area Network
LBR	Low Power PAN Border Router
LDPC	Low Density Parity Check
LEACH	Low Energy Adaptive Clustering Hierarchy
LECIM	Low Energy, Critical Infrastructure Monitoring
LFU	Least Frequently Used
LHIP	Lightweight HIP
LIDR	Light Detection And Ranging
LKM	Link Margin
LLC	Link Layer Control
LLCP	Logical Link Control Protocol
LLLN	Low Power, Low Rate and Lossy Network
LLN	Low Power and Lossy Network
LOAD	Lightweight On-demand Ad-hoc Distance Vector
LOADng	LOAD Next Generation
LoRa	Long Range
LPWAN	Low Power Wide Area Network
LS	Link State
LSB	Least Significant Bit
LTE	Long Term Evolution
LTN	Low Throughput Networks
M2M	Machine to Machine
M2P	Multipoint to Point
MAC	Media Access Control
Mbps	Megabits per second
MCTA	Management Channel Time Allocation
mDNS	Multicast DNS
MED	Minimal End Device
MEIP	Media over IP
MEMS	Micro Electro Mechanical Systems
MeshCoP	Mesh Commissioning Protocol
MFR	MAC Footer
MHR	MAC Header
MIB	Master Information Block
MIC	Message Integrity Code
MIMO	Multiple Input Multiple Output
MLE	Mesh Link Establishment
MMC	Modified Miller Code
MME	Mobility Management Entity
MPDU	MAC Protocol Data Unit
MPL	Multicast Protocol for Low Power and Lossy Networks
MQTT	Message Queue Telemetry Transport
MQTT-SN	MQTT for Sensor Networks
MS/TP	Master Slave/Token Passing
MSB	Most Significant Bit
MSDU	MAC Service Data Unit
MSF	Minimal Scheduling Function
MSSIM	Mean Structural Similarity
MTC	Machine Type Communication
MTR	Multi-Topology Routing
MTU	Master Slave/Token Passing

NA	Neighbor Advertisement
NACK	Negative Acknowledgment
NAS	Non-Access Stratum
NAT	Network Address Translation
NB	Narrow Band
NB-Fi	Narrowband Fidelity
NB-IoT	Narrow Band IoT
ND	Neighbor Discovery
NFC	Near Field Communication
NHC	Next Header Compression
NPBCH	Narrowband Physical Broadcast Channel
NPCW	Normal Priority Contention Window
NPDCCH	Narrowband Physical Downlink Control Channel
NPDSCH	Narrowband Physical Downlink Shared Channel
NPRACH	Narrowband Physical Random Access Channel
NPSS	Narrowband Primary Synchronization Signal
NPSSS	Narrowband Secondary Synchronization Signal
NPUSCH	Narrowband Physical Uplink Shared Channel
NR	New Radio
NRS	Narrowband Reference Signal
NS	Neighbor Solicitation
NSEC	Next Secure
NUD	Neighbor Unreachability Detection
NZR	Unipolar Nonreturn to Zero
OFDM	Orthogonal Frequency Division Multiplexing
OOK	On-Off Keying
OPC UA	Open Platform Communications United Architecture
OPCODE	Operation Code
OQPSK	Offset QPSK
OS	Operative System
OSI	Open Systems Interconnection
P2P	Point to Point
PAM	Pulse Amplitude Modulation
PAN ID	PAN Identifier
PAN	Personal Area Network
PCA	Priority Channel Access
PCF	Point Coordination Function
PDR	Packet Delivery Ratio
PDU	Packet Data Unit
PEGASIS	Power-Efficient Gathering in Sensor Information Systems
P-GW	Packet Data Node Gateway
PID	Proportional Integral Derivative
PINGREQ	Ping Request
PINGRESP	Ping Response
PKI	Public Key Infrastructure
PLC	Power Line Communication
PNC	Piconet Coordinator
PRB	Physical Resource Block
PS	Packet Switching
PSK	Phase Shift Keying
PSKc	Pre-Shared Key for the Commissioner
PSM	Power Saving Mode
PSNR	Peak Signal to Noise Ratio

PTYPE	PDU Type
PUBACK	Publish Acknowledgment
PUBCOMP	Publish Complete
PUBREC	Publish Received
PUBREL	Publish Release
QAM	Quaddrature Amplitude Modulation
QDCOUNT	Question Count
QoS	Quality of Service
QPSK	Quaddrature PSK
QR	Query/Response
RA	Recursion Available
RA	Router Advertisement
RADAR	Radio Detection And Ranging
RAP	Random Access Procedure
RCODE	Response Code
RD	Recursion Desired
REED	Router Eligible End Device
RERR	Route Error
REST	Representational State Transfer
RF	Radio Frequency
RFD	Reduced Function Devices
RFID	Radio Frequency Identification
RFTDMA	Random Frequency and Time Division Multiple Access
RIP	Routing Information Protocol
RIPng	RIP Next Generation
RISF	Response Interface Spacing
ROLL	Routing over Low Power and Lossy Networks
RPL	Routing for Low Power
RPMA	Random Phase Multiple Access
RR	Receiver Report
RR	Resource Record
RRC	Radio Resource Control
RREP	Route Reply
RREP-ACK	RREP Acknowledgment
RREQ	Route Request
RS 232	Recommended Standard 232
RS 485	Recommended Standard 485
RS	Router Solicitation
RSSI	Received Signal Strength Indicator
rt	resource type
RTC	Real Time Communication
RTCP	Real Time Control Protocol
RTOS	Real Time Operating System
RTP	Real Time Transport Protocol
RTS	Request to Send
RU	Resource Unit
RZ	Unipolar Return to Zero
SA	Source Address
SAA	Stateless Address Autoconfiguration
SAE	Source Address Encoding
SAM	Source Address Mode
SAP	Service Access Point
SCADA	Supervisory Control And Data Acquisition

SCEF	Service Capability Exposure Function
SC-FDMA	Single Carrier FDMA
SCHC	Static Context Header Compression
SCTP	Stream Control Transmission Protocol
SD-DNS	Service Discovery DNS
SDES	Source Description
SDN	Software Defined Network
SDP	Session Description Protocol
SDR	Software Defined Radio
SDW	Software Defined WAN
SF	Scheduling Function
SFD	Start Frame Delimiter
S-GW	Serving Gateway
SIB	System Information Block
SIFS	Short Inter Frame Spacing
SigComp	Signal Compression
SIP	Session Initialization Protocol
SNAP	Subnetwork Access Protocol
SNOW	Sensor Network Over White Spaces
SNR	Signal to Noise Ratio
SOA	Service Oriented Architecture
SOAP	Simple Object Access Protocol
SoC	System on Chip
SoF	Start of Frame
SoM	System on Module
SONAR	Sound Navigation And Ranging
SPI	Serial Peripheral Interface
SPIN	Sensor Protocols for Information via Negotiation
SPIN-BC	SPIN Broadcast
SPIN-EC	SPIN Energy Conservation
SR	Sender Report
SS	Spread Spectrum
SSAP	Source Service Access Point
SSDP	Simple Service Discovery Protocol
SSIM	Structural Similarity
SSRC	Synchronization Source
SUBACK	Subscribe Acknowledgment
SWD	Serial Wire Debug
SWoT	Social Web of Things
sz	maximum size
TC	Truncation
TCP	Transport Control Protocol
TDD	Time Division Duplex
TDMA	Time Division Multiple Access
TG4g	IEEE 802.15.4g Task Group
TG4k	IEEE 802.15.4k Task Group
TLS	Transport Layer Security
TLV	Type-Length-Value
TMF	Thread Management Framework
TMSP	Time Synchronized Mesh Protocol
TPC	Transmit Power Control
TSCH	Time Slotted Channel Hopping
TTL	Time to Live

TTN	The Things Network
UAC	User Agent Client
UART	Universal Asynchronous Receiver Transmitter
UAS	User Agent Server
UAV	Unmanned Aerial Vehicle
UDP	User Datagram Protocol
UE	User Equipment
ULA	Unique Local Unicast Addresses
UNB	Ultra Narrow Band
UNSUBACK	Unsubscribe Acknowledgment
UPnP	Universal Plug and Play
URI	Uniform Resource Identifier
URL	Uniform Resource Locator
V2I	Vehicle to Infrastructure
V2V	Vehicle to Vehicle
V2X	Vehicle to Everything
VAD	Voice Activity Detection
VoIP	Voice over IP
VoLTE	Voice over LTE
VPN	Virtual Private Network
WAN	Wide Area Network
WAVE	Wireless Access in Vehicular Environment
Weightless SIG	Weightless Special Interest Group
WEP	Wired Equivalent Privacy
Wi-Fi	Wireless Fidelity
Wi-SUN	Wireless Smart Utility Networks
WLAN	Wireless Local Area Network
WPA	Wi-Fi Protected Access
WPA2	Wi-Fi Protected Access II
WPAN	Wireless Personal Area Network
WSN	Wireless Sensor Network
WUR	Wake Up Radio
XEP	XMPP Extension Protocol
XLEACH	eXtended LEACH
XMPP	eXtensible Messaging and Presence Protocol
ZCL	ZigBee Cluster Library
ZDP	ZigBee Device Profile

Part I
Understanding IoT Communications

This part of this book, that includes two chapters, deals with the bases of IoT communications that are the building blocks needed to understand architectures and topologies. Chapter 1 provides an introduction that addresses the evolution of IoT, discussing components and the main differences with traditional networking schemes. The chapter also introduces a use case that explores some of the technologies and mechanisms that are presented later in this book. Chapter 2 focuses on the topologies and the interaction between the different players involved in IoT networking. It describes *wireless sensor networks* (WSNs) and their importance in the context of modern IoT networks.

1 Introduction

1.1 M2M and IoT

Understanding IoT communication requires understanding its origins and progression throughout the years. IoT results from the evolution of several other technologies like *wireless sensor networks* (WSNs) and *machine-to-machine* (M2M) communication. M2M is heavily linked to IoT due to their many similarities. Although the distinction between M2M and IoT has been an area of great debate, there is a consensus on what their main differences are [14].

M2M communication systems rely on devices and applications interacting, without any human intervention, over wireless and wireline channels. Each of these entities transmits and receives data and supports basic functionality that includes source encoding, channel encoding, and modulation. In general, each full duplex bidirectional link between two devices is composed of two independent half duplex unidirectional links. In a typical M2M scenario, where a device and an application communicate with each other, the latter sends requests to the former to sense some meaningful environment parameter or asset that can be used, in turn, by the application to make an automated decision. For example, the device could be an engine, while the application could be a piece of software that based on the device internal temperature would decide on the flow rate of coolant injected in the engine. This scheme represents a traditional communication system that is integrated by the series of components that are shown in Fig. 1.1. An analog signal generated by the temperature sensor is converted to the digital domain, compressed by means of a *source encoder* (SE) and then processed through a *channel encoder* (CE). Essentially, the SE converts the analog temperature signal into digital readouts by relying on an *analog-to-digital* converter (ADC) that performs sampling and quantization. The CE, in turn, adds controlled redundancy to the readouts in order to improve reliability when they are transmitted over the noisy channel. By virtue of being digital, readouts and controlled redundancy only exist in the context of a processor or computer memory. However, for them to be transmitted to the application, they must be first sent over the analog channel. This implies the presence of an additional stage where these digital values are converted, through a *modulator* (M), back into an analog signal. This signal, known as a modulated wave, traverses the analog channel, and when it arrives at the far end, it is demodulated by means of a *demodulator* (DM). The DM converts the analog signal into digital temperature readouts and associated controlled redundancy. The controlled redundancy is removed by the *channel decoder* (CD) that restores the digital temperature readouts. Since the application is an algorithm that is run by a piece of software, there is no need to convert the information any further. At this point, the application uses the temperature readouts as samples that can be processed by a generic *Proportional Integral Derivative* (PID) controller algorithm. The application decides and increases or decreases the flow of coolant by transmitting an actuation flow rate. Since this value is already digital, it is processed by a CE that introduces controlled redundancy to improve reliability. The digital rate and its associated redundancy are modulated and converted into a modulated wave that is transmitted over the channel. When the signal arrives at the engine, it is demodulated and processed by the CD. Because the engine is an analog device, the digital flow rate is converted by a *source decoder* (SD) into an analog value that is applied to the coolant flow regulator of the engine. The same way that source encoding is performed by an ADC, source decoding is carried out by a *digital-to-analog* converter (DAC).

For most communication systems [8], not just M2M scenarios, the functionality provided by SE, SD, CE, CD, M, and DM is typically mapped in layers that ease software and firmware implementation and deployment. More details of these layered architectures in the context of not only M2M but also IoT are discussed in Sect. 1.2 and further presented in Chap. 2.

R. Herrero, *Fundamentals of IoT Communication Technologies*, Textbooks in Telecommunication Engineering, https://doi.org/10.1007/978-3-030-70080-5_1

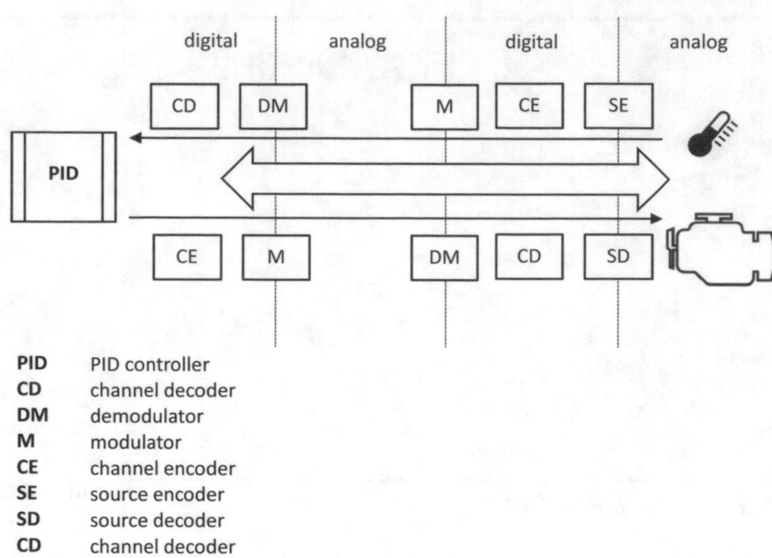

Fig. 1.1 M2M communication system

As it can be inferred from the example above, M2M usually involves proprietary mechanisms where source, channel encoding, as well as modulation are not standardized. Moreover, many legacy M2M scenarios are not even necessarily digital, and, in many cases, just plain modulation and demodulation are good enough to provide some basic functionality. Telemetry, which represents the quintessential M2M application, relies on the remote measurement of a parameter that is processed at some monitoring station. An early example of analog telemetry includes the synchro which is a variable coupling transformer where the magnitude of the magnetic coupling between the primary and secondary varies in accordance with the position of a rotable element. The Panama Canal, completed in 1914, relied on the extensive use of synchros to monitor locks and water levels. This and other telemetry mechanisms evolved for decades until 1968 when the digital *Supervisory Control And Data Acquisition* (SCADA) system became that standard technology for distributed measurement and control. Although initially SCADA systems were built from mainframe computers and required intensive human supervision to operate, they eventually became more automated and therefore more compatible with the M2M paradigm [18].

Another example of well-known M2M systems are the programmable logic controllers that consist of an industrial computer with several digital input and output ports that can be used for sensing and actuation [6]. These devices usually include built-in wireline communication ports that rely on physical and data link mechanisms ranging from low-rate RS-232 to high-rate Ethernet.

Because most M2M applications involve devices that are remotely monitored and controlled by other devices that make decisions, these architectures tend to be vertical in nature. Specifically, a sensor interacts with the environment by sampling and generating a readout that is transmitted over a communication channel to an application. The application, in turn, processes the readout and makes a decision that may or may not involve actuation. If it does involve actuation, the application transmits over the channel a command or some other relevant piece of information that is used by the actuator to perform a specific action against the environment.

Typical M2M applications just deal with one type of parameter (i.e. temperature, humidity, speed, inventory level) where a simple decision is made (i.e. open or close a valve). Lack of standardization, which manifests by the lack of communication protocols, many times plays against M2M applications and devices attempting to interact with each other.

Because single-device scenarios limit the level of automation that a particular system can have, it is just natural to try to expand them by incorporating multiple devices. A more complex solution with multiple heterogeneous devices associated with multiple heterogeneous parameters implies a horizontal approach to the machine communication problem. A horizontal approach, in turn, requires a standard mechanism to enable communication between dissimilar devices. Traditional M2M frameworks, unfortunately, do not usually provide such a mechanism [14]. Luckily, the *Internet Protocol* (IP) can be used as the fundamental building block to enable this integration. Specifically, IP serves as the glue that connects different devices with different applications and leads to what it has been called the *Internet of Things* (IoT). IoT is consequently a superset of traditional M2M schemes that is only possible due to standardization by means of IP. Figure 1.2 illustrates the difference between M2M and IoT deployments. Each scenario shows three *cyber-physical systems* (CPSs) where in every CPS a device (D_1, D_2, or D_3) interacts with a single asset (i.e. temperature, humidity, speed, inventory level). The sensor data generated by each CPS reaches three separate applications (A_1, A_2, and A_3).

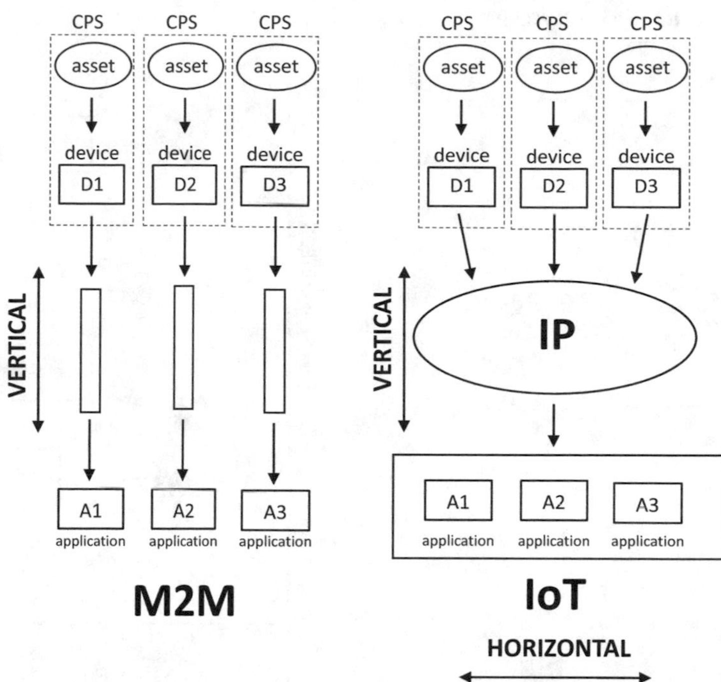

Fig. 1.2 M2M vs IoT

Under traditional vertical M2M communication, applications neither share data nor make global decisions that affect more than one CPS. On the other hand, under horizontal IoT, a single super application that encompasses A_1, A_2, and A_3 makes decisions that affect all three CPSs. Moreover, sensor traffic from the different CPSs reaches the super application by means of IP communication.

In the long run, IoT is intended to rely on the automated IP interaction of billions of devices all over the globe. However, IoT relies on IP version 6 (IPv6) addresses that overcome the well-known IP version 4 (IPv4) address shortage. Specifically, IPv4 relies on 32-bit addresses that can only support up to $2^{32} \approx 4300 \times 10^6$ addresses. On the other hand, IPv6 relies on 128-bit addresses that account for up to $2^{128} \approx 3000 \times 10^{35}$ addresses. IPv4 *network address translation* (NAT) mechanisms that have a goal to overcome the address shortage require network infrastructure changes that make them impractical. Moreover, since NATting relies on using transport layer ports to provide additional network layer, addressing it constitutes a violation of the traditional IP layered architecture described in Sect. 1.2.

Using IPv6 "as is" on IoT applications is not usually possible because IPv6 exhibits some characteristics that make it incompatible with most IoT scenarios. Specifically, IoT is typically associated with devices that produce and consume small chunks of data, so it is advantageous to keep the network overhead as negligible as possible. Long 128-bit addresses and IPv6 limitations on the minimum size of packets put heavy restrictions on certain IoT wireless and wireline technologies. These problems are minimized by introducing IoT-specific adaptation mechanisms, like *IPv6 over low-power wireless personal area networks* (6LoWPAN), that, among many things, compress fields and lift some of the restrictions on the size of packets [20].

1.2 Layered Architectures

M2M and IoT communication systems represented by the functional blocks described in Sect. 1.1 are difficult to implement, take too long to develop, and are very expensive to maintain. A more efficient approach consists in mapping these functions into multiple layers that can be independently updated. For example, while modulation and source encoding are, respectively, mapped into physical and application layers, channel encoding is distributed over multiple layers that provide great functional granularity and enable additional mechanisms like security and networking support.

Under a layered architecture, each layer is a black box with an input that processes the output of the layer underneath, and an output becomes the input of the layer above. The exchange of data between layers is performed by means of standardized interfaces supported by the input and output facilities [15]. Granularity and standardization ease implementation and make system maintenance a lot simpler and cheaper. Essentially, network software updates apply only to those layers that oversee the specific functionality that needs to be upgraded.

In a layered architecture, layers can be made "shorter" or "taller" by, respectively, removing and adding more functions. To an extreme, a flat architecture is one where all

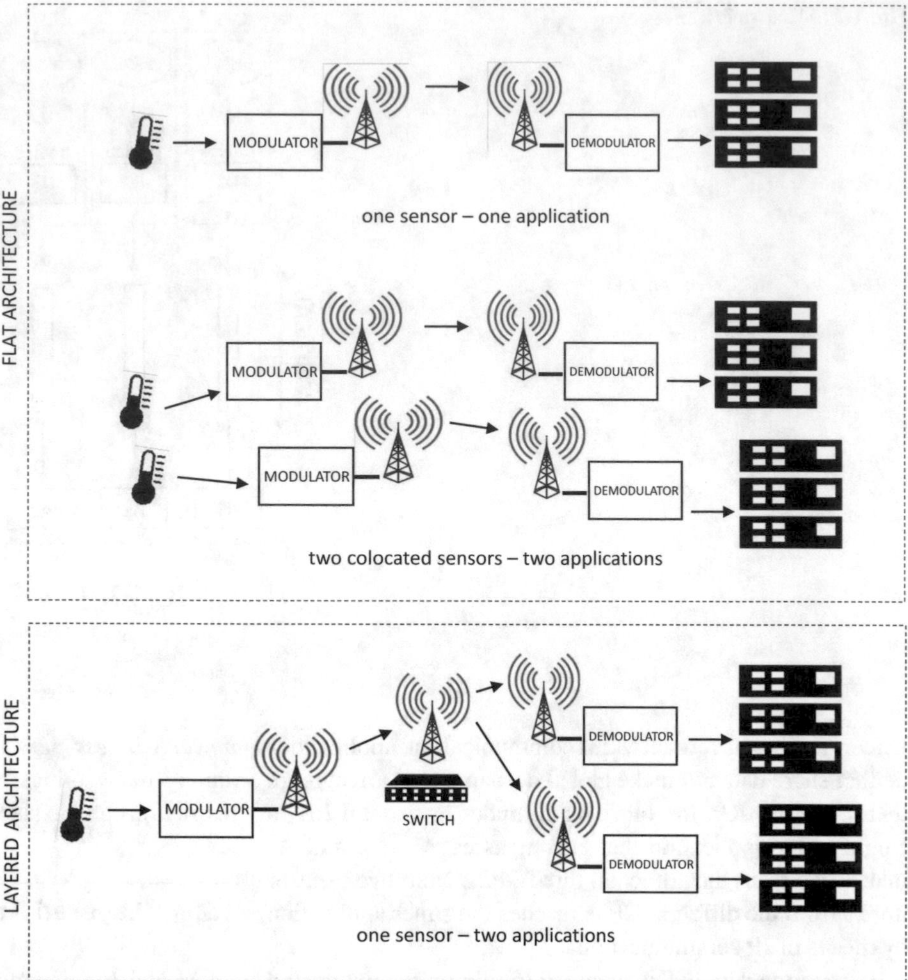

Fig. 1.3 Flat vs layered architectures

functions are performed by a single layer. Too many layers can also be a problem due to the extra overhead associated with handling them. To understand the advantages of a layered architecture when compared to a flat one, consider the example in Fig. 1.3. The system consists of a temperature sensor transmitting readouts to multiple applications. In a flat architecture, the support of two applications requires two colocated sensors transmitting readouts since every device and application relies on a single layer providing all functionality. In general, each additional application requires changes of the sensing infrastructure. Flat architectures are inflexible with simple topology changes requiring complex scenario modifications. This results in higher costs and longer deployment times that can affect the overall feasibility of the solution. On the other hand, a layered architecture supports packet switching that enables a single application to simultaneously communicate with multiple endpoints. Essentially, each layer processes information that has been segmented in units of packets that can be queued and transmitted throughout the network. In a layered IoT architecture, the support of additional endpoints does not require dramatic changes of the network infrastructure.

	OSI
layer 7	application
layer 6	presentation
layer 5	session
layer 4	transport
layer 3	network
layer 2	link control
layer 1	physical

IETF	
application	layer 5
transport	layer 4
network	layer 3
link control	layer 2
physical	layer 1

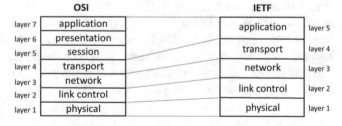

Fig. 1.4 IETF vs OSI stacks

Although there are several models that can be used to represent layered architectures, the two most important ones are the *Open Systems Interconnection* (OSI) and the *Internet Engineering Task Force* (IETF) models. The stacks of layers of these two models are comparatively shown in Fig. 1.4.

The OSI model introduces seven layers. First, the physical layer is dedicated to the transmission and reception of data over the channel including the conversion of information between the digital and the analog domains. This layer involves physical phenomena including traditional signal modulation and demodulation. Second, the link layer provides error control mechanisms for reliable transmission of information over the channel. Third, the network layer ensures that information packets are delivered to the destination. Fourth, the transport layer provides support of multiplexing of traffic from different applications. Fifth, the session layer oversees managing multiple sessions between applications. Sixth, the presentation layer provides formatting of information for further processing including security extensions for encryption and decryption. Seventh, the application layer involves application-specific services as well as the conversion of information between the digital and analog domain. Packets at each layer receive different names; they are called messages at the application layer, they are called segments at the transport layer, they are called datagrams at the network layer, and they are called frames at the link layer. The five-layer IEFT model is similar to the OSI model with the main difference being that OSI layers 5 through 7 are mapped into a single application layer. Since IoT is based on Internet technologies originally specified by IETF, this book follows the IETF layered architecture model as the main mechanism to build IoT protocols, frameworks, and solutions [17].

1.3 System Components

Two additional and very important components are needed for IoT communication to work, a device and an application. Summarizing, in its simplest format and as illustrated in Fig. 1.5, an IoT scenario typically involves three main fundamental components: (1) devices, (2) a communication channel, and (3) an application performing analytics. These components interact with the "outside world"; devices interact with assets, while applications interact with enterprise processes. Devices are sensor, controllers, and actuators typically running on small constrained embedded computers based on *system-on-chip* (SoC) and *system-on-module* (SoM) hardware. A SoC is a chip that provides a *central processing unit* (CPU), memory, storage, input/output digital and analog interfaces, and *radio-frequency* (RF) signal processing. A SoC may also include a *graphical processing unit* (GPU) and support for several peripherals. On the other hand, a SoM is a circuit board that includes a SoC in addition to some other discrete chips that provide additional functionality.

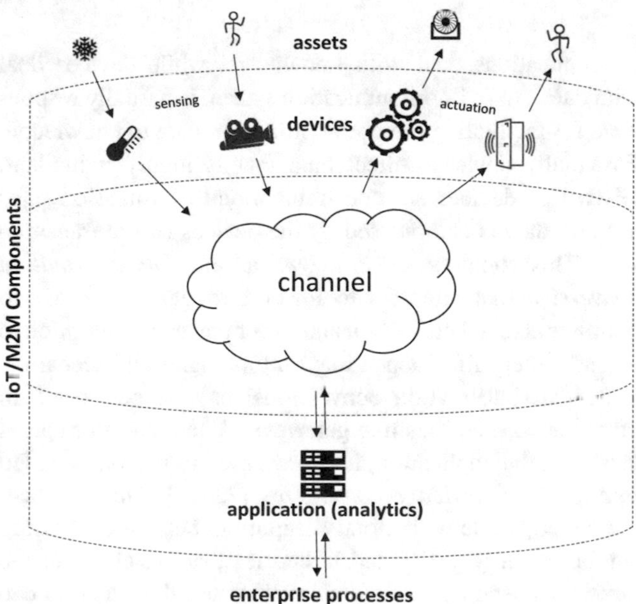

Fig. 1.5 M2M/IoT components

The communication channel is the medium that enables the propagation of signals between devices and applications. Because, in a traditional sensor-application scenario, signal propagation is usually from multiple devices to a single application, it is called *multipoint-to-point* (M2P). In a layered architecture, simultaneous communication between multiple devices and an application is efficiently carried out by means of network packet switching. Moreover, being an IoT system a superset of several interacting M2M subsystems, the channel is a critical component that integrates devices and application and where the standardization of communication protocols is key.

Embedded devices are resource constrained; they have very little memory and exhibit very low computational capacity in order to minimize power consumption to extend battery life. Low power consumption also limits transmission rates and signal coverage. The communication between a device and an application may involve signals traversing different channels that affect the very nature of the communication protocol stacks and the network itself. Specifically, the conversion between different protocols at each layer, as signals traverse channels, is performed by IoT gateways. Gateways, depending on the protocol complexity, range from simple embedded devices to more complex specialized equipment [19].

Although IoT is associated with several types of networks, there are two main types: (1) *wireless personal area networks* (WPANs) and *low-power wide-area networks* (LPWANs). The difference between them is the trade-off between coverage and transmission rate. LPWAN is associated with very low transmission rates (up to 50 Kbps) over long distances (in the order of kilometers). On the other hand, WPAN comparatively provides higher transmission rates (in the order of Mbps) but supports shorter distances (in the order of hundreds of meters). Both technologies rely on devices communicating with gateways that provide access to applications through the Internet core. Under LPWAN, this connectivity is one-hop with direct communication between a device and the gateway. Under WPAN, the connectivity is multi-hop with a single device communicating with the gateway by relying on intermediate devices to propagate its data. This latter scenario

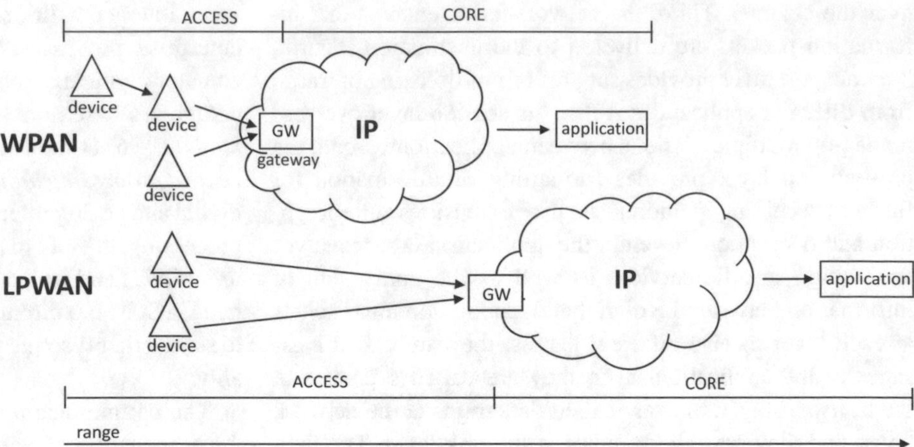

Fig. 1.6 WPAN vs LPWAN

is called mesh connectivity and enables WPAN to operate over long distances at a cost of the additional latency introduced by the packet forwarding of the intermediate devices. WPANs support end-to-end IP connectivity by relying on gateways. These gateways are fairly simple components that just convert lower layer traffic (physical and link layers). LPWANs do not typically support end-to-end IP connectivity, and their gateways are more complex as they convert full stacks of protocols.

Figure 1.6 shows the difference between a WPAN and an LPWAN. As it can be seen, the portion of the network between the devices and the gateway is called the access side, while everything else is called the core side or backbone. Note that the WPAN supports multiple devices connected in mesh to increase access side coverage. Additionally, the WPAN gateway is fully IP compliant, while the LPWAN gateway, on the other hand, only supports IP on one interface [1].

In the context of IoT, applications typically perform operational analytics to extract knowledge from the data generated by devices. This knowledge is used, in turn, to make automated actuation decisions and to provide visualization for human interaction. Knowledge extraction is a complex subject that is outside the scope of this book. It relies on several techniques ranging from traditional statistical and predictive analysis to machine learning and data mining.

► **Transmission Rate vs Bandwidth vs Throughput** One important distinction in communication systems is the difference among transmission rate, bandwidth, and throughput. The transmission rate represents how much digital information is being transmitted over the communication path. It is measured in units of *bits per second* (bps). It typically includes the redundancy added by the CE. Note that the transmission rate is sometimes referred to as *network bandwidth*. The digital bits are converted into a signal in the analog domain by means of the modulator. This signal, which is transmitted over the channel, is characterized by its bandwidth. The bandwidth is measured in units of Hertz (Hz). Through the modulation technique, a larger analog bandwidth is usually associated with a higher digital transmission rate. The demodulated analog wave is converted back into digital information that arrives at the receiver at a rate known as throughput. Throughput, as the transmission rate, is measured in units of bps. Because of network impairments like network packet loss, throughput is usually lower than the transmission rate. The relationship between these three parameters is shown below:

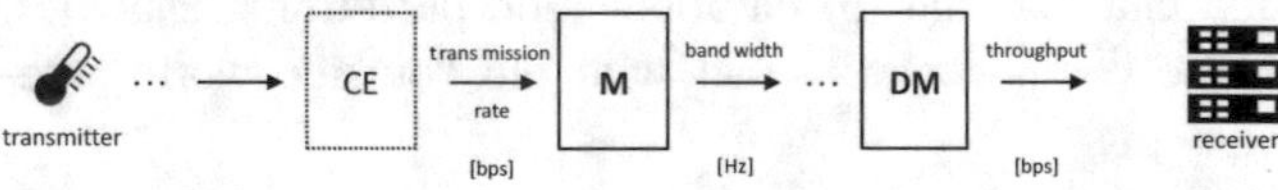

Applications deal with knowledge, while devices deal with data. An IoT communication system is partially responsible for facilitating the conversion from data to knowledge. Essentially devices collect data that is highly redundant. Given that devices are constrained, and transmission rates are low, data is compressed by the devices before transmission. This compression is known as the *data-information conversion* that attempts to lower throughput in order to optimize channel utilization and lower power consumption to extend battery life. Depending on topologies and scenarios, more data-information conversions may be performed in other network entities like gateways. When the information arrives at the application, it is processed in what it is called *information-knowledge conversion* [22, 16]. This is a scenario compatible with cloud computing, but this processing can be partially performed in entities that are closer to the device. Depending on how close they are, the situation can be representative of mist computing and fog computing scenarios.

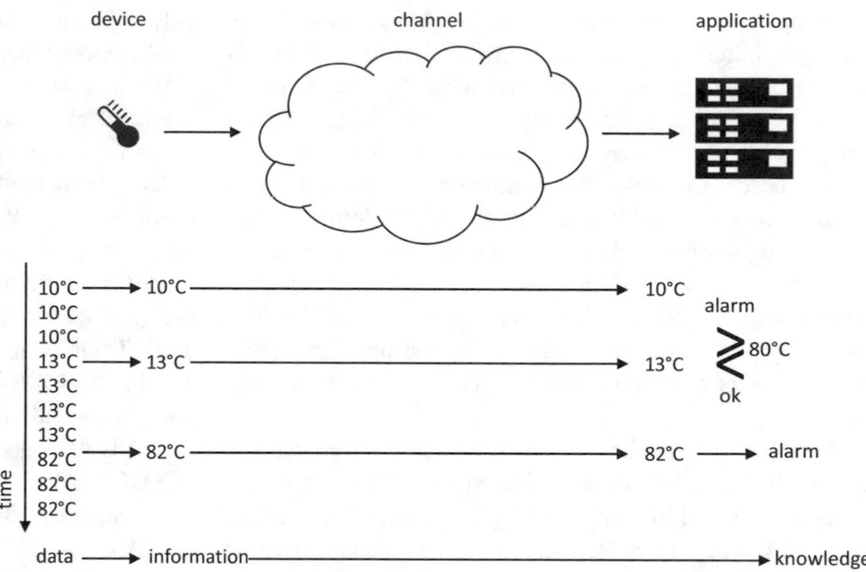

Fig. 1.7 Data transformation example

Figure 1.7 illustrates an example of data to knowledge transformation in an IoT scheme. A temperature sensor periodically transmits readouts to an application. Since these readouts are highly redundant, the device just sends temperature values whenever a change is detected. This data-information conversion reduces the amount of network traffic, decreases the transmission rate, and lowers power consumption in order to prolong battery life. The information is used by the application to determine whether the temperature at the sensor location is too high. Its goal is to convert information into knowledge by means of a threshold comparison that triggers an alarm that can be used, in turn, to perform actuation. Of course, this is a very simple example of analytics being performed by the application. Moreover, due to its simplicity, the procedure could be easily performed on the device. In more generic scenarios, however, the information-knowledge conversion is complex enough that can only be performed by an application.

Example 1.1 Consider the communication system shown in Fig. 1.1. What components are responsible for the *data-information* and the *information-knowledge* conversions?

Solution The data-information conversion is performed at the SE since it converts an analog temperature signal into digital readouts that result from ADC sampling and quantization. The SE also includes an embedded processor that performs data compression and removes redundancy in a controlled way in order to generate information. The overall goal is to lower

(continued)

(continued)

the transmission rate. The information then arrives at the application that runs an algorithm to convert it into knowledge in order to make a decision that results in actuation.

1.4 Use Case: IoT Applied to UAVs

Let's focus now on an introductory IoT use case that involves several different protocols and mechanisms that fall under the umbrella of the topologies and architectures discussed in the previous sections. *Unmanned aerial vehicles* (UAVs) are excellent candidates of IoT automation since they can be used for remote sensing and more specifically to capture hyperspectral images [12]. Hyperspectral images are made of data cubes, which, including a few dozens of spectral bands, support different applications ranging from the detection of land vegetation to the monitoring of atmospheric products derived from the processing of lower-level radiance information.

Essentially UAVs can capture these images and perform on-board compression as part of the data-information conversion. Information is then transmitted to an application that by means of real-time image processing makes flight path decisions in order to zoom in and obtain other images of interest. The processing performed by the application is part of the information-knowledge conversion. To summarize, raw images are converted at each UAV into information by means of compression techniques and then transmitted to the application that transforms the information into knowledge through image processing mechanisms. Knowledge is used, in turn, to change the flight path in order to support zooming and resolution changes that enable additional fine-tuning required for analysis. Compression at the UAV is performed by means of a hyperspectral image *coder/decoder* (codec) that takes a raw image as input and produces a lossy compressed image as output. Lossy compression is a term used to describe the compression mechanism that removes non-essential information in order to lower its size. Lossy compression leads to noise due to the distortion introduced by the removal of such information. In general, codec-based compression is a lot less computationally complex than the image classification performed by the application and therefore a lot more suitable for an embedded device like a UAV.

The autonomy of the UAVs is such that it covers an area divided into multiple subareas known as cells of size A_{cell}. Cells are scanned by UAVs by taking hyperspectral pictures of the different parts of the cell. The size of the area covered by a single image is given by A_{image}. The application, in the context of this scenario, is called mission planner. The planner resides at a ground station and calculates flight paths of UAVs to efficiently coordinate the mission by considering environmental parameters like wind and weather conditions as well as terrain information and other inputs from the UAVs. Moreover, the mission planner controls the movement of the UAVs and receives in return location information, such as *Global Positioning System* (GPS) positions, and other sensory information from the UAVs.

Communication between the planner and the UAVs is through a hybrid network, shown in Fig. 1.8, that combines a high-rate WPAN and a low-rate LPWAN. UAVs are powered by batteries such that energy consumption is a limiting factor that conditions high-rate WPAN links to exhibit short-distance coverage. Long-distance coverage is only possible by means of multi-hop communications where a UAV communicates with the planner through intermediate UAVs that forward packets. The drawback of this scheme is overhead and latency. Similarly, power limitations condition long-range LPWAN links to be low rate.

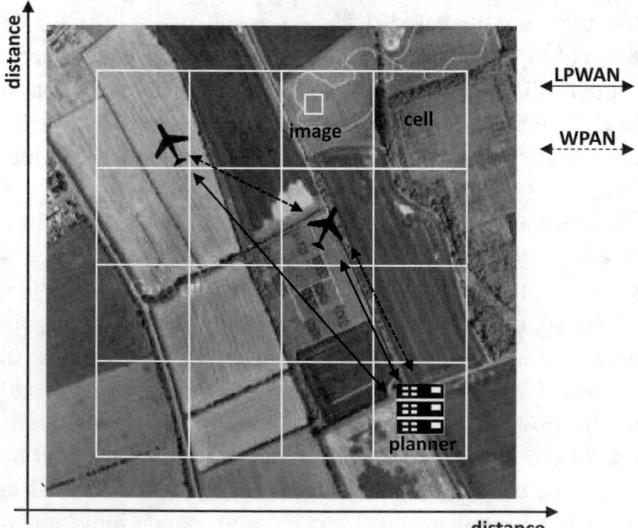

Fig. 1.8 Hybrid network

The idea behind a hybrid network architecture is to rely on the high-latency WPAN for the transmission of hyperspectral images and rely on the low-latency LPWAN for transmission of sensor and flight control information. In general, due to the nature of each network, different technologies are typically used. Traditionally, WPAN technologies range from IEEE 802.11 (Wi-Fi) to IEEE 802.15.4. Similarly, LPWAN technologies include a myriad of mechanisms like *narrowband IoT* (NB-IoT) and *LoRa*. Note that within the family of WPAN protocols, some technologies like IEEE 802.15.4, when compared to others like IEEE 802.11, exhibit certain LPWAN features like lower rate and larger coverage. This is particularly important in the context of this use case where IEEE 802.15.4 and IEEE 802.11n, respectively, provide low-rate LPWAN and high-rate WPAN communications.

The planner extracts knowledge by means of image processing- and machine learning-based classification. Note that these tasks cannot be performed at a UAV since the required computational complexity is too high. Moreover, because computational complexity is related to power consumption, even if computation were possible, the extra processing could lead to battery drain that would render the solution impractical. Therefore, UAVs are only in charge of less intensive hyperspectral image compression.

Each cell is scanned by one UAV using a mounted camera with resolution of $R_{\text{px}} \times R_{\text{py}}$ (pixel). The resolution results from a trade-off between processing capabilities associated with lossy compression and transmission rates. In fact, depending on the codec and the compression algorithm under consideration, the compressed image results in a stream of M_{image} bits. Each image covers an area A_{image} that includes multiple spectral bands. Additionally, each band is rectangular with a k aspect ratio given by the ratio of the width to the height of the image and a *field of view* (FOV) given by the length of the diagonal of the rectangle. By changing altitude, UAVs can zoom in and zoom out by, respectively, decreasing and increasing the FOV. Note that the covered area is calculated as

$$A_{\text{image}} = \left(k \times \frac{\text{FOV}}{\sqrt{k^2+1}}\right) \times \left(\frac{\text{FOV}}{\sqrt{k^2+1}}\right)$$

where the relationship between k and FOV is shown in Fig. 1.9. The total amount of data transmitted on a given UAV-to-UAV link, defined as M_{cell}, is given by

$$M_{\text{cell}} = n \times M_{\text{image}}$$

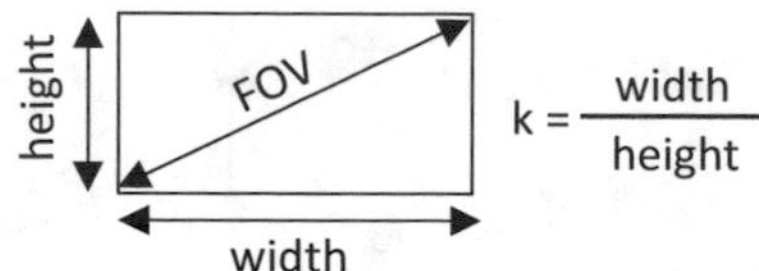

Fig. 1.9 Aspect ratio k vs FOV

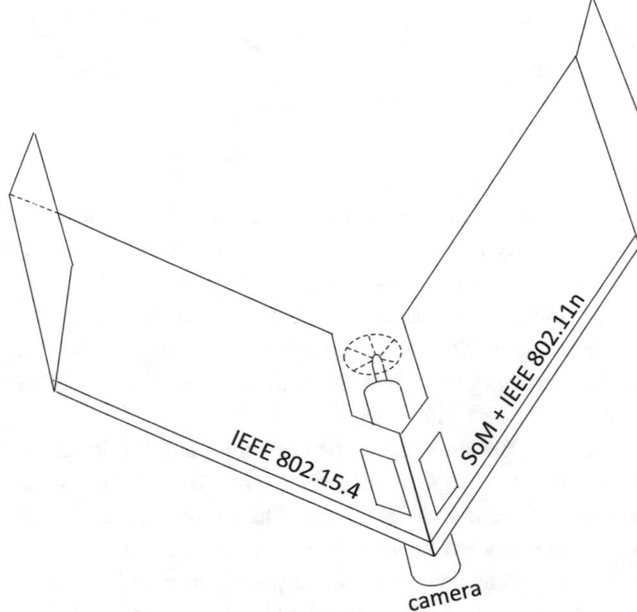

Fig. 1.10 UAV winglet

where $n = \frac{A_{\text{cell}}}{A_{\text{image}}}$ is the number of images scanned in each cell.

The UAV is a winglet, illustrated in Fig. 1.10, that has a wingspan of 0.8 m and a weight of 1/2 kg including a hyperspectral camera. A single electrical motor drives a propeller that enables the UAV to fly. A lithium polymer battery gives the UAV an autonomy of half an hour. The average linear speed is about 10 m/s, and the maximum altitude is around 400 m high. Each UAV has an autopilot that provides, among other things, a GPS unit and pressure as well as inertial sensors. The autopilot also gives UAVs automated takeoff and landing capabilities as well as mechanisms to navigate through a set of waypoints dynamically programmed by the planner. In this solution, the low-rate communication between UAVs and planner is by means of a IEEE 802.15.4 radio that enables a 250 Kbps peak transmission and a 1 km coverage. Although this low rate is good enough for transmission of actuation and sensor traffic that includes control messages and status information, it is not high enough to support compressed hyperspectral images. To that end, the UAV also has a multi-antenna *Multiple Input-Multiple Output* (MIMO) IEEE 802.11n radio that provides high throughput with a peak transmission of 600 Mbps and a coverage of 100 m. Because an IEEE 802.11n radio supports multiple spectral bands in addition to those supported by the IEEE 802.15.4 radio, interference can be greatly minimized. More details of these and other technologies are presented in Chap. 3.

The brain of the UAV is a SoM that interacts with the autopilot and enables both IEEE 802.11n and IEEE 802.15.4 connectivity. This SoM runs Ubuntu Core, a well-known Linux distribution intended for IoT systems, that supports the upper layer protocol stacks, including drivers and applications, for interaction with the planner. By keeping the weight of the UAVs low, many advantages can be accomplished: (1) battery consumption is reduced, (2) UAVs can fly higher, (3) UAVs can fly longer, (4) UAVs can fly faster, and (5) in case of malfunctioning, liability due to physical and personal damage can be greatly minimized. Unfortunately, low-weight systems are also computationally less complex and support slower communications because of smaller batteries. Additionally, low-weight systems support fewer add-on modules and are more likely to have UAV flight instability issues due to the presence of the hyperspectral camera and the external antennas. The hyperspectral camera is a fast 60 g 80 mm×97 mm×159 mm camera that captures 25,650×990 spectral bands that can be used, among other things, for land vegetation detection.

Leaving IP connectivity and routing issues aside, discussed in detail in Chap. 4, we can focus on the mechanisms to transmit low-rate sensor and flight control information as well as high-rate hyperspectral images [9]. Low-rate traffic transmitted over IEEE 802.15.4 relies on the *Constrained Application Protocol* (CoAP) [21]. CoAP is a lightweight and highly efficient protocol that enables the management of IoT sessions. More importantly, CoAP supports *Representational State Transfer* (REST) transactions that optimize the interaction of the entities involved in the communication. Figure 1.11 illustrates the flow of the CoAP transactions used for the transmission of sensor and flight control information [10]. First, the planner transmits a CoAP GET request to observe the sensor readouts originated at the UAV. Observation is a feature associated with REST and adopted by certain protocols like CoAP that enables an observed device to transmit readouts whenever parameter changes are detected. Observation removes the need of an application to be constantly polling a device. In that figure, dash lines identify both the original GET request and the UAV CoAP 2.05 Content updates that carry readouts from sensors. These updates are real-time processed and used, in turn, by the application to send CoAP POST requests with lists of waypoints that define the flightpath. Solid lines identify each POST request and its corresponding acknowledgment carried in an empty 2.05 Content message. More details of CoAP are presented in Chap. 5.

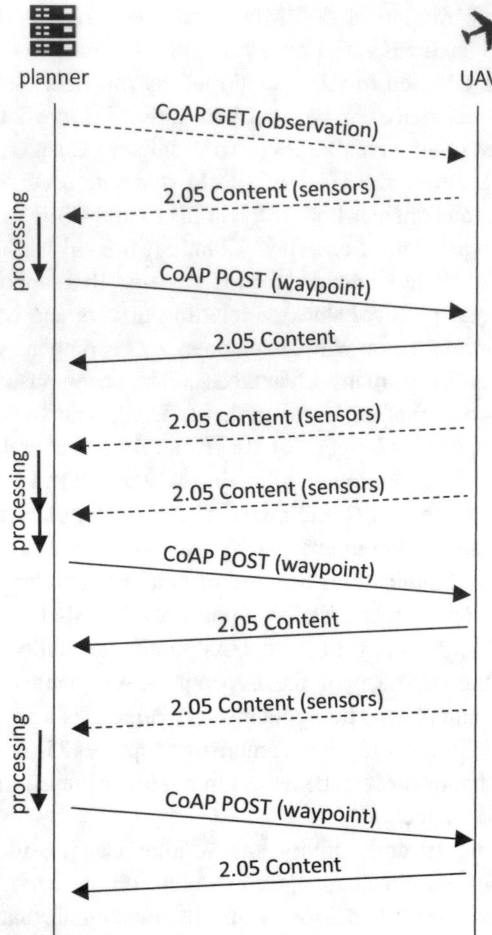

Fig. 1.11 CoAP traffic

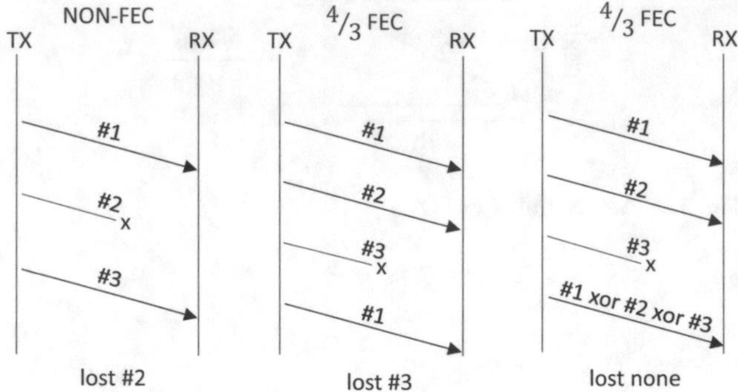

Fig. 1.12 FEC

On critical issue with this solution is how to efficiently transmit hyperspectral images from the UAVs to the planner. Since hyperspectral images are responsible for high-throughput traffic, well-known *real-time communication* (RTC) mechanisms similar to those used in *Voice over IP* (VoIP) solutions can be applied to IoT. The goal is to rely and extend traditional RTC protocols for transmission of compressed hyperspectral images. In this context, RTC consists of two different components: (1) signaling that refers to the exchange of information between entities in order to establish a session and (2) media that is about the transmission of the hyperspectral images themselves. Signaling provides extensions for negotiation of hyperspectral image codecs by means of the *Session Description Protocol* (SDP) that runs in the context of the more generic *Session Initialization Protocol* (SIP). Media, on the other hand, is transmitted over the *Real-Time Transport Protocol* (RTP) that runs on top of *User Datagram Protocol* (UDP). UDP provides some minimal sequence and timing control but lacks any data integrity protection. This causes RTP traffic to be affected by network loss that leads to media quality issues. Specifically, the communication links between UAVs and the planer rely on radio propagation that is mainly by the way of scattering over surfaces and diffraction over and around them in a situation typical of a multipath scenario that usually results in signal *fading* that causes network packet loss. Note that SIP, RTP, and SDP are discussed in detail in Chap. 5.

Network packet loss can be prevented and minimized by the intensive use of *forward error correction* (FEC), *negative acknowledgment* (NACK), as well as a combination of both approaches known as *hybrid error correction* (HEC) [11]. All these mechanisms have been widely used in the context of conventional media. In this solution, however, FEC is applied due to two main reasons: (1) it is computationally simple and (2) it exhibits very low latency. Moreover, FEC is independently applied to both uplink and downlink media streams. The IETF RFC 5109 standardizes a mechanism to encode generic FEC information via separate RTP FEC streams negotiated by means of SIP. Since the standard does not recommend any concrete FEC mechanism and just provides the general guidelines for correct signaling of FEC information, a generic bidimensional parity code is used. Specifically, the payloads of two consecutive hyperspectral data cube RTP packets are XORed together into a single FEC RTP payload.

To understand the implications of FEC, we can look at Fig. 1.12 that shows an example of traffic being transmitted with and without FEC. If FEC is disabled, lost network packets result in lost packets at the application level. This is shown in the left flow in the figure. If the packet #2 is lost, the application is not able to process it. On the other hand, if FEC is enabled, redundant packets are transmitted to compensate for those packets dropped by the network layer. The relationship between redundant and original packets is given by the code rate defined as the ratio k/n where n is the total number of transmitted packets (redundant and original) and k is the number of original packets. A $3/4$ code rate indicates that every four transmitted packets, three are original and one is redundant. There are many ways to specify how

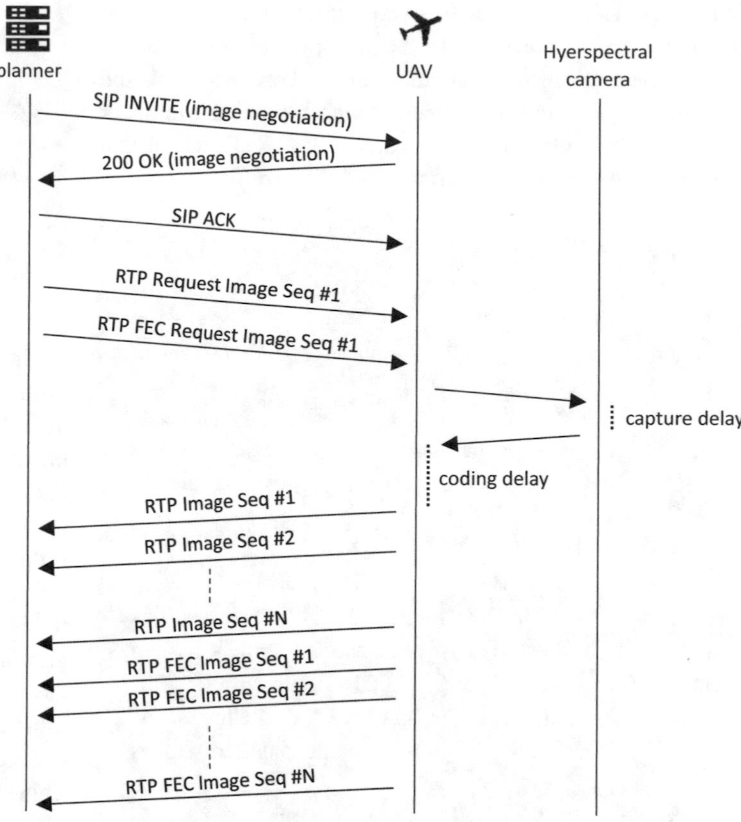

Fig. 1.13 SIP and RTP traffic

the redundant packet should be generated. One possibility is randomly selecting one out of the three original packets and retransmitting it. This leads to application layer packet loss if the network drops any of the other two packets that do not have redundant copies. This is shown in the middle flow in Fig. 1.12. If packet #3 is lost, the application is not able to process it since it has no redundant packet associated with it. Alternatively, by XORing all three data packets into a single redundant FEC packet, it is possible to recover up to one of the lost original packets. Specifically, and as shown in the figure, if the packet #3 is lost, this packet can be recovered from the redundant packet by XORing all other received packets; that is, $\#1 \oplus \#2 \oplus (\#1 \oplus \#2 \oplus \#3) = \#3$. Therefore, the XOR operation ($\oplus$) provides a simple mechanism to encode information about all three original packets in a single redundant packet. Because image compression payloads are typically of variable length, XORing requires padding packets with zeros to make them all the same size.

Each hyperspectral image is well over 80 MBytes of raw data that can be compressed by means of the *mblpc* hyperspectral codec [12]. In order to minimize throughput and network load, the codec is configured to provide lossy compression as a trade-off between rate and distortion. The rate is inversely related to the distortion, such that higher transmission rates are associated with lower image quality. Because different applications require different quality levels, they also provide different average transmission rates.

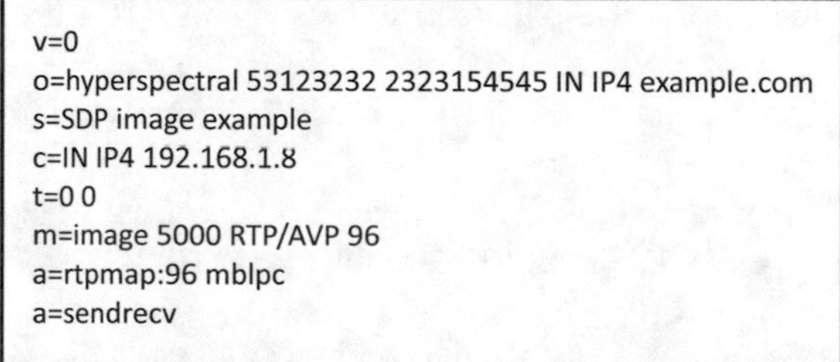

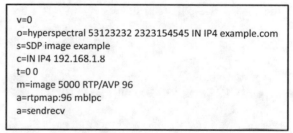

```
v=0
o=hyperspectral 53123232 2323154545 IN IP4 example.com
s=SDP image example
c=IN IP4 192.168.1.8
t=0 0
m=image 5000 RTP/AVP 96
a=rtpmap:96 mblpc
a=sendrecv
```

Fig. 1.14 SDP offering

Figure 1.13 shows a media session between the planner and a UAV to request a single hyperspectral image. Note that although communication is from a UAV to a planner, traffic may flow through multiple intermediate UAVs. A hyperspectral image is transmitted through the interaction of SIP, RTP, and SDP. First, the planner requests a new session by transmitting a SIP INVITE message that includes an SDP section, shown in Fig. 1.14, that identifies the offered hyperspectral codec as well as transport parameters associated with the RTP traffic. The UAV then accepts the request and issues a 200 OK response that also includes an SDP section that specifies the negotiated codec and transport parameters. The planner then sends a SIP ACK message to terminate the three-way handshake. At this point, the planner transmits a sequence of RTP packets (in accordance with the negotiated session) to request a partial or full hyperspectral image.

When the UAV receives this request, its camera captures an image (taking around 25 ms per spectral band) that is compressed, packetized (chunked into several packets), and transmitted back to the planner. Reliability against network impairments is introduced by means of FEC RTP packets that are transmitted alongside the original RTP traffic.

Example 1.2 Consider the SIP flow in Fig. 1.13, and assume that the distance between the UAV and the planner is 200 m. How long does it take for the session to be established when assuming a realistic Wi-Fi 5 Mbps rate in this scenario? What happens if 250 Kbps IEEE 802.15.4 networking were used instead? Assume average packet sizes of 320 bytes. Consider only propagation and transmission delays.

Solution Since there is no packet loss, the transmission rate and throughput are identical. The delay specifies how fast the first RTP packet can be sent. This can be done as soon as the SIP ACK is transmitted. The propagation delay is given by how long it takes for the SIP INVITE and SIP 200 OK to be propagated. This is given by $d_{\text{prop}} = 2 \times \frac{d}{c} = 1.34\ \mu\text{s}$ where c is the speed of light given by $c = 300 \times 10^6$ m/s and $d = 200$ m is the distance between UAV and planner. Note that using the speed of light in this calculation is a best-case scenario assumption as an electromagnetic wave propagates slightly slower in regular (non-vacuum) channels.

The transmission delay is given by how long it takes for the SIP INVITE, SIP 200 OK, and SIP ACK to be transmitted. This is given by $d_{\text{trans}} = 3 \times \frac{L}{R} = 1.53$ ms where $L = 8 \times 320$ bytes $= 2560$ bits is the packet size and $R = 5 \times 10^6$ bps is the Wi-Fi transmission

(continued)

(continued)

rate. Note that because $d_{\text{trans}} \gg d_{\text{prop}}$, the propagation delay can be typically ignored and the delay can be considered as $d \approx d_{\text{trans}} = 1.53$ ms. This is even more noticeable when considering IoT-friendly IEEE 802.15.4 where a $R' = 0.25 \times 10^6$ bps transmission rate leads to a $d \approx 3 \times \frac{L}{R'} = 30.72$ ms delay.

The payload of each RTP packet includes specific information that applies to both uplink and downlink. For those originated at the planner and sent to a UAV, each packet includes control information that specifies the list of spectral bands that are requested. Figure 1.15 shows the specific fields included in each packet. Specifically, an 8-bit sequence number that specifies the hyperspectral image being requested is followed by a variable number of groups of spectral bands. These groups are indicated by an 8-bit start and end band fields that specify the band numbers associated with each group. Downlink packets originated at a UAV and sent to the planner include the compressed data cubes encoded as illustrated in Fig. 1.16. Each response includes an 8-bit sequence number that specifies the data cube being transmitted. This sequence number matches that of the incoming request. The response includes an 8-bit band number that specifies the spectral band number being encoded as well as a 16-bit index that represents the incremental index number of the portion of spectral band included in the payload. Additionally, the payload includes a 16-bit size field that indicates the length of the portion of hyperspectral band data along with the compressed portion of the band itself.

One issue with wireless IoT communications is the problem caused by the characteristics of the fading channel. This is a well-known phenomenon that can be modeled by the two-state Markov process shown in Fig. 1.17 [13]. The model

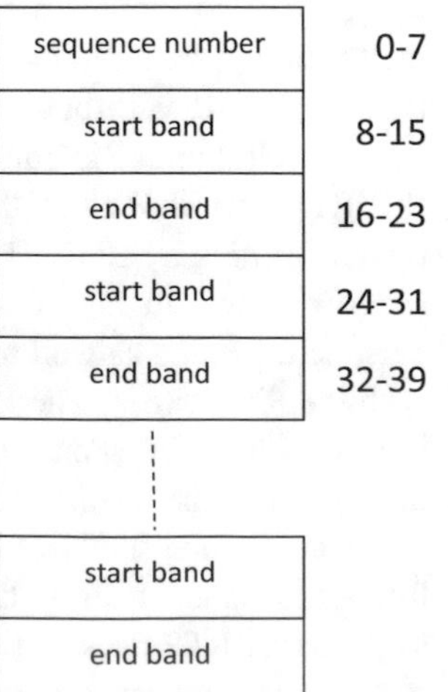

Fig. 1.15 Packetization of control information

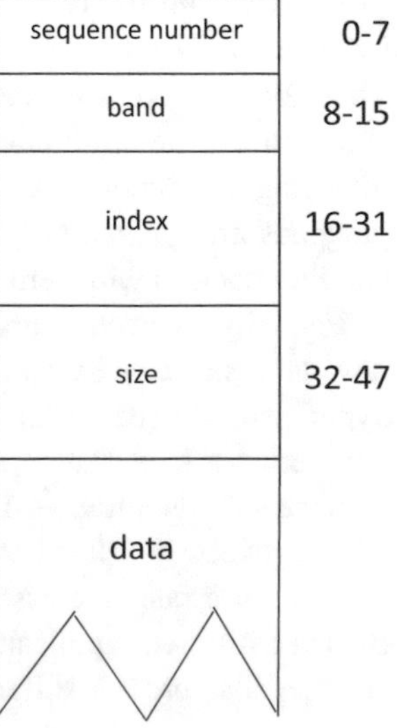

Fig. 1.16 Packetization of hyperspectral images

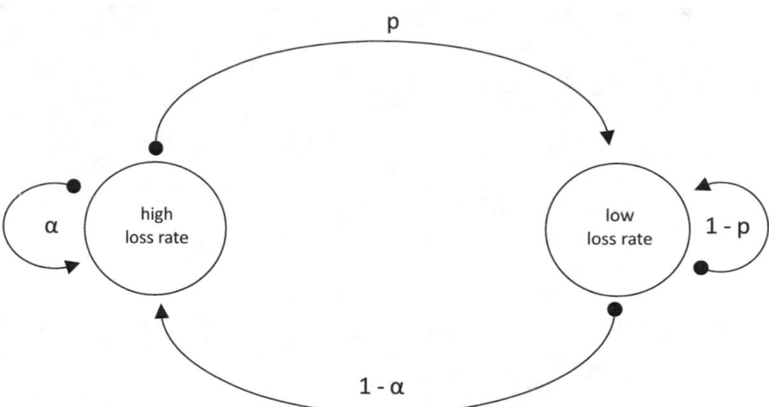

Fig. 1.17 Fading channel

relies on two states such that the channel is either in a low-loss state or in a high-loss state. While in low-loss state packets may be dropped, the loss probability is lower than when the channel is in the high-loss state. Transitions are given by two main parameters: (1) p, the transition probability from low- to high-loss states, and (2) α, the probability that the channel stays in a high-loss state. The assumption is that all packets are dropped when the channel is in a high packet loss state, while no packets are lost if the channel is in a low packet loss state. From this model, it is clear that the nature of the channel is such that bursty packet loss is likely to occur because when the channel is in a high-loss state, several packets may be dropped before the channel transitions back to the low-loss state.

One way to assess the efficiency of the solution, evaluating its resilience against fading and other performance problems, is to check the quality of the transmitted hyperspectral images. Although there are many algorithms that can be used to assess image quality, there are only a few that are effective when considering the effects of network impairments. To this end, the evaluation of data cube quality can be performed by means of different technologies including *peak signal-to-noise ratio* (PSNR) and *structural similarity* (SSIM). These mechanisms provide quality scores that can be used to understand the degradation introduced by the encoding and packetization in a lossy environment. PSNR provides the ratio between the energy of the hyperspectral and the energy of the error between the image and its reference. SSIM is used to measure the similarity between images by means of comparing local patterns of pixels and considering the strong interdependency of sections that are spatially close [23]. SSIM scores range from −1 to 1 where 1 is only achievable when a received image is identical to its reference. The SSIM scores that result from all spectral bands in an image are averaged to obtain a *mean structural similarity* (MSSIM) score that represents the quality of a data cube.

Table 1.1 Effects of FEC in transmission of hyperspectral images

		PSNR (dB)		MSSIM	
α	p	No FEC	FEC	No FEC	FEC
0.01	0.01	42.91	45.28	0.90	0.94
0.01	0.05	39.11	43.83	0.86	0.92
0.01	0.10	35.74	41.71	0.79	0.89
0.05	0.01	40.39	44.16	0.86	0.92
0.05	0.05	36.09	41.49	0.81	0.89
0.05	0.10	32.75	40.10	0.77	0.86
0.10	0.01	37.17	41.07	0.83	0.88
0.10	0.05	31.66	39.49	0.74	0.86
0.10	0.10	27.48	37.63	0.68	0.83
0.15	0.01	32.71	39.85	0.76	0.86
0.15	0.05	25.11	37.12	0.69	0.82
0.15	0.10	21.05	33.52	0.64	0.79

Table 1.1 shows PSNR values as well as MSSIM scores for transmission of hyperspectral images with and without FEC-assisted RTP transport. These values incorporate the effects of lossy compression due to the mblpc codec configured to provide a compression rate of around 1 *bits per pixel band* (bppb). The measured transmission rates are 1.608 Mbps and 2.575 Mbps for RTP and FEC-assisted RTP, respectively. Note that, as expected, there is high correlation between PSNR values and MSSIM scores and that the best results are obtained when FEC is enabled. In this latter case, the maximum PSNR and MSSIM improvements are 12 dB and 20%, respectively. As the bursty packet loss intensifies, when both α and p go up, the overall improvement becomes more significant. This can be better seen with a graphical example; Fig. 1.18 shows PSNR as a function of the network loss for both low and high loss burstiness. It is important to mention that although this scheme relies on basic XOR FEC, other more efficient alternatives like *Reed-Solomon* (RS) codes, *turbo codes* (TC), and *low-density parity-check* (LDPC) codes can be used instead.

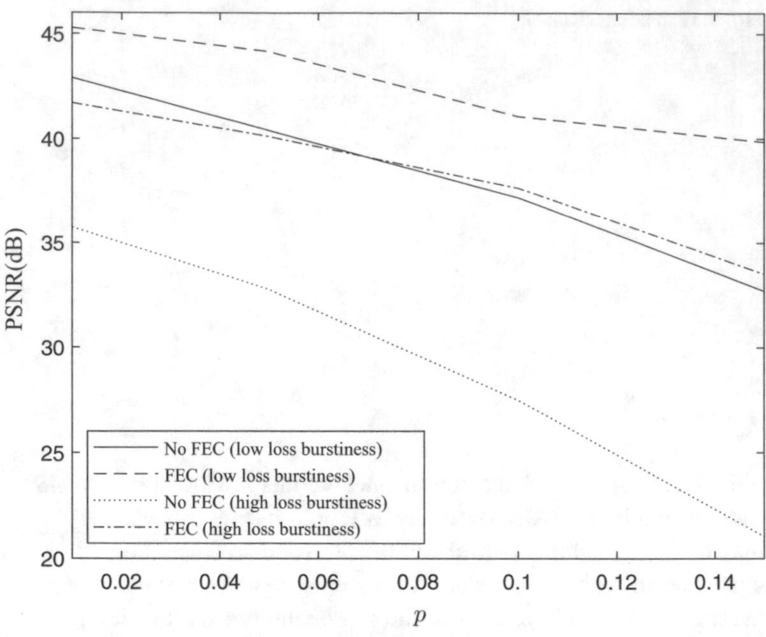

Fig. 1.18 PSNR vs loss

1.5 Why Now?

One of the main questions is *why now?*, that is, why has IoT become so popular in the last few years? Although remote sensing and actuation have been around for well over a century at this point, there are specific technological and scientific developments, shown in Fig. 1.19, that have led to the IoT explosion. On the device side: material science improvements including *microelectromechanical systems* (MEMS) used to build micro-sized sensors and actuators, printable electronics that enable fast and cost-effective development of embedded devices, and smart textiles for the creation of next-generation wearables [5]. Energy production and storage advancements that include smart grids to generate electricity through affordable photovoltaic panels and energy harvesting that relies on miniaturized ultracapacitors and batteries [4]. Embedded processing developments that result from cheaper and computationally more complex SoM and SoC platforms that are essential to sensors and actuators that include lightweight, low-power 32-bit and 64-bit *Advanced RISC Machine* (ARM) processors and the support of several *Input/Output* (I/O) and networking interfaces and ports in a very small footprint [7].

On the application and analytics side: software architecture evolution toward Open *Application Program Interfaces* (APIs) and REST interfaces that based on the web paradigm extend the *Service-Oriented Architecture* (SOA) to IoT. Specifically, SOA is a system design approach where applications use the services that are available in the network. Decision-making support by means of knowledge frameworks supported by *artificial intelligence* (AI) technologies like machine learning that rely on huge data sets accessible through big data distributed data mining mechanisms. Virtualization that provides multiple independent execution environments to run different applications on the same hardware and enable the conversion from hardware-specific systems to inexpensive software-based solutions [3].

From the perspective of this book, however, the most relevant breakthroughs are on the communication and networking sides. This includes ubiquitous Internet access that is instantaneously available everywhere and provides cheap end-to-end connectivity. Networking technologies that are standardized through different mechanisms ranging from physical layer wireless IEEE 802.15.4 and wireline *Power Line Communication* (PLC) to application layer *Message Queue Telemetry Transport* (MQTT) and the aforementioned CoAP. This book focuses on these and many other protocols and technologies, detailing their interaction as well as their most important features and characteristics including security, resilience in a constrained environment to support power, and transmission efficiency.

1.6 Applications

IoT applications span over several industries relying on multiple overlapping technologies. Of all industries, however, consumer electronics is probably the one that has evolved the most with the introduction of IoT. Some common consumer electronics applications are connected gadgets, wearables, robotics, and participatory sensing that tied to the *Social Web of Things* (SWoT) enables users to use their sensors' readouts in order to share relevant information like traffic and pollution conditions. IoT applications in the retail bank-

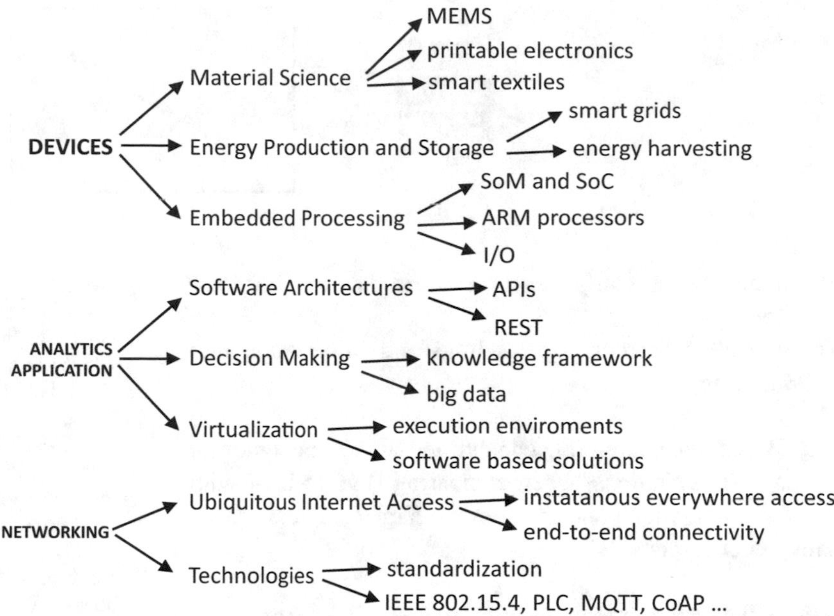

Fig. 1.19 Developments that lead to IoT

ing industry involve micro payments, logistics, product lifecycle information, and shopping assistance. The automobile industry is also being affected by IoT through autonomous cars and multi-modal transport that enables the coordination for end-to-end delivery of goods. Another sector to consider is agriculture where IoT applications include forestry, farming, livestock monitoring, and urban agriculture that enables small-scale farming in cities and other highly populated zones [14].

The environmental industry initially by means of WSNs and now with IoT has improved applications intended to sense pollution and monitor air, water, soil, and weather. The infrastructure industry, on the other hand, relies on IoT advancements to provide home and building automation including access control as well as monitoring of roads and railroads. Similarly, public utility companies are also starting to take advantage of IoT and enhancing their services by means of smart grids; water, gas, as well as oil management; and heating and cooling control and monitoring. The health industry is another sector that has been heavily disrupted by IoT; this includes remote monitoring of health, assisted living, behavioral change, treatment compliance, and fitness applications. A whole new industry that is based on the smart city concept has resulted from IoT; some areas of interest are integrated environments, sustainability, and inclusive living. IoT has also resulted in improved industrial processes with applications ranging from simple automation and remote operations to resource management and robotics including manufacturing and control of heavy machinery. This has led to IIoT that is a key factor in the context of the Fourth Industrial Revolution also known as Industry 4.0 [2]. Other critical elements of the IoT evolution include the integration of newer and more efficient LPWAN technologies that natively support IPv6 with *Protocol Stack Virtualization* (PSV) techniques that enable the deployment of *massive IoT* solutions.

Summary

IoT is driven by several scientific and technological breakthroughs and developments that provide an evolution that starts with traditional M2M and WSN schemes. This chapter explored this evolution by introducing an overview of the main concepts needed to understand IoT. It detailed the convenience of the layered architecture as a model for design, development, and deployment of IoT communication technologies. The chapter presented a quick overview of the main IoT components and how they lead to the two most popular IoT network topology families: WPANs and LPWANs. These topologies are put in the context of a use case that supports IoT communication with UAVs. Other relevant IoT applications ranging from home automation to IIoT are also discussed.

Homework Problems and Questions

1.1 In an M2M scenario, an application performs image processing of a video stream captured by a camera and unlocks a door when the right person is identified:

(a) Describe the different elements of the communication system

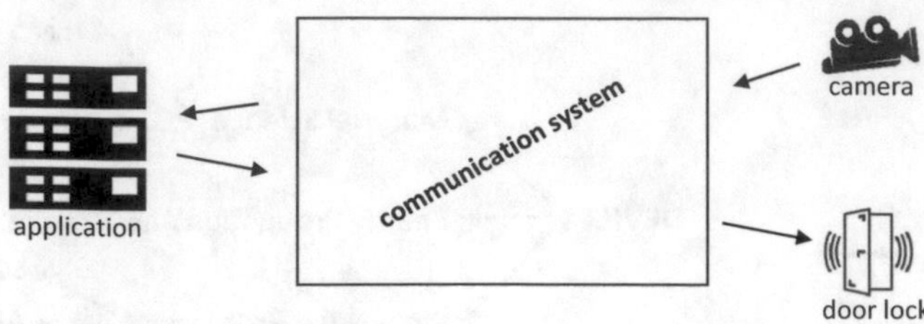

(b) Indicate what components transform data into information
(c) Indicate what components transform information into knowledge

1.2 Based on addressing capabilities alone and ignoring other NAT limitations, when comparing IPv6 to IPv4 with NAT support, which one provides the largest address space? How much larger is it?

1.3 In the transformation example of Fig. 1.7, each temperature readout is a 1-byte number, and the extracted knowledge can be encoded as a 1-bit number. What are the compression rates for each transformation? The compression rate is given by the ratio between the uncompressed and compressed values.

1.4 In Fig. 1.6, why is the WPAN gateway inside the IP cloud, while the LPWAN one is not?

1.5 What are the implications of combining OSI session, presentation, and application layers into a single IETF application layer?

1.6 What components are responsible for the data-information and information-knowledge conversion stages in the use case in Sect. 1.4? What type of conversion does each stage involve?

1.7 In Figs. 1.13 and 1.11, respectively, consider an average RTP image packet size of 1000 bytes transmitted every 12 ms and an average 2.05 Content packet size of 10 bytes transmitted every 1 ms. Ignoring any additional overhead due to lower layers, what channel capacity is minimally expected for WPAN and LPWAN networks?

References

1. Al-Sarawi, S., Anbar, M., Alieyan, K., Alzubaidi, M.: Internet of things (IoT) communication protocols: review. In: 2017 8th International Conference on Information Technology (ICIT), pp. 685–690 (2017)
2. Conway, J.: The industrial internet of things: an evolution to a smart manufacturing enterprise. White paper (2015)
3. Geng, H.: Internet of Things and Data Analytics in the Cloud with Innovation and Sustainability, pp. 1–28 (2017)
4. Geng, H.: Internet of Things and Smart Grid Standardization, pp. 495–512 (2017)
5. Geng, H.: MEMS, pp. 147–166 (2017)
6. Geng, H.: Scada Fundamentals and Applications in the IoT, pp. 283–293 (2017)
7. Hassan, Q.F.: A Tutorial Introduction to IoT Design and Prototyping with Examples, pp. 153–190 (2018)
8. Haykin, S.: Communication Systems, 5th edn. Wiley, New York (2009)
9. Herrero, R.: Real time transmission of hyperspectral images onboard UAVs. In: 4th AETOS International Workshop on "Research Challenges for Future RPAS/UAV Systems"
10. Herrero, R.: MQTT-SN, CoAP, and RTP in wireless IoT real-time communications. Multimed. Sys. (2020). https://doi.org/10.1007/s00530-020-00674-5
11. Herrero, R., Cadirola, M.: Effect of FEC mechanisms in the performance of low bit rate codecs in lossy mobile environments. In: Proceedings of the Conference on Principles, Systems and Applications of IP Telecommunications, IPTComm '14. Association for Computing Machinery, New York (2014). https://doi.org/10.1145/2670386.2670387
12. Herrero, R., Cadirola, M., Ingle, V.K.: Preprocessing and compression of Hyperspectral images captured onboard UAVs. In: Carapezza, E.M., Datskos, P.G., Tsamis, C., Laycock, L., White, H.J. (eds.) Unmanned/Unattended Sensors and Sensor Networks XI; and Advanced Free-Space Optical Communication Techniques and Applications, vol. 9647, pp. 8–16. International Society for Optics and Photonics, SPIE (2015). https://doi.org/10.1117/12.2186169
13. Hohlfeld, O., Geib, R., Hasslinger, G.: Packet loss in real-time services: Markovian models generating QoE impairments. In: 2008 16th International Workshop on Quality of Service, pp. 239–248 (2008). https://doi.org/10.1109/IWQOS.2008.33
14. Holler, J., Tsiatsis, V., Mulligan, C., Avesand, S., Karnouskos, S., Boyle, D.: From Machine-to-Machine to the Internet of Things: Introduction to a New Age of Intelligence, 1st edn. Academic, New York (2014)
15. Kurose, J.F., Ross, K.W.: Computer Networking: A Top-Down Approach, 6th edn. Pearson, London (2012)
16. Lucero, S., et al.: IoT platforms: enabling the internet of things. White paper (2016)
17. Palattella, M.R., Accettura, N., Vilajosana, X., Watteyne, T., Grieco, L.A., Boggia, G., Dohler, M.: Standardized protocol stack for the internet of (important) things. IEEE Commun. Surv. Tutorials **15**(3), 1389–1406 (2013)
18. Ravi Kumar, K.S., Gottapu, J., Mandarapu, C., Rohit, M.S., Murthy, T.V.N., Mavuri, M.: Introduction to M2M communication in smart grid. In: 2015 International Conference on Control, Instrumentation, Communication and Computational Technologies (ICCICCT), pp. 250–254 (2015)
19. Sethi, P., Sarangi, S.R.: Internet of things: architectures, protocols, and applications. J. Electr. Comput. Eng. **2017**, 9324035 (2017). https://doi.org/10.1155/2017/9324035

20. Shelby, Z., Bormann, C.: 6LoWPAN: The Wireless Embedded Internet. Wiley, Chichester (2010)
21. Shelby, Z., Hartke, K., Bormann, C.: The Constrained Application Protocol (CoAP). RFC 7252 (2014). https://doi.org/10.17487/RFC7252. https://rfc-editor.org/rfc/rfc7252.txt
22. Siow, E., Tiropanis, T., Hall, W.: Analytics for the internet of things: a survey. ACM Comput. Surv. **51**(4) (2018). https://doi.org/10.1145/3204947
23. Wang, Z., Bovik, A.C., Sheikh, H.R., Simoncelli, E.P.: Image quality assessment: from error visibility to structural similarity. Trans. Img. Proc. **13**(4), 600–612 (2004). https://doi.org/10.1109/TIP.2003.819861

Concepts of IoT Networking

2

2.1 IoT Networking

From a functional perspective, an IoT network, like most packet switched networks, is made of two types of devices: endpoints that are known as hosts and are source or destination of messages and routers that assist in the propagation of messages throughout the network. Both, hosts and routers, form communication systems with transmitters and receivers connected to channels by means of links. Each router supports multiple hosts that, in turn, are connected to sources and sinks as illustrated in Fig. 2.1 [13]. By means of packet switching, the network provides an alternative to dedicated links between endpoints as routers allow for resource sharing in a highly efficient manner. In the context of IoT, hosts are typically sensors, actuators, controllers, and devices in general as well as applications like those performing complex decision-making. Routers, on the other hand, can be dedicated equipment or as in the case of capillary networks, described in the next section, other devices. By giving plain devices, like sensors and actuators, routing capabilities, it is possible to lower deployment times and costs by maximizing hardware reutilization.

Links, depending on the nature of the channel, can be wireless associated with free propagation or wireline associated with guided propagation. Guided propagation typically involves (1) copper conductors that are twisted at a rate of several twists per centimeter in order to mitigate *electromagnetic interference* (EMI); (2) coax cables that are made of inner and outer conductors, separated by a dielectric insulating material with stronger immunity to EMI and comparatively higher bandwidth that enables higher transmission rates; and (3) optical fibers that exhibit no EMI and support much higher transmission rates [16]. On the other hand, free propagation occurs between corresponding antennas in the Earth's atmosphere, underwater, as well as in free space. Examples of free propagation include wireless broadcast and satellite channels. In the context of IoT, the decision between relying on a wireless and relying on a wireline solution is related to device deployment costs and times. Specifically, in order to support a massive number of devices, wireline solutions usually require huge infrastructure changes that are too expensive and take too long to implement. Wireless architectures with battery-powered devices are the most common type of IoT deployment. Alternatively, wireline scenarios that take advantage of preexistent power wiring for communications are also popular.

Depending on the needs and requirements of a given scenario, limiting factors, including transmission power, channel bandwidth, and deployment costs, favor one solution over the others. One important consideration is that a transmitted signal is affected by the channel noise. By the time the signal arrives at the receiver, it has been attenuated and affected by channel noise leading a specific *signal-to-noise* (SNR). Because the channel capacity theorem states that the maximum achievable transmission rate is a direct function of SNR, higher SNR typically means higher transmission rates [4,15]. Note that the transmission rate is error-free if the right channel encoding mechanism is used. In IoT networks, this can be challenging as preserving battery life usually implies low signal power and low SNR that translates into low transmission rates. More details of this interaction are described in Chap. 3.

Because IoT devices, like sensors, interact with the physical environment, they monitor infinite precision analog assets like temperature, humidity, or light intensity that cannot be packetized and transmitted without certain transformations. This was previously described as part of the example of Fig. 1.1 that shows the communication path between a temperature sensor and an application. To put it in the context of a layered architecture, first, the analog variable is converted to a digital number represented by a sequence of bits generated

R. Herrero, *Fundamentals of IoT Communication Technologies*, Textbooks in Telecommunication Engineering, https://doi.org/10.1007/978-3-030-70080-5_2

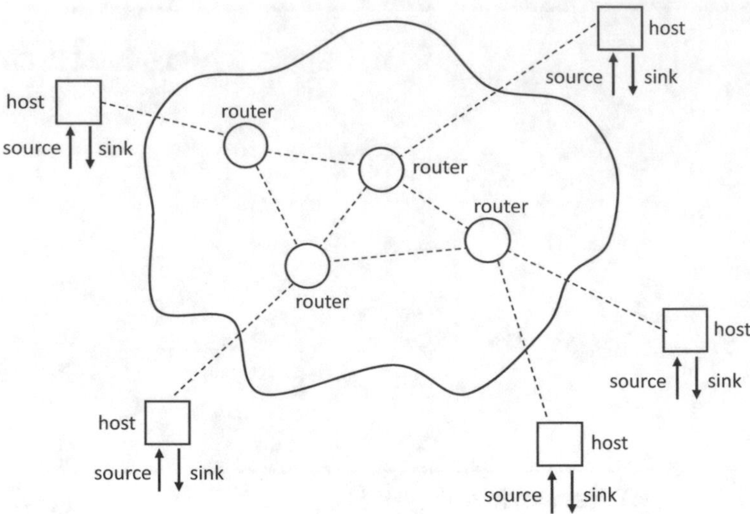

Fig. 2.1 Communication network

by source encoding at an ADC. Note that source encoding can be thought of as being performed at the application layer. The converted digital value representing the asset can then be prepared for transmission by adding reliability, addressing and additional routing information as part of channel encoding [9]. This is typically done by appending headers and other fields to the converted digital value in order to build a packet in a way that is consistent with transport, network, and link layers. For example, a checksum field that provides FEC is an example of channel encoding. The resulting packet can then be transmitted over the channel as a modulated wave.

Note that modulation is performed by the physical layer. Since the channel is analog because it exists in the physical world, modulation performs one more conversion. Essentially, the modulator converts the packet into electrical signals that can be transmitted through wires or propagated through antennas. As the analog signal traverses the channel, it is attenuated and becomes affected by noise that lowers the SNR and limits performance. When the signal arrives at the receiver, the demodulator, at the physical layer, restores the digital packet by converting the signal into a stream of bits. Subsequently, the channel decoder removes any address fields and additional reliability information performed at link, network, and transport layers, and it forwards the payload to the application layer. In sensing scenarios, the consumer of the payload is an application that makes automated decisions. In actuation scenarios, however, digital data is generated by an application and transmitted through the channel to a device that performs actuation. In this case, since the consumer of the digital payload is analog, source decoding converts it into an analog signal by means of a DAC. This is consistent with the communication path between the application and the engine in the example of Fig. 1.1.

2.2 Types of Networks

As indicated in Fig. 2.2, IoT networks sit in between applications and devices that, in turn, interact with assets. The IoT system is the set made of devices, applications, and networks, while the asset is external to the IoT system. In all cases, IoT networks usually involve wireless and wireline links that use different physical and link layers but that they all rely on IPv6 connectivity. In the IoT domain, there are two common scenarios for communication between two endpoints: (1) one-hop communication (i.e. sensor directly talking to an application) and (2) multi-hop communication (i.e. sensor indirectly talking to an application by relying on intermediate devices to forward packets). This latter scenario is representative of capillary networks initially introduced as part of WSNs and now widely extended to many IoT technologies [10]. Figure 2.3 illustrates a capillary network where *sensor 1* communicates with the application through *sensor 2* and *sensor 3*. Alternatively, and due to routing, *sensor 1* can talk to the application through *sensor 4*. Similarly, Fig. 2.4 shows the same devices and application but under a one-hop topology.

Example 2.1 Consider the following scenario where a sensor transmits a 200-byte packet over three different paths based on specific routing objectives. Each path is associated with a number of hops needed to reach the destination. More hops imply shorter links that support higher transmission rates as signal degradation is lower. In this particular scenario, links A, B, and C support nominal transmission rates of 240 Kbps, 64 Kbps, and 152 Kbps, respectively. How long does it

(continued)

Fig. 2.2 IoT networks

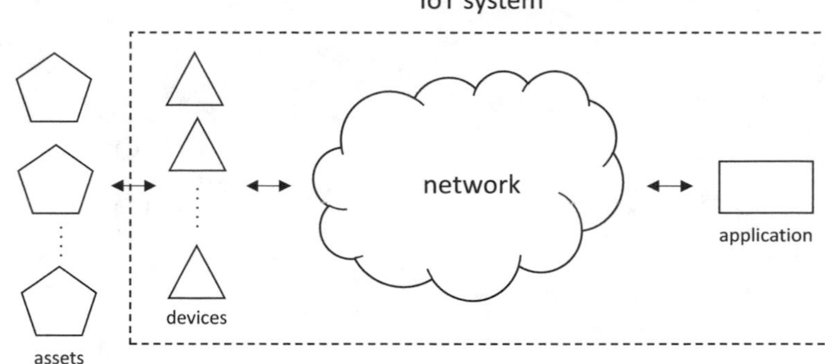

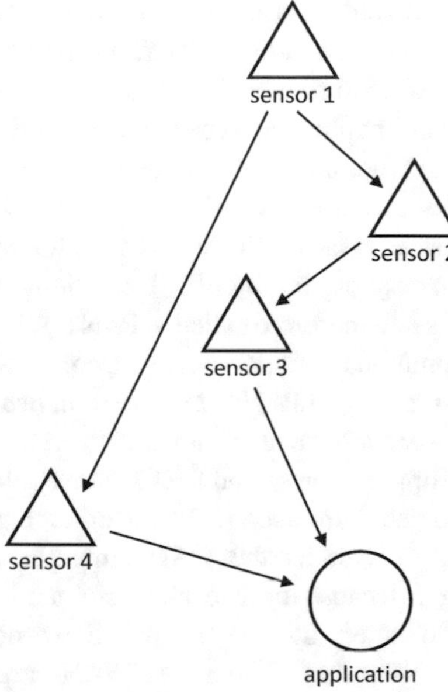

Fig. 2.3 Multi-hop capillary network

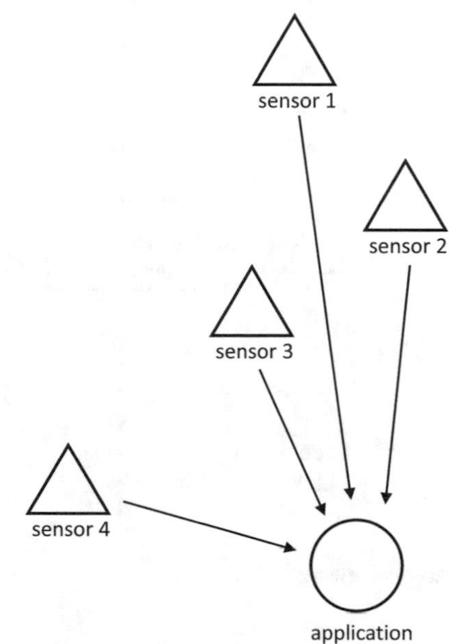

Fig. 2.4 Single-hop network

(continued)

take for a packet to reach the gateway over the three different paths (d_A, d_B, and d_C)?

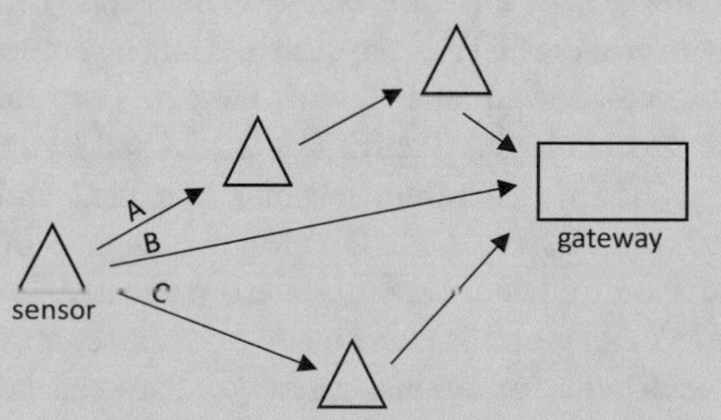

Solution The delay is given by the number of hops per path multiplied by the delay over a single link. Specifically the corresponding delays are $d_A = 3 \times$

(continued)

(continued)

$^{200\times8\text{ bits}}/_{240000\text{ bps}} = 20$ ms, $d_B = {}^{200\times8\text{ bits}}/_{64000\text{ bps}} = 25$ ms, and $d_C = 2 \times {}^{200\times8\text{ bits}}/_{152000\text{ bps}} = 21.05$ ms. Note that although path A traffic is retransmitted by two intermediate nodes, it exhibits lower delay when compared to paths B and C because of its higher transmission rate.

Networks have been traditionally classified based on the area they cover; as shown in Fig. 2.5, the three main types are *local area networks* (LANs), *metropolitan area networks* (MANs), and *wide-area networks* (WANs). A LAN involves an office, a lab, and a building, and it does not rely on third-party communication infrastructure to provide high-speed service. A WAN, on the other hand, involves a very large geographical region and therefore relies on third-party infrastructure to function. The throughput of a WAN is typically much lower than that of a LAN. The size of a MAN falls somewhere in between that of a LAN and WAN. As a

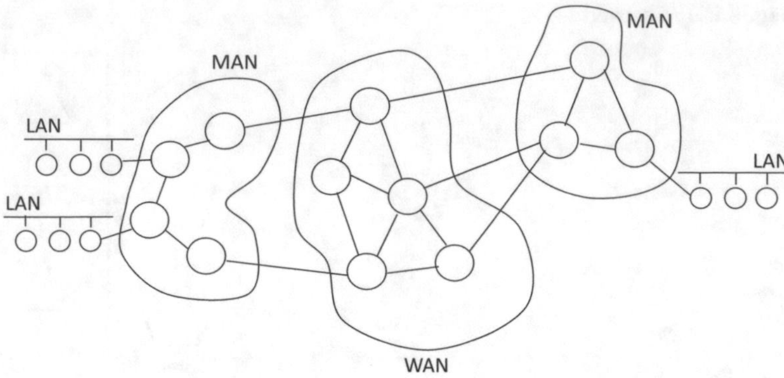

Fig. 2.5 Network classification

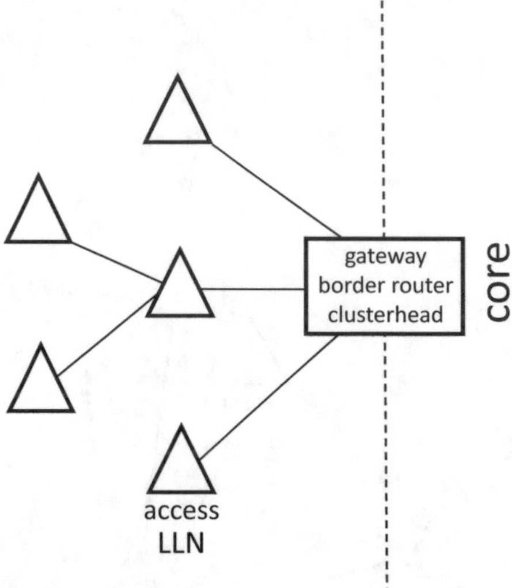

Fig. 2.6 IoT access network

WAN, a MAN also relies on third-party communications but operates at higher speeds typically linking LANs and WANs. Larger than WANs, an *Internet area network* (IAN) connects endpoints within a cloud environment.

Under IoT, networking typically refers to the access network where the devices are located. Figure 2.6 illustrates a generic IoT access network. In general, the overall topology is such that devices interact with an IoT gateway, border router, or clusterhead that acts as the boundary between access and core sides. The core network is usually a mainstream IP network that provides global connectivity to applications. These applications, in turn, many times reside in the cloud infrastructure.

Recent developments of virtualization have led to the integration of WANs and *software-defined networks* (SDNs) into *software-defined WAN* (SD-WANs) that provide reliable long-distance WAN access at a reduced cost by relying on IP tunnels transmitted over the public Internet.

All these network types can be deployed as wireless or wireline topologies. For example, WWAN and WLAN, respectively, indicate the wireless versions of WAN and LAN. Essentially, a "W" prefix indicates that the network type is wireless. Similarly, if no prefix "W" is included, it is assumed to be a plain wireline-based network type.

Most IoT access side networks, where constrained devices are usually deployed, classify as low-rate short-range LAN subtypes. A *home area network* (HAN) is one such subtype that supports IoT communication at home and building levels. Another very popular subtype that is of particular importance in the IoT domain is that of *personal area networks* (PANs). PANs provide a small coverage at relatively low transmission rates in order to extend battery life. As indicated in Sect. 1.3, a very big family of IoT solutions are based on wireless PANs and therefore called WPANs. WPANs are one of the most common network types in the context of IoT. With a smaller coverage than PANs, *body area networks* (BANs) are made of *wearables* and *implants* as well as other small devices that support fitness and health-care applications. As it is also indicated in Sect. 1.3, the other big family of IoT networks falls under the WAN umbrella due to their large coverage. Because these technologies are low power in order to extend battery life, they are called low-power WANs or LPWANs. Note that although LPWANs are inherently wireless, they are called LPWAN and not LPWWAN [3].

Both WPANs and LPWANs, by virtue of being wireless, rely on batteries. Moreover, in order to extend battery life, power consumption must be minimized. Low power, unfortunately, implies weak signals and therefore low SNR that is responsible for high packet loss and low transmission rates. Most IoT technologies, as shown in Fig. 2.6, fall under what it is called *low-power and lossy networks* (LLNs) or sometimes called *low-power, low-rate, and lossy networks* (LLLNs) [2].

There are several PAN physical and link layer mechanisms that can co-exist in an IoT environment. They range from wireless technologies like IEEE 802.15.4 and *Bluetooth Low Energy* (BLE) to wireline schemes like PLC and *Master-Slave/Token-Passing* (MS/TP). More complete description of these and other technologies are presented in Chap. 3. Since IoT PANs and WPANs involve devices communicating with each other or against gateways, they can be laid out in different topologies including buses, rings, and master-slave schemes. In most cases, these technologies are part of proprietary network stacks that do not rely on the Internet for communications. The IoT revolution has brought into existence several network layer protocols that can be used with these heterogeneous mechanisms to provide broad IPv6 communication. Specifically, 6LoWPAN and other similar mechanisms provide IPv6 connectivity on top of physical and layer protocols like IEEE 802.15.4 to uniformly support IoT applications. Details of the IoT network layers including 6LoWPAN support are presented in Chap. 4. There are also several LPWAN stacks that co-exist in several full and hybrid IoT environments. They are mostly wireless technologies that include LoRa, SigFox, NB-IoT, and many more presented in Chap. 8.

2.3 Devices

IoT involves, by definition, interaction with the physical environment that is performed by means of logical devices that can be controllers, actuators, sensors, and gateways [17]. These logical devices are supported by hardware provided by physical devices running on constrained embedded computers and systems, first introduced in Sect. 1.3. They are embedded because they are typically part of a bigger CPS. They are constrained because they have limited memory and computational complexity. In general, these computers are characterized by a power source, a networking stack, and a processor.

The power sources can be traditional *alternating current* (AC) and *direct current* (DC) electric power lines, batteries, or hybrid schemes that support energy harvesting by means of, for example, solar panels [6]. In regard to the networking stacks, and depending on the hardware and software capabilities, there are different protocols associated with layers and technologies as diverse as IEEE 802.15.4 and CoAP. As stated before, the main goal of this book is to explore these IoT communication technologies.

The processors are in most cases low-power constrained embedded computers with limited computational complexity and small *instruction set architectures* (ISAs). Because of the IoT interaction with the physical environment, these devices are particularly reliable when it comes to control over timing. Therefore, in many cases, embedded processing is a synonym of real-time processing. The simplest devices rely on microprocessors with basic *central processing units* (CPUs) that combine several peripheral devices like memories, I/O interfaces, and timers. They are usually 8-bit processors that consume extremely small amounts of energy and rely on power cycles as well as sleep modes to minimize energy consumption and extend battery life. These small embedded devices are well known to operate on small batteries for several years.

More advanced devices rely on 32-bit and 64-bit ARM processors that comparatively consume more power but provide a lot higher computational complexity including, sometimes, support of *digital signal processing* (DSP) capabilities. Many not-too-complex embedded processors rely on co-processors that offload complex functionality like signal and network processing. In general, embedded processors, regardless of their complexity, include several I/O interfaces that are accessible to system designers by means of pins on SoCs and SoMs. These interfaces provide basic communication between peripherals within a device by supporting point-to-point and bus infrastructures [8]. The *Serial Peripheral Interface* (SPI) enables very simple, fast serial communication between a master embedded processor and multiple slaves. Similarly, the *Inter-Integrated Circuit* (I^2C) interface provides a more complex, but a bit slower, bus architecture that supports multiple co-existent bidirectional master and slave peripheral. A *universal asynchronous receiver-transmitter* (UART) interface supports a generic and reliable mechanism for the transmission of data over serial link. Both, RS-232 and RS-485, are well-known examples of UART serial interfaces that despite being old and slow are widely used in the industry. As opposed to SPI and I^2C, UART-based serial interfaces are sometimes used by some IoT technologies. For example, wireline MS/TP relies on RS-485 for its physical layer.

Another serial interface that is widely supported by most embedded processors is known as *Joint Test Action Group* (JTAG). JTAG, standardized as IEEE 1149.1, is used to perform a boundary scan of an integrated circuit to enable circuit testing in situations where using electrical probes is virtually impossible. It also enables the examination and control of the state of an embedded processor. A newer mechanism is Serial Wire Debug (SWD) that provides similar functionality with fewer pins.

In the context of IoT devices, some embedded processors also include ADC and DAC interfaces for conversion of signals from the analog to the digital and from the digital to the analog domains, respectively. Last but not least, most embedded processors include *general-purpose input and output* (GPIO) ports that can be used to read and write two-level digital signals for interaction with sensors and actuators [14].

In order to run complex software, complex hardware is needed. Constrained 8-bit embedded processors are a lot weaker candidates than 64-bit ARM processors to run complex *operating systems* (OSs). Based on levels of complexity, OSs can be classified as main-loop, event-based, embedded, or full-featured. A main-loop OS consists of a simple bootloader that executes a single threaded process that continuously polls sensors and performs actuation in response. An event-based OS is a bit more sophisticated and relies on hardware interrupts to report events to an application. An embedded OS, usually called real-time OS (RTOS), is lightweight but includes all basic building blocks of traditional OSs including threading, sockets, and contention mechanisms that provide concurrency and real-time functionality [5]. Full-feature OSs, on the other hand, include all the components and modules that belong to commercial-grade OSs distributed into *kernel* and *user space* elements. In many scenarios, highly constrained devices run as bare-metal devices that do not rely on any OS support, and their firmware provides all functionality.

Based on hardware and software capabilities, devices can be simple or complex. A simple device relies on a main-loop or event-based OS running on a battery-powered constrained embedded processor. Examples of simple devices are basic sensors or actuators. Simple device communication is low rate and may or may not rely on IP networking. When a simple device does not natively include an IP interface, Internet connectivity is provided by means of an IoT gateway. Specifically, many simple devices talk to a gateway that converts non-IP into IP traffic. LPWANs are quintessential examples of this scenario. A complex device, on the other hand, relies on an RTOS or on a full-featured OS with fully compliant IP stacks. These stacks provide PAN, LAN, and WAN access that provide direct communication to the Internet. Complex devices, such as gateways and stand-alone sensors, rely on external power lines that enable higher transmission rates. Keep in mind that the topic of software/hardware interaction and capabilities of embedded devices in the context of IoT is quite complex and it falls outside the scope of this book.

► **RTOSs** RTOSs give devices capabilities that are not typically available in bare-metal environments. Specifically, they provide very efficient task concurrency, communication, and scheduling. RTOSs are typically as limited as the devices they run on. A list of some of the most popular ones is shown below.

Contiki	BSD-licensed open-source, lightweight OS.
	Implemented in C and ported to Atmel AVR, MSP430, and ARM M processors.
	Very low memory requirements (order of kilobytes).
Open WSN	Open-source OS created by UC Berkeley.
	Implemented in C.
TinyOS	BSD-licensed open-source OS.
	It runs on devices with 16 KB of memory.
	Applications are written in nesC a version of C that optimizes resources.
RIOT	LGPLv2-licensed open-source OS.
	Implemented in C and C++.
	Minimum memory requirement of 1.5 KB.
Zephyr	Apache 2.0-licensed open-source OS.
	Implemented in C and ported to ARM M, x86, and RISC-V 32 processors.
	High-end support includes POSIX APIs.
FreeRTOS	MIT-licensed open-source OS.
	Implemented in C and ported to 35 microcontrollers and processors.
	Full kernel support.
MBed	Apache 2.0-licensed open-source OS.
	Implemented in C and C++. Ported to ARM M processors.

IoT devices, and more specifically sensors, controllers, and actuators, can be self-configured and self-organized. The idea is that they can be deployed in a network such that once they are powered up, they can be automatically provisioned and configured to become functional right away. Devices, at this point, have all the information that enables them to communicate with gateways and applications. Moreover, certain devices can also self-propel and support mobility that allows them to deploy themselves in unaccessible remote areas while preserving connectivity. In general, a highly desired property of IoT devices is reliability such that they can operate for years without any human interaction.

2.3.1 Sensors

Sensors are logical devices that sense or measure an asset of the physical environment. Examples of assets include not only physical parameters like temperature, humidity, and light intensity but also other measurable quantities like inventory and population sizes. The sensors in each case retrieve temperature, relative humidity, and light intensity as values measured in Centigrade degrees, percentage, and Lux, respectively [24]. Sensors can be classified in accordance to their size; they can be nanoscopic in the order of 1–100 nm, mesoscopic between 100 and 10,000 nm, microscopic in the range between 10 and 1000 μm, and macroscopic when above 1 mm. Moreover, depending on size and logistics, sensors are sometimes deployed as an array of devices specially in the context of WSNs.

Depending on the complexity of the embedded processor, a sensor may perform some local processing in order to remove redundancy in a controlled way. An example of this removal is source encoding where sensor readouts are digitized and compressed. Compression can be lossless or lossy depending upon whether the original samples can be recovered or not. Specifically, through source encoding, data is converted into information that can be transmitted at lower rates reducing the channel bandwidth requirements and improving power consumption. This data-information conversion is part of the transformation shown in Fig. 1.7 [21]. If more complex computational capabilities are available, more efficient processing can be done at the sensor. In special cases, this additional processing is performed under the umbrella of an information-knowledge conversion executed at the sensor. Examples of these more complex technologies include machine learning classification as well as numerical predictions by means of linear regression and other statistical mechanisms.

In general, a sensor is active if it is required to emit sounds or generate electromagnetic waves that can be detected by means of external observation. A sensor that is not active is passive. For example, temperature and humidity sensors as well as cameras are examples of passive sensors since it is impossible to tell whether they are sensing by just looking at them. On the other hand, *radio detection and ranging* (radar), *sound navigation and ranging* (sonar), and *light detection and ranging* (lidar) are examples of active sensors as they emit electromagnetic waves or they generate sounds that can be easily detected by observing them. Because they generate energy, active sensors consume a lot more power, and, therefore, they cannot rely on batteries to function. In opposition, passive sensors are more likely to rely on batteries and perform external energy harvesting by means of, for example, solar panels.

Battery life can be extended by means of *power duty cycles* where devices sleep by dramatically reducing power consumption at preprogrammed intervals. Specifically, while a device sleeps, it minimizes power consumption by only enabling basic functionality including a wake-up interrupt or notification. In order to minimize network throughput and preserve the power consumption of all devices to extend the network lifetime, it is preferable that duty cycles are coordinated throughout the network [6]. This is particularly important when considering capillary networks that rely on multi-hop communication where intermediate sensors act as routers. If a sensor does not know whether transitional devices are sleeping at any given time, it may waste energy to transmit data to an inactive router that is not able to propagate packets to destination.

From a networking perspective, depending on the use of the sensor readouts, transmission reliability is important. If sensor readouts are to be used to make real-time decisions like to change the flight path of a UAV, latency and packet loss must be as low as possible. On the other hand, if those readouts are to be used to perform offline data visualization, latency and packet loss requirements are a lot less restrictive. In general, application-specific *quality of service* QoS goals lead to different application latency and packet loss levels that tell how reliable sensor data transmission must be. The trade-off, shown in Fig. 2.7, between QoS and reliability has power consumption and, therefore, battery life implications.

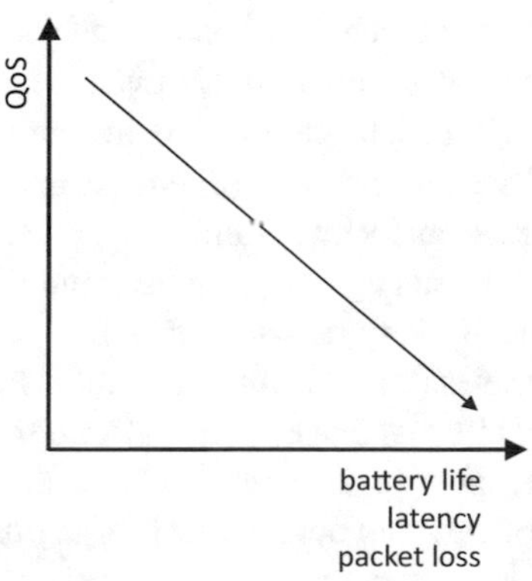

Fig. 2.7 QoS vs battery life, loss, and latency

2.3.2 Actuators and Controllers

Actuators are logical devices that perform some *external* change of an asset of the physical environment. An example of actuation is the activation of a fan to lower the temperature of a room. In this case, the actuator is the fan, and the asset is the temperature. Another example of actuation is the servos that can be used to change the flight path of a UAV. Similarly, the servos are the actuators, and the flight path is the asset. Actuation is typically tied to sensing through feedback mechanisms where decision-making takes sensor and actuation data as input and output parameters, respectively. Because of this, if a given physical device has a logical actuator, it is quite likely that a logical sensor is also present [21]. The opposite, that is, the presence of an actuator given a sensor is present, is a lot less common.

Same way sensors are associated with source encoding and DACs, actuators are associated with source decoding and ADCs. Actuators, however, are a lot simpler as they do not rely on data-information and information-knowledge conversions. Usually the knowledge from the application results in a command being sent down to the actuator. As with sensors, actuators are affected by loss and latency that results in different QoS levels.

Controllers are logical devices that perform some *internal* change in the physical device to assist sensing or actuation [22]. This involves, for example, having a camera zoom in and out, replacing optical filters, or having a transmitter turn antennas around. For most cases, controllers are deployed along with sensors and actuators as logical devices on the same physical device. When assisting sensing, control is affected by the same application QoS requirements that are needed by the sensor.

2.3.3 Gateways

Gateways are logical devices that serve as an interface between access side IoT devices and core side applications. Access side IoT devices are the sensors, actuator, and controllers, while the core side applications rely on analytics to make real-time decisions. When compared to other devices, gateway is a bit more advanced, demanding higher computational complexity that requires more resourceful and powerful embedded processors fed by power lines. This complexity is also needed for the gateway to have enough "horsepower" to simultaneously interact with multiple sensors, actuators, and controllers. This does not prevent, in certain scenarios typically associated with multi-hop communications, simpler devices like sensors and actuators from providing basic gateway functionality. Specifically, sometimes networks can rely on sensors and actuators taking turns in becoming temporary gateways that aggregate and forward packets to uplink appli-

Table 2.1 Device comparison

Device	Complexity	Networking	Form factor
Sensor	Low	WPAN/LPWAN	Small
Actuator	Low	WPAN/LPWAN	Medium
Controller	Low	WPAN/LPWAN	Medium
Gateway	High	WPAN/LPWAN WAN	+Large

cations. Of course, this is contingent on device computational complexity and battery life. Many times, gateways are critical in providing communication between devices, as they route all traffic up and down the network. This is especially true when the gateway acts as clusterheads that forward back and forth all packets in the access side.

In most IoT scenarios, gateways are known to provide interfaces between WPANs and LPWANs on the access side and mainstream WANs on the core side. In a more generic definition, gateways translate messages at different levels of the layered architecture [18]. They can (1) convert physical and link layer frames, for example, when forwarding them between wireline Ethernet and wireless IEEE 802.15.4; they can (2) convert network layer datagrams, for example, when forwarding them between IPv4 and 6LoWPAN/IPv6 layers; they can (3) convert transport layer segments, for example, when forwarding them between *Transport Control Protocol* (TCP) and UDP layers; and they can (4) convert application layer messages, for example, when forwarding them between *Hypertext Transfer Protocol* (HTTP) [1] and CoAP layers. Table 2.1 compares gateways against the other devices regarding computational complexity, networking capabilities, and hardware form factors.

Example 2.2 Consider the two gateways below: *gateway 1* (associated with a IoT scenario) only converts transport protocol 1 (T1) to transport protocol 2 (T2) and application protocol 1 (A1) to application protocol 2 (A2), while *gateway 2* (associated with a hybrid IoT scenario), respectively, converts physical protocol 1 (P1), link layer protocol 1 (L1), network protocol 1 (N1), transport protocol (T1), and application protocol 1 (A1) to physical protocol 2 (P2), link layer protocol 2 (L2), network protocol 2 (N2), transport protocol 2 (T2), and application protocol 2 (A2). Assuming that basic processing delay (i.e. address lookup) takes around 5 ms and more complex protocol conversion processing delay takes around 30 ms, how long does it take for a protocol 1 packet to be processed by both gateways?

(continued)

(continued)

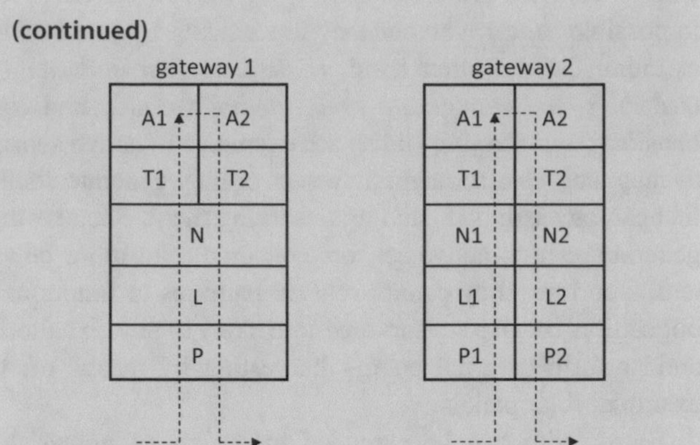

Solution The delay associated with *gateway 1* is given by $d_{\text{gw1}} = 3 \times 5$ ms $+ 2 \times 30$ ms $= 75$ ms. Similarly, the delay associated with *gateway 2* is given by $d_{\text{gw1}} = 5 \times 30$ ms $= 150$ ms. Clearly a hybrid IoT scenario where IP is only supported on the core side of the network requires gateways converting full stacks. Of course, this introduces processing delays that are a lot larger than those of IoT scenarios that support end-to-end IP connectivity.

2.4 Security

The security challenges in IoT networks are multiple [12,19]; (1) it is very important to minimize resource consumption, (2) it is critical to prevent link attacks ranging from passive eavesdropping to active interfering, and (3) wireless communication mechanisms associated with IoT networks render traditional security mechanisms many times unsuitable.

Security requirements are also important; (1) confidentiality implies that information must be encrypted to make sure that it is inaccessible to unauthorized users, (2) authentication dictates that only trusted sources can transmit data, (3) integrity means that the received messages are not altered in transit, (4) freshness guarantees that traffic does not include replayed datagrams, (5) availability ensures connectivity even when subjected to hostile attacks, (6) resilience provides consistent security levels even when devices are compromised, (7) resistance prevents intruders from getting full control of the network, and (8) energy efficiency implies that the network security must minimize power consumption.

Attacks and threats against both user and data in IoT networks are typically very destructive as IoT access networks are usually wireless and easily accessible. In this context, there are several threats that can be analyzed from the layered architecture perspective. The assumption is that an intruder is fully capable of carrying out an attack at any time but during deployment. IoT is special because it is highly susceptible to physical attacks where devices are destroyed or relocated. Because of these attacks, one or multiple devices can be put out of commission causing irreversible losses. During a physical attack, it is possible to extract keys and modify the behavior of the device by uploading a new firmware.

In IoT networks, *denial-of-service* (DoS) attacks are performed at different layers. Physical layer DoS attacks consist of an intruder generating interference that lowers the overall SNR of the modulated waves. Link layer DoS attacks are associated with forcing devices to deplete their energy. Specifically, attackers induce device media collisions that cause devices to retransmit frames that, in turn, reduce their battery life. In addition, frames with fake addresses in different WPANs can cause a device to transmit bogus replies that also drain its battery. This is further aggravated when a device relies on power duty cycles. In this case, the attacker attempts to keep the device busy in a scenario known as sleep deprivation torture. When the device that is being attacked is an IoT gateway, the results can be devastating as routing throughout the network is usually affected. Network layer DoS attacks are associated with an intruder flooding the network with many datagrams that fill queues and cause network performance degradation as well as throughput reduction. Replay protection is a mechanism that many networks put in place to make sure that attackers do not resend old datagrams in order to cause intentional congestion. To support this, datagrams include incrementing sequence numbers that are used by devices to keep track of the age of the datagrams. Devices drop stale datagrams that have sequence numbers that are lower than those of previously received datagrams. Unfortunately, this protection can be used to cause a type of DoS attack called replay protection attack. Essentially, an attacker sends bogus datagrams with high sequence numbers in order to prevent valid datagrams from being processed by a destination device [7].

In certain IoT protocols, for the sake of simplicity, acknowledgment messages associated with reliable transmissions are not protected by security. Attackers can take advantage of this situation by falsely acknowledging a transmitted message while producing enough interference for the original message to be dropped. The idea is to make the source believe that a message has been received even if it has not. In many IoT scenarios, cryptographic keys are periodically negotiated in order to make sure that confidentiality is preserved as time progresses. This is typically triggered by the IoT gateway in response to an unencrypted device request. An attacker can take advantage of this situation by pretending to be a valid device and forcing key renegotiation that leaves the device in-communicated.

Other network layer attacks include (1) routing information spoofing where the intruder spoofs, alters, or replays routing information in order to create loops that lead to datagram loss, (2) selective forwarding where an affected device only forwards certain datagrams in order to cause connectivity problems, (3) sinkhole attack where an affected device attempts to become the destination of all traffic in a certain location (this requires the device to broadcast low-cost routes to neighbors beforehand), (4) Sybil attack where a single device assumes multiple identities in order to become a destination of many other nodes in the network, (5) wormhole attack where the attacker captures datagrams in one location and retransmits them in another leading to routing and connectivity problems, and (6) neighbor discovery attack where neighbor discovery messages typically used in the context of WPANs are either dropped or corrupted in order to induce reachability issues.

2.5 Wireless Sensor Networks

WSNs have been briefly mentioned in previous sections, but because of their historical importance, it is very important to talk about them in the context of IoT [20]. WSN is an old legacy term that designates a system of a very large number of wireless sensors that interact with an application that monitors events and other phenomena occurring in the physical environment. Specifically, as indicated in Sect. 2.3.1, sensors measure assets like temperature and humidity and transmit these values to applications that perform data visualization and, in more advanced cases, knowledge extraction. Like IoT networks, WSNs rely on control for sensors to work efficiently. For example, if a sensor is a camera taking pictures, control by means of the camera zooming in and out leads to better sensing.

When, in addition to sensing and control, actuation is also performed by the devices, the network becomes a *wireless sensor and actuation network* (WSAN). In many cases, the term WS(A)N is used to designate a generic form of a WSN where actuation may or may not be also present. In general, WSANs are associated with decision-making, either automated or manual by means of human intervention, that rely on actuators to complete the feedback path. Note that, in most cases, sensing is always present regardless of whether actuation is in place or not. In general, if a network supports actuation, it is almost certain that it also supports sensing; the opposite is not always true.

WS(A)Ns are predecessors of modern IoT networks and, as such, incorporate all their basic components shown in Fig. 2.8. Clearly there are (1) assets, (2) sensors/actuators/controllers, (3) a communication network, (4) gateway(s), and (5) a central processor that serves as application [23]. There are differences in some of these components though. WS(A)Ns do not typically follow standards, like IP-based protocols, so communications are based on proprietary legacy mechanisms that prevent data

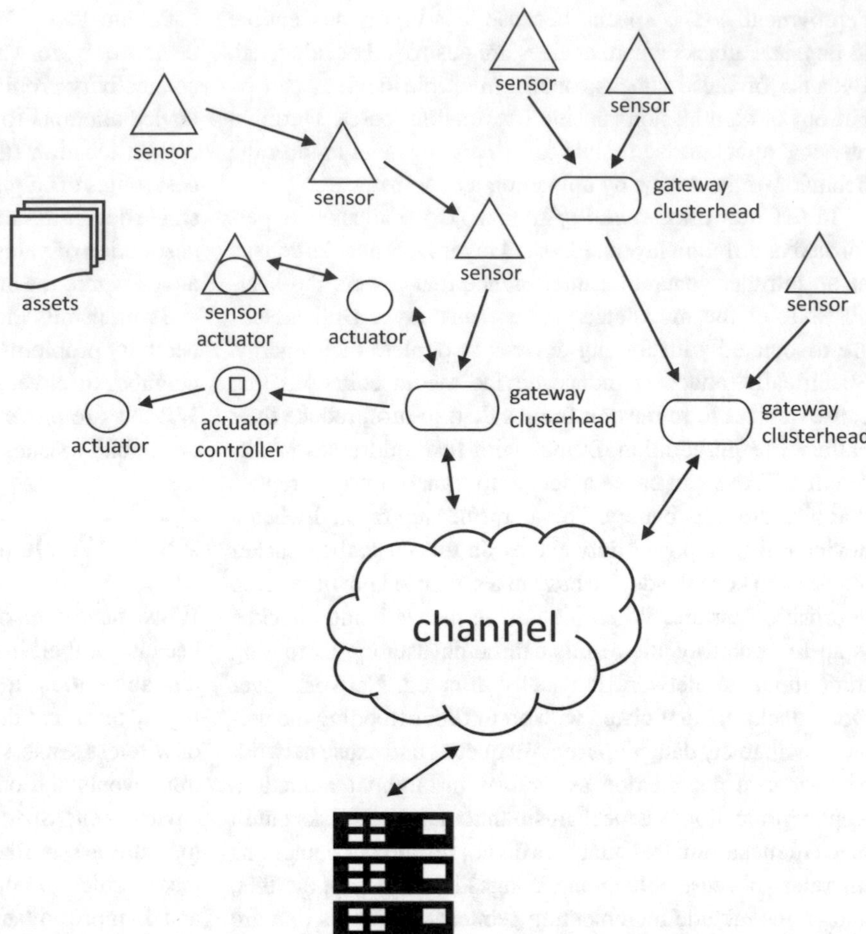

Fig. 2.8 WSAN

sharing and use of information in a wide sense. Sensors and actuators are wireless and only interact with gateways that collect packets and aggregate, transform, and forward them to the central processor.

Moreover, WS(A)N architectures, as illustrated in Fig. 2.8, are pyramidal with multiple devices communicating to multiple gateways that talk to a single application. Communication is therefore M2P and relies on clusters of devices that connect to other clusters by means of gateways that in turn communicate with other gateways in order to access the central processor. In many cases, the gateway functionality is carried out by sensors and actuators that behave as clusterheads. In fact, and in order to preserve battery life, devices typically take turns in becoming clusterheads as gateway communication requirements typically impose power consumption conditions that can lead to energy depletion. WS(A)Ns can also define mechanisms that reduce the overall battery consumption by allowing sensors to forward packets to other intermediate sensors that, in turn, retransmit these packets to reach the central processor. Since power consumption is exponentially affected by signal coverage, by minimizing the distance between transmitters and receivers, it is possible to improve the overall network energy efficiency.

In the context of WS(A)Ns, the central processor does not necessarily make real-time decisions, and in many cases, especially in WSN scenarios, information is stored for offline processing and visualization. This is particularly true for traditional WSNs that were built when analytics mechanisms like machine learning and data mining were not as developed as they are today.

Limited battery life at devices leads to a trade-off between computational complexity and transmission power. If more processing is performed at a sensor or actuator, more energy is needed to execute more complex algorithms. Similarly, if more traffic is to be generated by a device, more energy is needed for a higher transmission rate [11]. Fortunately, more processing results in fewer packets being transmitted and therefore lower transmission rate requirements. Figure 2.9 shows this trade-off; essentially, battery life is presented as a function of both processing and transmission rate. Specifically, battery life decreases as both processing and transmission rate increase, but more processing implies a lower transmission rate, and a higher transmission rate implies less

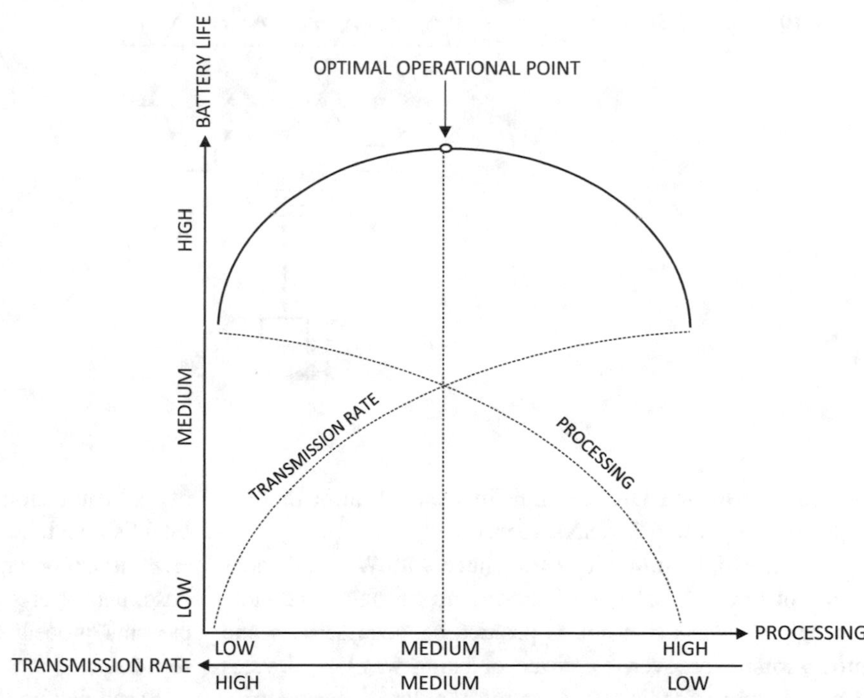

Fig. 2.9 Operational point

processing. When considering their accumulated effect, the optimum operational point that maximizes battery life results from a trade-off between transmission rate and processing. In general, maximizing the battery life of a device leads to finding an operational point that determines computational requirements and protocol stacks.

Most WSN applications produce traffic that is lightweight in nature since it involves the transmission of packets containing simple commands and events. This relates to both communication and processing requirements. On the other hand, some WSN applications rely on sensors that capture video by means of cameras. Video, as opposed to traditional sensor readouts, is highly demanding from transmission rate and computational complexity perspectives. Video codec selection at the device enables a WSN application to decide the operational point that represents the best trade-off between those two parameters. For example, transmitting raw uncompressed video minimizes processing but maximizes throughput, while video compression by means of the state-of-the-art *International Telecommunication Union Telecommunication Standardization Sector* (ITU-T) H.265 codec maximizes processing but minimizes throughput. Other less efficient codecs can be selected to find the best trade-off for a given WSN scenario.

When comparing IoT networks to traditional WS(A)Ns, one major difference is that the latter consist of networks with many sensors (and actuators) highly densely deployed. Specifically, because these devices are wireless and battery-powered, duty cycles are used to preserve battery life. In fact, most sensors affected by power duty cycles are installed in clusters where highly redundant sensors take turns to go to "sleep" in order to guarantee that sensor readouts are always transmitted uplink. Again, this depends on overall network coordination including traffic routing protocols. Summarizing, in the context of WSNs, battery life preservation is usually the result of several interrelated trade-off mechanisms: (1) low transmission rates at a cost of higher processing, (2) power duty cycles to minimize device processing, (3) multi-hop routing to lower transmission power, and (4) low computational complexity at a cost of higher transmission rates [20].

WSNs are classified depending on their size and more importantly on whether they support multi-hop communication. Specifically, category 1 WSNs are those that include a very large and highly dense deployment of devices. Transmission is multi-hop with communication from devices to gateways relying on intermediate sensors and actuators forwarding and aggregating packets. Under category 1 structures, multiple devices talk to other devices that forward traffic to gateways that, in turn, talk to other gateways that end up communicating with a single gateway that provides communication to the central processor. On the other hand, category 2 WSNs are simpler, with fewer devices directly connected to a single gateway that communicates with the central processor. The penalty to pay for the extended coverage provided by category 1 WSNs is lower transmission rates and higher latency when compared to category 2 WSNs. Figure 2.10 shows examples of both category 1 and 2 WSNs. In many solutions, category 1 and 2 WSNs overlap in order to provide different functionalities. As such, each network type relies on different physical and link layer technologies. Leaving other differences aside, the low- and high-throughput networks

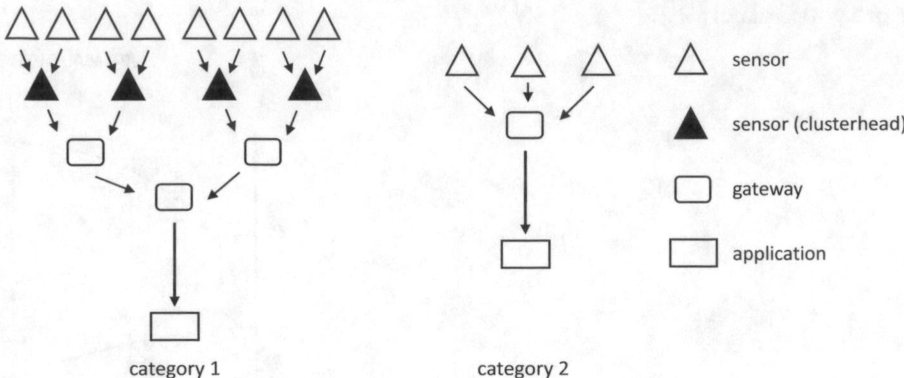

Fig. 2.10 WSN classification

presented in the IoT UAV example in Sect. 1.4 can be thought of as category 1 and 2 WSNs, respectively.

Because of the limitations associated with WSNs, devices may not have global identification and be only accessible through gateways that act as proxies between sensors and processors. Moreover, by virtue of being wireless, devices can also move leading to dynamic topology changes triggered by internal and external position changes. Internal mobility is caused by self-powered changes of the device position by means of control. External mobility results from an application or an operator changing the position of the device.

Mobile ad hoc networks (MANETs) are a well-known type of wireless networks that, from a traffic perspective, are similar in nature to WSNs. MANETs are representative of Wi-Fi (standardized by IEEE 802.11 and described in Sect. 3.3.1) networks configured in ad hoc mode as presented in Sect. 3.3.1. In MANETs, however, packets flow between a few endpoints *point-to-point* (P2P) as opposed to WSNs where transmission is mostly M2P from a large number of devices to a central processor [25]. WSNs are wireless but not always mobile, while MANETs are mobile. Moreover, traffic of WSNs can be highly redundant as it originates from close-by sensors that describe the same physical phenomena. This redundancy, when coordinated, can be exploited to increase the device battery life through the introduction of power cycles. In addition, as opposed to WSNs, MANETS are not deployed for long periods of time without human intervention and almost always do not run on batteries. The many differences between MANETs and WSNs make that routing protocols that would normally work in one case do not work in the other.

Due to the sheer size of WSNs, both network organization and device tracking are very important. Because traditional WSNs do not rely on standardized mechanisms, technologies to accomplish these goals include proprietary group management that provides device self-organization by means of authentication, association, registration, and session establishment. Additionally, and depending on the WSN, asset tracking enables asset detection and classification. Also, part of group management, dynamic device allocation presents procedures to provision, unprovision, and query sensors.

Some factors that are relevant to WSNs are (1) network coverage that indicates the topological coverage of a given WSN, (2) detectability that provides the probability that a given WSN is able to detect a specific asset or physical phenomenon, (3) node coverage that presents the level of device redundancy available to capture data if a sensor failure occurs, (4) node assessment that introduces provisioning capabilities to add additional sensors and actuators to accomplish redundancy goals whenever needed, and (5) control capability that enables orientation and mobility sensor changes that maximize coverage.

Most WSN devices perform in-network processing that relies on several techniques: (1) data aggregation where readouts from different sensors are collected by a sensor closer to the gateway or clusterhead, (2) data fusion where redundancy from readouts of different sensors is removed by another sensor, (3) data analysis where a single sensor converts data into information though signal processing and lightweight machine learning techniques, and (4) grid and hierarchical computing where data analysis is performed by those sensors that have the best context information to do it. Central processors complement in-network processing by taking the pre-preprocessed information and converting it to useful knowledge as mentioned in previous paragraphs.

Like IoT, under WSN, controller side data management is critical. The main aspects of data management include database management that can be used to set, query, and store device configuration and provisioning parameters as well as sensor readouts and actuation information. In some scenarios, high availability is provided by distributing databases over multiple entities including several controllers and devices supporting in-network processing. Data in WSN is multidimensional for both storage and retrieval, and therefore efficient spatial and temporal searching is mandatory.

When designing WSNs, it is important to consider hardware and power constraints that limit physical layer parameters like transmission rate. Moreover, deploying a WSN requires understanding of network topologies and how to deal with fault tolerance, mobility, and scalability. As previously mentioned, WSNs are typically self-configuring systems that must be able to deal with unpredictable states and situations. Other less flexible scenarios that rely on static or semi-static topologies are pre-configured and do not support self-configuration.

Desired characteristics of WSNs are low network and computational latency, power efficiency, and low transmission rates that enable efficient use of the wireless channel. The idea is to minimize the amount of traffic generated by each device in order to create enough room for all sensors to share the medium. In-network processing is a good mechanism to accomplish this goal at a cost of increased energy consumption that results from the trade-off between computational complexity and transmission power [20]. Traditional WSNs due to their proprietary nature have not been very efficient at addressing these issues. Standardization by means of IoT protocols and technologies give WSNs a new way to overcome these limitations.

Summary

Most IoT communication solutions are deployed as a combination of access and core networks where devices interact with applications that extract knowledge in order to make smart decisions. This chapter started by delving into concepts associated with different IoT networking topologies to then focus on the multiple elements of connectivity including security aspects. This led to presenting the roles supported by actuators, sensors, controllers, and gateways in the context of IoT networks. The chapter also investigated the main differences between modern IoT scenarios and traditional WSNs.

Homework Problems and Questions

2.1 Draw the topology of a capillary network with four sensors, three actuators, and a simple gateway that provides connectivity to the Internet. Assume that only one of the sensors and two of the actuators support routing capabilities.

2.2 For the use case in Sect. 1.4:

(a) what mechanism plays the role of controller (if any)?
(b) what mechanism plays the role of IoT gateway (if any)?
(c) what element is the asset?

2.3 In Fig. 2.9, the trade-off between embedded processing and transmission rate is shown. How can the figure be modified to consider throughput instead of transmission rate?

2.4 From a power consumption perspective, what are the differences between passive and active sensors?

2.5 The end-to-end delay of a communication system is defined as $(N - 1 + P)\,\frac{L}{R}$ where N and P are the number of links and transmitted packets, respectively, L is the average packet size, and R is the transmission rate. What is the extra latency due to transforming a high-power single-hop access IoT network into a low-power four-hop one? Assume that in both scenarios the packet size and the transmission rate is the same.

2.6 What are some of the drawbacks of replay protection? How do they relate to power consumption?

2.7 The efficiency of a communication system η is given by $\eta = \frac{R}{C}$ where R is the transmission rate and C is the channel capacity. In an IoT M2P scenario, the access interface of a gateway has an 85% efficiency. What is the efficiency of the core interface if the access and core side channel capacities are 250 Kbps and 1 Gbps, respectively? What are the implications of these results?

2.8 In a WSN scenario, several sensors are to be used to transmit readouts of an asset that is 20 km away. If the coverage of each sensor is 500 m, what category of WSN is the best deployment option? Why?

References

1. Belshe, M., Peon, R., Thomson, M.: Hypertext Transfer Protocol Version 2 (HTTP/2). RFC 7540 (2015). https://doi.org/10.17487/RFC7540. https://rfc-editor.org/rfc/rfc7540.txt
2. Bormann, C., Ersue, M., Keranen, A.: Terminology for Constrained-Node Networks. RFC 7228 (2014). https://rfc-editor.org/rfc/rfc7228.txt
3. Chew, D.: Protocols of the Wireless Internet of Things, pp. 21–45 (2019)
4. Cover, T.M., Thomas, J.A.: Elements of Information Theory (Wiley Series in Telecommunications and Signal Processing). Wiley-Interscience, New York (2006)
5. Elk, K.: Embedded Software for the IoT: The Basics, Best Practices and Technologies, 2nd edn. CreateSpace Independent Publishing Platform, North Charleston, SC (2017)

6. Geng, H.: Internet of Things and Smart Grid Standardization, pp. 495–512 (2017)
7. Geng, H.: Security of IoT Data, pp. 399–406 (2017)
8. Hassan, Q.F.: A Tutorial Introduction to IoT Design and Prototyping with Examples, pp. 153–190 (2018)
9. Haykin, S.: Communication Systems, 5th edn. Wiley, New York (2009)
10. Holler, J., Tsiatsis, V., Mulligan, C., Avesand, S., Karnouskos, S., Boyle, D.: From Machine-to-Machine to the Internet of Things: Introduction to a New Age of Intelligence, 1st edn. Academic, New York (2014)
11. Huang, Y.M., Hsieh, M.Y., Sandnes, F.E.: Wireless Sensor Networks and Applications, pp. 199–219. Springer, Berlin (2008)
12. Jurcut, A.D., Ranaweera, P., Xu, L.: Introduction to IoT Security, pp. 27–64 (2020)
13. Kurose, J.F., Ross, K.W.: Computer Networking: A Top-Down Approach, 6th edn. Pearson, London (2012)
14. Lee, E.A., Seshia, S.A.: Introduction to Embedded Systems: A Cyber-Physical Systems Approach, 2nd edn. The MIT Press, Cambridge (2016)
15. Popovski, P.: Information-Theoretic View on Wireless Channel Capacity, pp. 201–234 (2020)
16. Robertazzi, T., Shi, L.: Networking and Computation: Technology, Modeling and Performance (2020)
17. Sehrawat, D., Gill, N.S.: Smart sensors: analysis of different types of IoT sensors. In: 2019 3rd International Conference on Trends in Electronics and Informatics (ICOEI), pp. 523–528 (2019)
18. Sethi, P., Sarangi, S.R.: Internet of things: architectures, protocols, and applications. J. Electr. Comput. Eng. **2017**, 9324035 (2017). https://doi.org/10.1155/2017/9324035
19. Shrobe, H., Shrier, D.L., Pentland, A.: Data Security and Privacy in the Age of IoT, Chap 13, pp. 379–401 (2018)
20. Sohraby, K., Minoli, D., Znati, T.: Wireless Sensor Networks: Technology, Protocols, and Applications. Wiley, New York (2007)
21. Stanley, M., Lee, J., Spanias, A.: Sensor Analysis for the Internet of Things (2018)
22. Xia, F., Kong, X., Xu, Z.: Cyber-physical control over wireless sensor and actuator networks with packet loss (2010)
23. Yang, S.H.: Wireless Sensor Networks: Principles, Design and Applications. Springer Publishing Company, Incorporated (2013)
24. Yasuura, H.: Introduction, pp. 1–6. Springer International Publishing, Cham (2017)
25. Zheng, J., Jamalipour, A.: Wireless Sensor Networks: A Networking Perspective. Wiley/IEEE Press, Hoboken/Piscataway (2009)

Part II
IoT Protocol Layers

This part of this book, that includes three chapters, explores the different communication layers of standard IoT solutions. Chapter 3 introduces the physical and link layers of wireless and wireline IoT technologies. The chapter analyzes them from a perspective of their most important characteristics like transmission rate and power efficiency. Chapter 4 explores the network and transport layers including adaptation mechanisms that enable IoT physical and link layer protocols to support IPv6. Chapter 5 deals with the application layers by describing the most relevant methods for session management and the transmission of device and application-generated media and data.

3 Physical and Link Layers

3.1 About Physical and Link Layers...

This section provides a brief introduction to some of the fundamentals of physical and link layers. As such, a reader familiar with these mechanisms can skip this section. The link layer provides basic connectivity for adjacent endpoints to interact with each other over a single link. In order to accomplish this goal, the link layer relies on several mechanisms: framing, channel coding, link access, and reliable delivery. Framing is a generic technique that consists of adding fields and special synchronization markers to data propagated down from upper layers. Framing in the context of the link layer implies adding headers and delimiters to network layer datagrams in order to make frames. On the other hand, channel coding is an optional mechanism that embeds FEC information in one of these headers. This information, in turn, can be used by the receiver to determine whether a given frame has been corrupted by channel noise, attenuation, and other interference. Traditional methods of FEC rely on block and convolutional codes that map frame fields into codewords that enable error detection and, sometimes, correction. These two functionalities, framing and channel coding, are part of the *Link Layer Control* (LLC) sublayer [22].

Link access is provided by a set of rules, known as *media access control* (MAC), that determine how frames are received and transmitted over the physical channel. If multiple endpoints share the same channel, their transmission is restricted to the contention that exists when trying to send frames simultaneously. MAC, in this case, provides the mechanisms that describe how the channel is shared and how it becomes accessible in an efficient way. As part of this sublayer, MAC addresses are needed to indicate source and destination endpoints and sometimes intermediate waypoints. Generally, MAC must guarantee endpoints fair access to the channel based on priorities and QoS goals. For example, if a single device transmits a very long frame, it prevents other devices from also sending traffic causing the overall system latency to increase. MAC is, therefore, responsible for balancing access latency to benefit all users in accordance with network requirements. Link access does not guarantee that frames are not corrupted by the channel when they are transmitted. If channel coding is in place, receivers can detect corrupted frames and avoid their processing. Since the corrupted frames are dropped from the transmission path, transmitters can resend them to guarantee reliability. Specifically, reliable delivery is an optional mechanism that provides the infrastructure to signal and support these retransmissions. It typically involves some feedback technique by which receivers indicate whether a given frame has been received (i.e. by means of an acknowledgment) or not (i.e. by means of a negative acknowledgment). Moreover, it sometimes includes timers that are reset on successful reception of acknowledgments. When these timers expire, retransmissions and connectivity loss events are triggered. In a similar way to framing and channel coding that are grouped together in the LLC sublayer, link access and reliable delivery are part of the greater MAC sublayer. All these four sublayers are shown in Fig. 3.1.

The channel capacity theorem provides the maximum transmission rate as a function of the two main characteristics of a communication system, namely, channel bandwidth and SNR at the receiver for a channel subjected to *additive white Gaussian noise* (AWGN) [19]. Note that this is a simple model that does not consider, among many things, the fading characteristics of wireless channels. However, the model is good enough to understand some of the effects of system parameters in its performance. In general, the larger the bandwidth and SNR are, the higher the allowable transmission rate is. Similarly, the larger the SNR is, the lower the *bit error rate* (BER) is. BER is defined as the probability that a received frame bit has been corrupted by channel noise, attenuation, and interference. Note that SNR at the receiver is given by the ratio between the transmitted signal power and the channel noise power. Because both channel noise and bandwidth are typically fixed, the only element that can be adjusted is the

R. Herrero, *Fundamentals of IoT Communication Technologies*, Textbooks in Telecommunication Engineering, https://doi.org/10.1007/978-3-030-70080-5_3

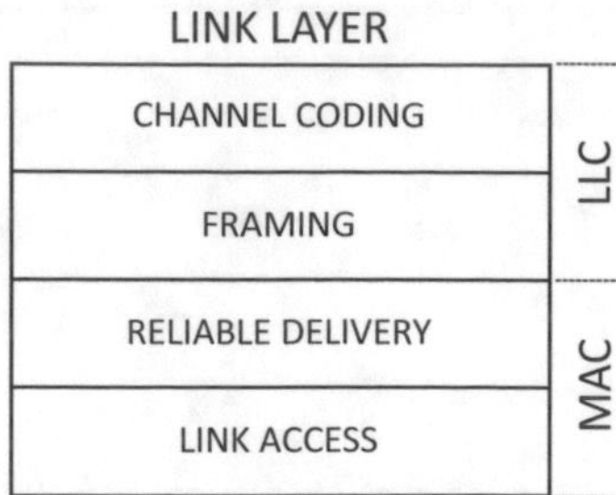

Fig. 3.1 Link layer

transmission power at the transmitter. If transmission power increases, then transmission rate also increases, while BER lowers as SNR gets larger.

Unfortunately, in the context of a wireless scheme, increasing transmission power leads to depleting batteries a lot faster. Essentially, in order to extend battery life, power consumption must be minimized. This idea is illustrated in Fig. 3.2. For fixed channel bandwidth and noise, long battery life leads to low power consumption that translates into low SNR at the receiver. Low SNR, in turn, implies both low channel capacity and high BER. Low channel capacity limits the transmission rate, and high BER causes high packet loss. High packet loss and low transmission rates are representative of LLNs discussed in Sect. 2.2. Note that packet loss can be minimized by means of reliable delivery and channel coding techniques that enable the retransmission of missing and corrupted packets at a cost of a larger transmission rate. Again, this is not always possible as channel capacity typically limits transmission rates.

Low transmission rates associated with LLNs greatly affect how MAC works. Since the MAC sublayer must guarantee fair access to the channel, a single user cannot transmit for too long as it prevents the other users from accessing the media. In this case, transmitting traffic as short bursts at low transmission rates translates into limiting the frame size to a maximum allowable size known as the *maximum transmission unit* (MTU) size.

The physical layer is responsible for processing the link layer frame by subjecting it to both transmission and propagation. It is propagation that defines the characteristic of the transmission, and it is the channel that defines the characteristics of the propagation. As indicated in Chap. 1, depending on the channel characteristics, the physical layer can support guided propagation or free propagation. Guided propagation occurs in man-made guiding channels like waveguides, coaxial cables, twisted pair cables, and optical fibers that are examples of traditional wireline communications. On the other hand, free propagation occurs between corresponding antennas in the Earth's atmosphere and underwater. Free propagation is, of course, the basis of wireless broadcasting.

FEC is the generic mechanism under which, through channel encoding, error control codes are propagated from the transmitter to the receiver. Since the channel encoder adds controlled redundancy in the form of additional bits in the frames sent from the transmitter to the receiver, the transmission rate is slightly increased. On the receiver, the channel decoder removes the controlled redundancy and restores the error controlled source frame. To accommodate the transmission rate increase, the modulation scheme is typically modified to increase either the bandwidth or the number of bits per symbol. Because channel bandwidth is a limiting factor and a modulated wave with higher bandwidth cannot be always used, changing the modulation scheme is usually the most common option. When block codes are used for error control, a payload is partitioned into chunks of data called messages, and each message, when controlled redundancy is added, becomes a codeword. A block code is identified by the tuple (n, k) that, as indicated in Fig. 3.3, implies that out of n bits, k bits are message bits and $n-k$ are controlled redundancy bits. Note that n is the block length, while k is the message length [17]. For every k message bits entering the block channel encoder, n code bits are generated leading to a code rate $r = \frac{k}{n}$ with $0 \leq r \leq 1$. An alternative to block codes are convolutional codes that operate on continuous streams of bits instead of on blocks of bits.

Transmission in wireline scenarios is performed by means of line codes [16]. Line coding is used to modulate digital bits of a link layer frame into an electrical signal that can be transmitted over the channel. Different mechanisms exist to fulfill this objective, but they all attempt to accomplish two goals that affect the transmitted signal; (1) minimize its DC component, and (2) maximize its transitions. Minimizing the DC component increases the propagation range of a signal and the distance between repeaters. On the other hand, maximizing signal transitions enables the receiver to better synchronize with the transmitter.

Figure 3.4 shows the modulation of the codeword *101010* by means of *Unipolar Nonreturn-to-Zero* (NZR) where bit *1* is modulated by transmitting a square pulse of positive amplitude for the duration of the bit T_b, while bit *0* is modulated by not transmitting any signal during that period. Because the DC component of any waveform modulated by this mechanism is always positive, NZR is not a good option for modulation over wireline channels. Also, if an all ones or an all zeros sequence is transmitted, the resulting modulated wave exhibits no transitions, and therefore no synchronization information can be extracted.

Figure 3.5 shows the modulation of the codeword *101010* by means of *Polar NZR* where bits *1*' and *0* are respectively represented by transmitting a positive or negative square pulse for the duration of the bit T_b. Unless the incoming sequence has an even distribution of zeros and ones, the

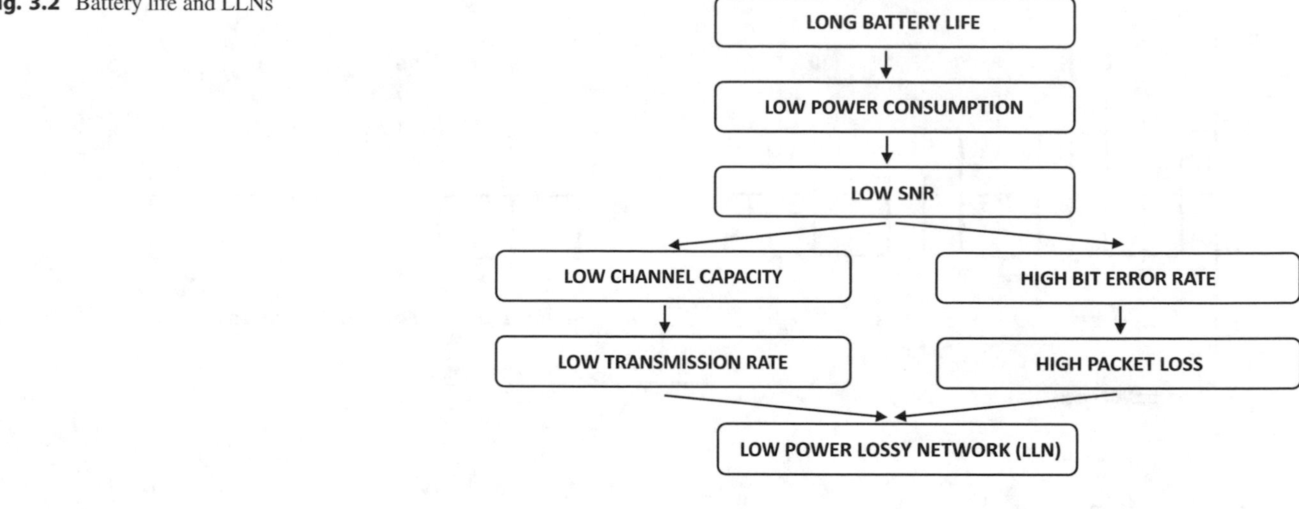

Fig. 3.2 Battery life and LLNs

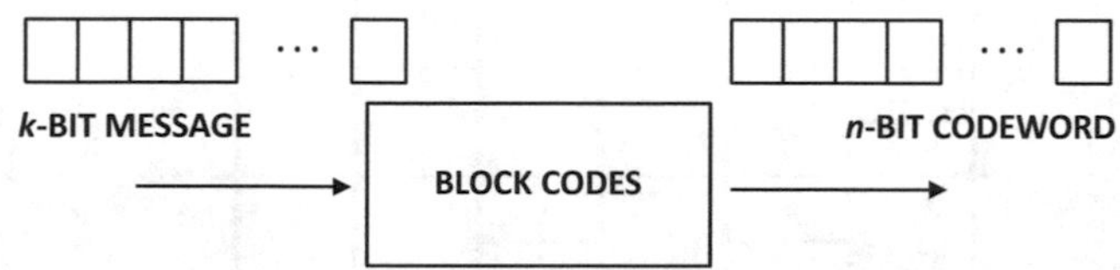

Fig. 3.3 Block codes

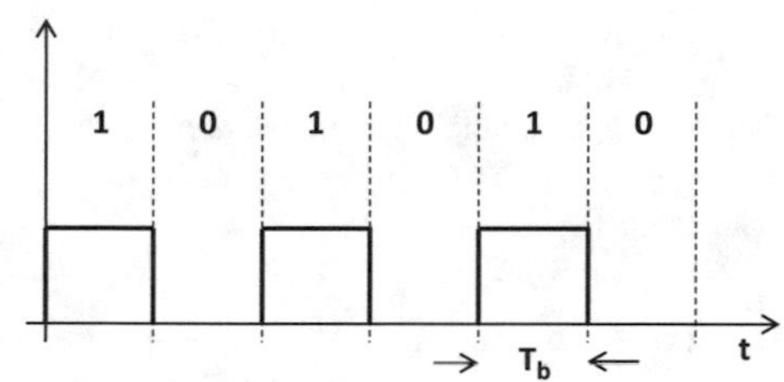

Fig. 3.4 Unipolar NRZ

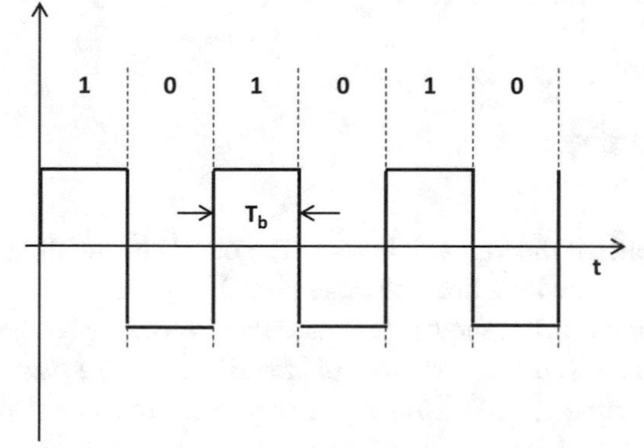

Fig. 3.5 Polar NRZ

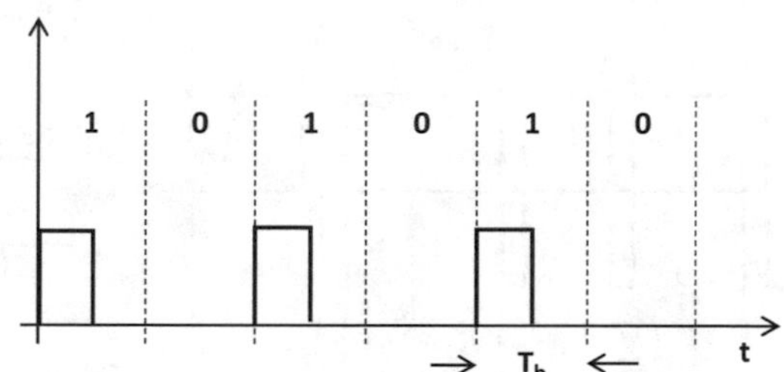

Fig. 3.6 Unipolar RZ

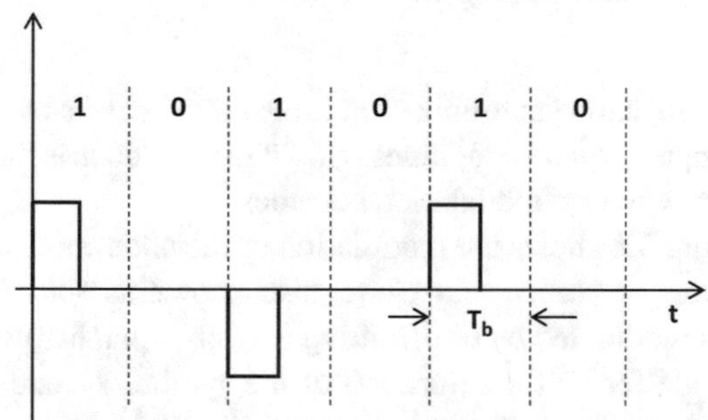

Fig. 3.7 Bipolar RZ (AMI)

modulated wave exhibits a positive DC component, and no synchronization information can be extracted deeming this modulation impractical in most wireline channels.

Figure 3.6 shows the modulation of the codeword *101010* by means of *Unipolar Return-to-Zero* (RZ). Bit "1" is modulated by transmitting a square pulse of positive amplitude for half the duration of the bit T_b, and bit "0" is modulated by not transmitting any signal during that period. Because the DC component of any waveform modulated by this mechanism is always positive, it is not a good alternative for modulation over wireline channels. However, as opposed to unipolar NRZ, clock information can be extracted if an all-one sequence is received.

Figure 3.7 shows the modulation of the codeword *101010* by means of *Bipolar RZ*, also known as *Alternate Mark Inversion* (AMI). Under this scheme bit *1* is modulated by transmitting a square pulse of alternate positive and negative amplitudes for half the duration of the bit T_b, while bit *0* is modulated by not transmitting any signal during that period. The DC component of a bipolar RZ modulated wave is al-

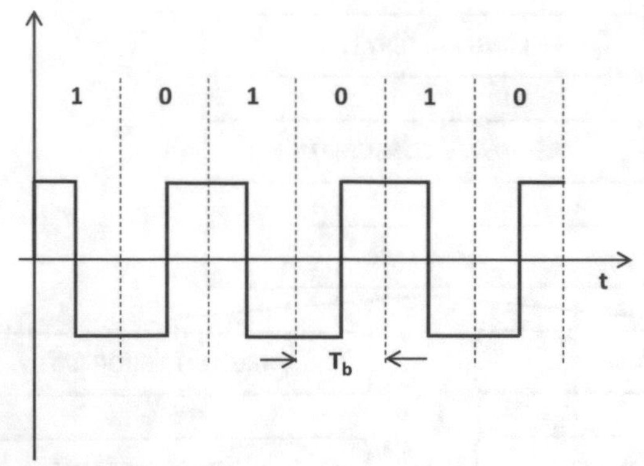

Fig. 3.8 Manchester code

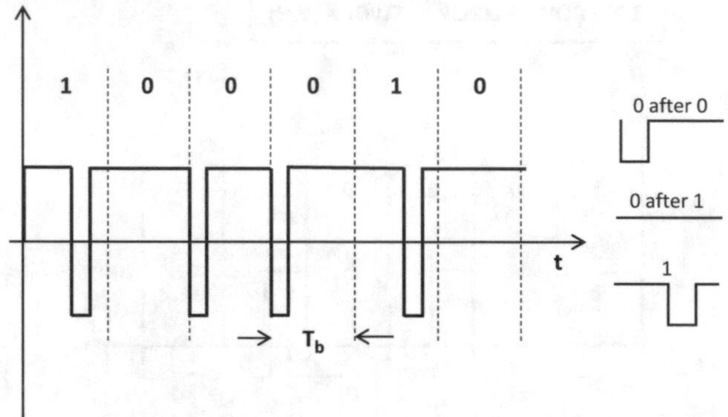

Fig. 3.9 Modified Miller code

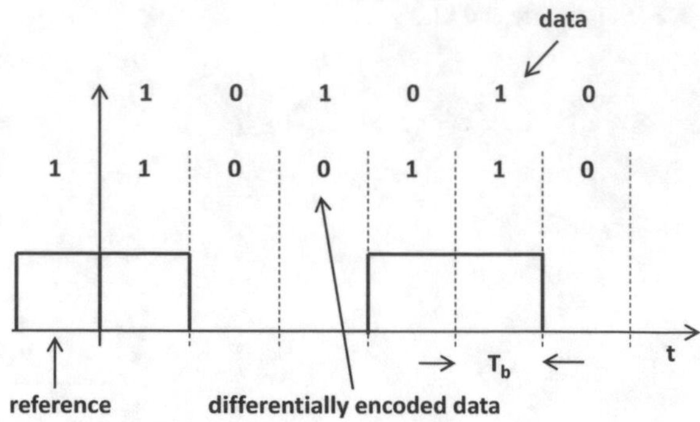

Fig. 3.10 Differential encoding

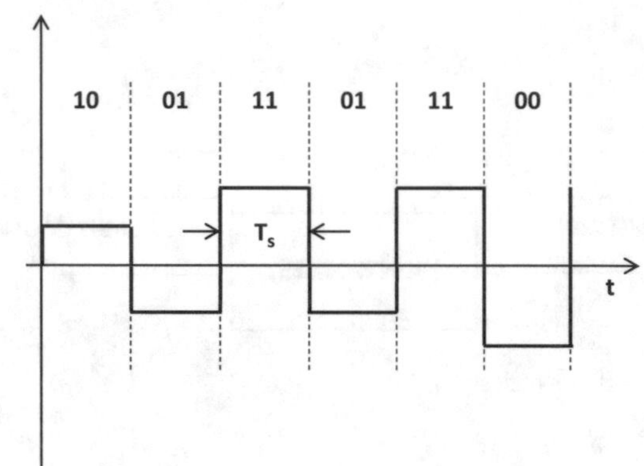

Fig. 3.11 Multilevel pulse-amplitude modulation

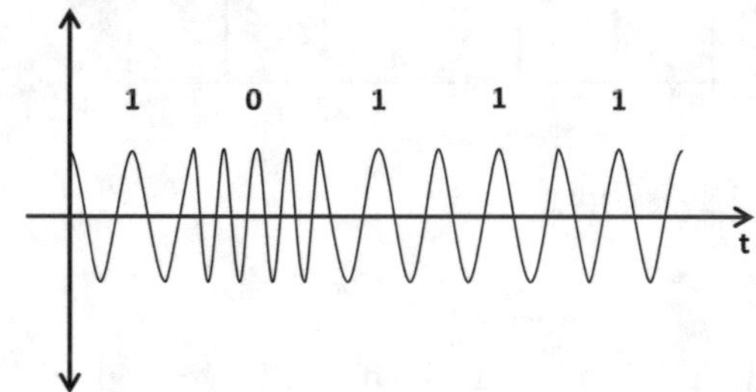

Fig. 3.12 FSK

ways zero, and synchronization information can be extracted if the input sequence includes ones. This line code is the base of many wireline modulation schemes.

Figure 3.8 shows the modulation of the codeword *101010* by means of *Manchester Code*, also known as *Split Phase*. Bit *1* is modulated by transmitting a symbol that has positive amplitude for half the duration of the symbol T_b, and it has negative amplitude for the other half. Similarly, bit *0* is modulated by transmitting a symbol that has negative amplitude for half the duration of the symbol T_b, and it has positive amplitude for the other half. The modulated wave does not have a DC component because the symbols themselves do not have DC components. Moreover, because each symbol transitions between negative and positive levels, synchronization information can be extracted from the modulated wave.

Figure 3.9 shows the modulation of the codeword *100010* by means of *modified Miller code* (MMC) where bit *1* is modulated by transmitting pulses and bit *0* is modulated by transmitting a pulse (or not) depending on the previous symbol. This mechanism attempts to introduce signal transitions even when multiple zeros are transmitted sequentially.

Figure 3.10 shows the modulation of the codeword *101010* by means of *differential encoding* where given a reference bit, the output bit is identical to this reference if the input bit is one and the output bit is the opposite of the reference if the input bit is zero. The output bitstream is then modulated with a line code like unipolar NRZ.

Figure 3.11 shows the modulation of the codeword *100111011100* by means of multilevel *pulse-amplitude modulation* (PAM). Under this scheme, groups of two bits are mapped into different levels of the square signal with pulses transmitted every T_s seconds. In general, if n bits are mapped into $M = \log_2(n)$ levels, the scheme is known as PAM-M. The scheme shown in Fig. 3.11 is, therefore, known as PAM-4.

Line codes are typically suitable for wireline communications, but they get filtered out when transmitted over wireless channels like radio and satellite links that are passband. Because continuous wave sinusoidal carriers are compatible with passband channels, they are the preferred mechanism

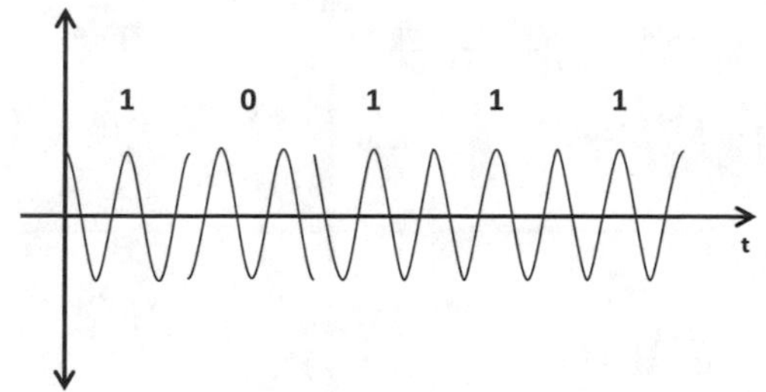

Fig. 3.13 PSK

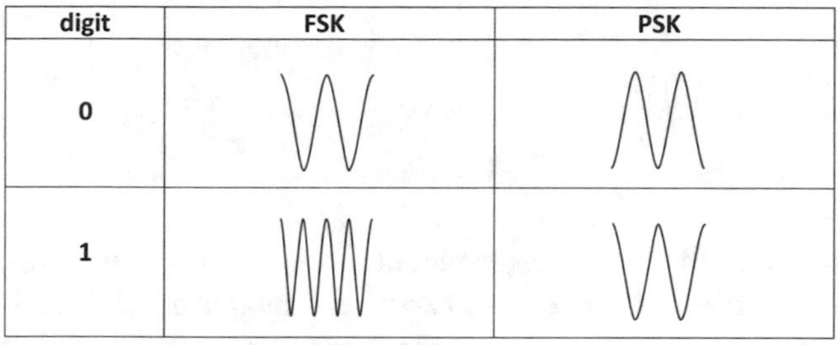

digit	FSK	PSK
0		
1		

Fig. 3.14 FSK and PSK digital-to-analog mapping

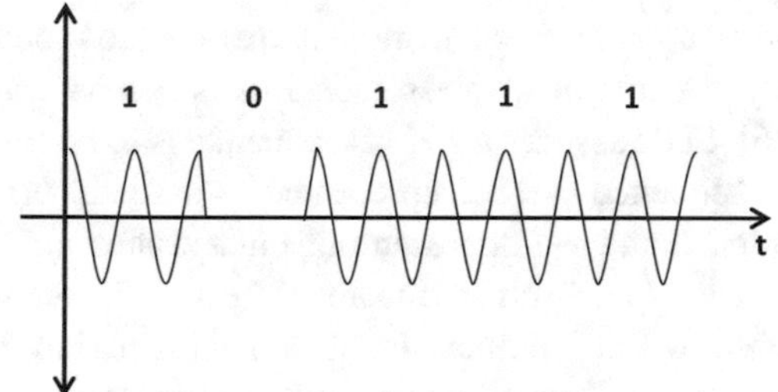

Fig. 3.15 ASK

for digital transmission in wireless environments [12]. The characteristics of the passband channel, that is, the lowest and highest frequencies, determine the sinusoidal carrier frequency to use in a digital passband modulation scheme. Depending on what carrier parameter is used to carry digital information, modulation can be *frequency-shift keying* (FSK) or *phase-shift keying* (PSK). FSK, shown in Fig. 3.12 and PSK, shown in Fig. 3.13, rely on changing the carrier frequency and phase, respectively. Both FSK and PSK are the digital equivalents of analog *frequency modulation* (FM) and *phase modulation* (PM) modulations and therefore exhibit constant envelopes that increase reliability against impairments in wireless channels.

Essentially FSK and PSK provide a one-to-one mapping between binary digits zero and one and the sinusoidal waveforms illustrated in Fig. 3.14. Another scheme, not as popular as FSK and PSK, that is analogous to analog *amplitude modulation* (AM) is *amplitude-shift keying* (ASK). ASK, shown in Fig. 3.15, relies on changing the amplitude of a sinusoidal carrier based on the input bitstream. Specifically, the amplitude of the signal is either maximum or zero,

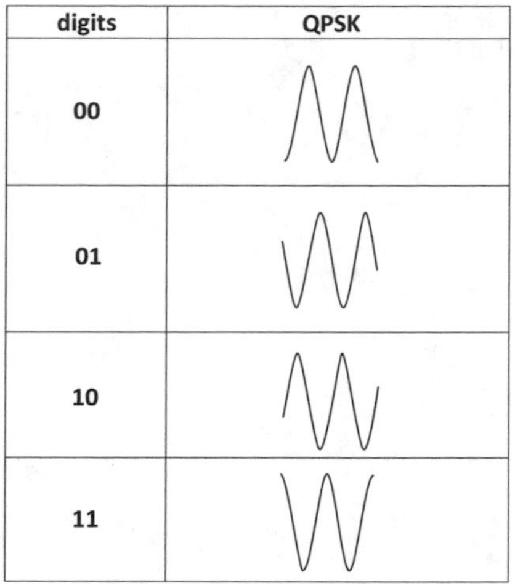

digits	QPSK
00	
01	
10	
11	

Fig. 3.16 QPSK digital-to-analog mapping

respectively, depending on whether a digital one or zero is transmitted.

In the case of PSK, the phase separation between each sinusoidal is 180°. This PSK scheme with 2 signals is also called *binary PSK* (BPSK) or 2-PSK. This idea can be further extended by grouping multiple digits into sinusoidal waves with different phase values. When mapping 2-bit digits into these signals, the phase separation is 90° instead. This scheme, shown in Fig. 3.16, is called *Quadrature PSK* (QPSK) or 4-PSK. One issue with QPSK is when transitioning from *00* to *11*. This situation introduces a phase shift of 180° that, as it can be seen, is responsible for bandwidth widening and additional noise. To prevent this, QPSK is modified as *Offset QPSK* (OQPSK) where 180° phase transitions are done in two 90°-transition stages. In addition, PSK can be further modified to rely on relative phase differences as opposed to absolute phase values. The idea is to improve the symbol detection capabilities at the receiver. This leads to a modulation scheme called *Differential PSK* (DPSK). Examples of DPSK are *DBPSK* when the PSK signal is binary.

If, besides the phase, the carrier amplitude is also modified, the modulation scheme is called *quadrature amplitude modulation* (QAM). When eight signals are used to map 3-bit digits, illustrated in Fig. 3.17, the modulation is called 8-QAM. In general, if M is the number of symbols, then $n = \log_2(M)$ is the number of bits per symbol. Note that this scenario leads to M-QAM, M PSK, and M-FSK modulation schemes. Because the receiver must detect the different symbols to perform inverse mapping, for fixed symbol energy, a larger number of symbols implies a higher likelihood that the receiver will fail to correctly demodulate them due to channel noise. Similarly, a larger number of symbols translates into a higher transmission rate as each symbol is transmitted on a fixed time interval. In general, as part of a trade-off known

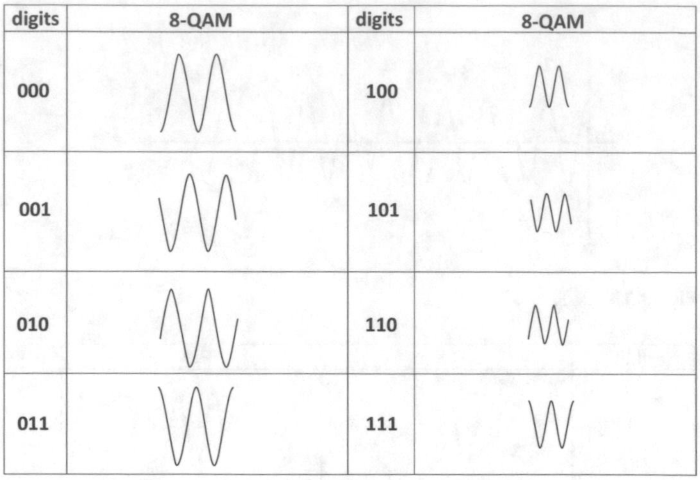

digits	8-QAM	digits	8-QAM
000		100	
001		101	
010		110	
011		111	

Fig. 3.17 8-QAM digital-to-analog mapping

as *rate-distortion*, higher transmission rates correspond to higher distortion.

Example 3.1 If a wireless sensor relies on a 8-PSK modulation scheme and the symbol duration is $T = 25\,\mu\text{s}$, what are the symbol (R_s) and transmission (R_b) rates?

Solution Symbol rate is $R_s = \frac{1}{T} = 40{,}000$ symbols per second. Transmission rate is $R_b = \log_2(8) \times R_s =$ 120 Kbps.

All passband modulations presented so far rely on a single carrier acting at any time. It is natural to think that this scheme can be expanded to simultaneously rely on multiple carriers each one using a particular modulation. The idea is to divide the available channel bandwidth W into a number of subchannels of equal bandwidth Δf such that, as illustrated in Fig. 3.18, around $K = W/\Delta f$ subchannels are available for modulation. When the carrier waves are orthogonal, that is

$$\int_0^T S_i(t)S_j(t)dt = 0 \qquad (1)$$

where S_n is one M of the symbols associated with the modulation, T is the symbol length and $i \neq j$, and the scheme is called *Orthogonal Frequency Division Multiplexing* (OFDM). Orthogonality improves resilience against packet loss caused by channel noise and multipath phenomenon. If $T = 1/\Delta f$, where Δf is the frequency difference between symbol sinusoidal waves, the carrier waves are orthogonal for any possible value of the carrier phase. Because orthogonality restricts the period of each symbol, it also imposes a new limitation of the transmission rate in each subchannel. In general, depending on SNR subchannel characteristics, different modulation schemes can be applied to each of them. For example, in a simplified scenario with two carriers ($K = 2$), one carrier can modulate QPSK and another 64-QAM such that the overall transmission rate is given by the aggregate of the rates over each subchannel, namely, $R = \frac{2+6}{T} = \frac{8}{T}$ bps.

Another popular mechanism that enables the modulation of binary streams in wireless scenarios is *spread spectrum* (SS) [26]. Let's assume a *101* bit sequence is to be transmitted, that sequence can be represented as a signal $b(t)$ with amplitudes 1 and -1 representing binary ones and zeros, respectively. If each bit is transmitted every T_b seconds, it is possible to find another signal $c(t)$ with period T_b that represents a pseudo-random sequence, known as code, that includes a number of binary digits, known as chips, transmitted every T_c seconds.

Example 3.2 An OFDM scheme consists of 16 QPSK subchannels with symbol duration $T = 3.2\,\mu\text{s}$, what is the transmission rate (R_b) ?

Solution Symbol rate is $R_s = \frac{1}{T} = 312{,}500$ symbols per second. Transmission rate is $R_b = 16 \times \log_2(4) \times R_s = 10$ Mbps.

Figure 3.19 shows both $b(t)$ and $c(t)$, where $c(t)$ is represented by the pseudo-random 15-chip *100011110101100* code. Of course, in this case, the code length is therefore 15 chips.

Under *direct-sequence SS* (DSSS), the sequence to be transmitted $b(t)$ is multiplied by the code $c(t)$ to produce the spread spectrum signal $m(t)$ shown in Fig. 3.20. When the destination endpoint receives $m(t)$, it can recover $b(t)$ by multiplying it by code $c(t)$ because $m(t) \times c(t) = b(t) \times c(t) \times c(t) = b(t)$ since $c(t)^2 = 1$. The requirement is for both, transmitter and receiver, to use the same code signal

Fig. 3.18 Discrete multicarrier modulation

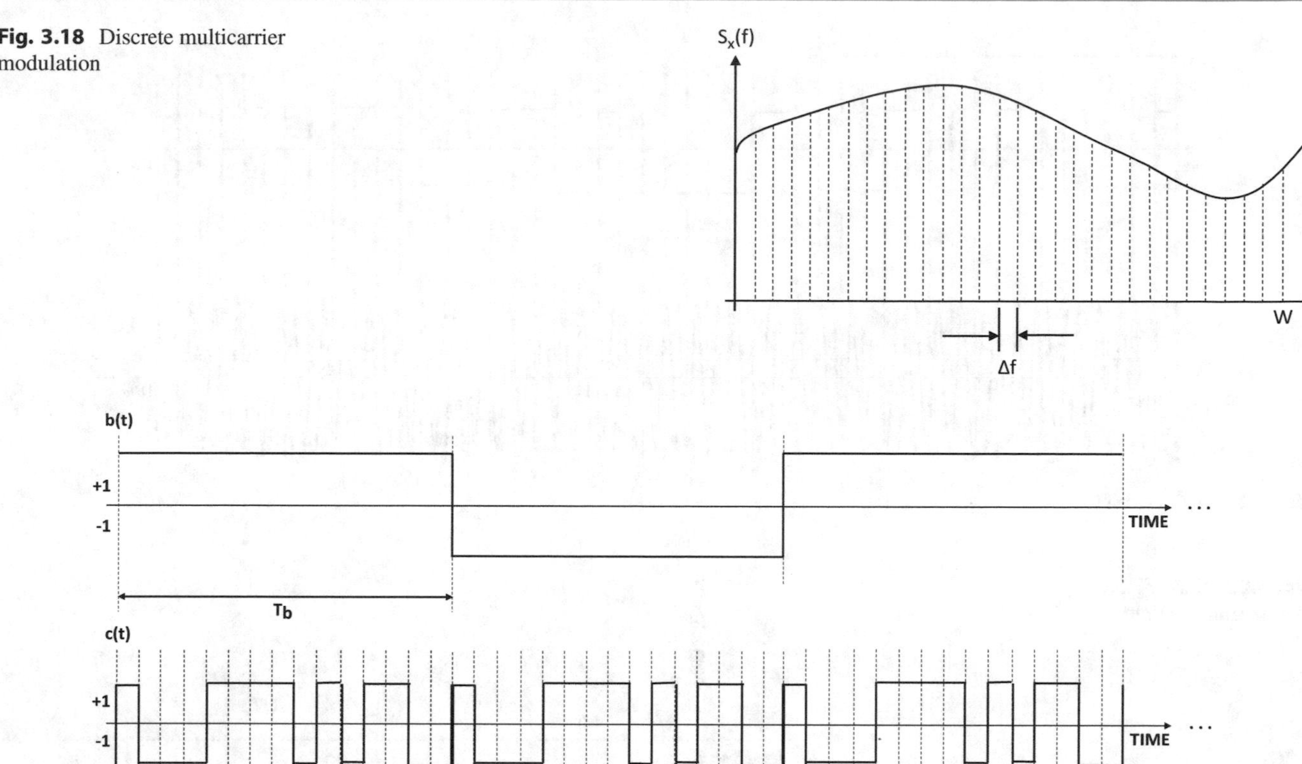

Fig. 3.19 Modulating wave $b(t)$ and code signal $c(t)$

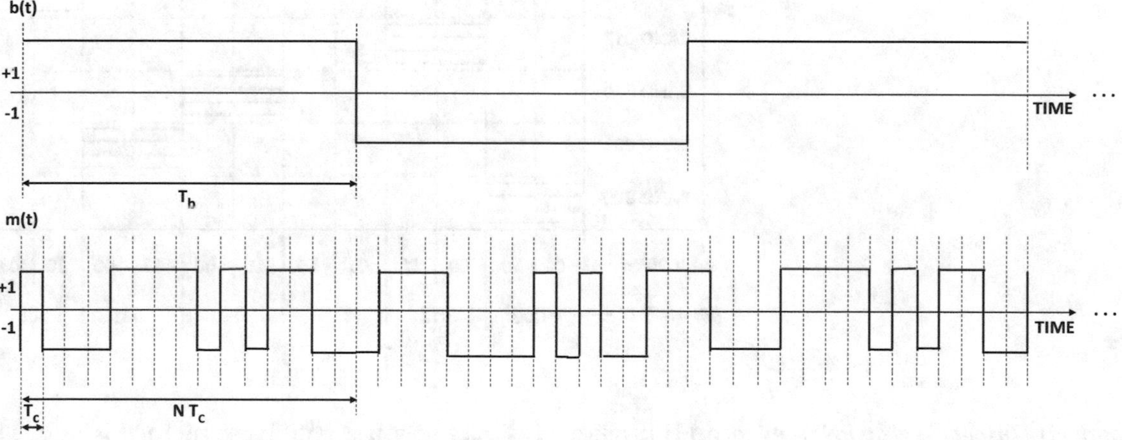

Fig. 3.20 $b(t)$ vs. $m(t)$

$c(t)$. In general, the longer the code, the more the resilience against channel noise and interference. Because each code identifies each communication path, multiple streams can share the same wireless channel. By relying on pulses, $m(t)$ is a line code that cannot be wirelessly transmitted. To provide wireless support, a DSSS signal is modulated by, for example, BPSK.

Figure 3.21 shows the $m(t)$ and the corresponding DSSS/BPSK modulated signal $x(t)$. The spectrum representation of $m(t)$ exhibits a bandwidth that is much larger than that of $b(t)$, thus the name spread spectrum. Under DSSS the ratio between the symbol width and the chip width is known as the spreading factor (or processing gain) SF $= \frac{T_b}{T_c}$. In general, the higher the spreading factor, the more resilience against channel noise and interference.

Another type of spread spectrum is called *Frequency Hopping SS* (FHSS) that relies on sequentially spreading the message signal spectrum. Based on the code generated from a

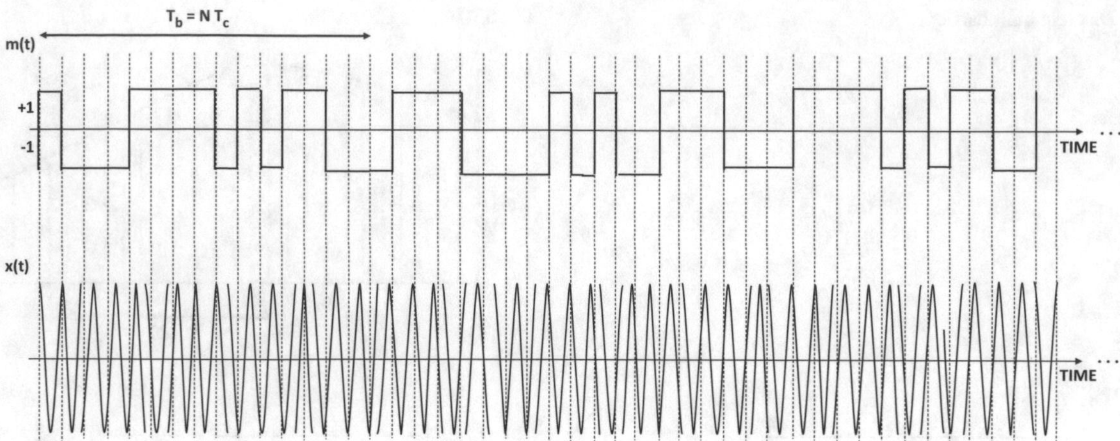

Fig. 3.21 $b(t)$ vs. $x(t)$

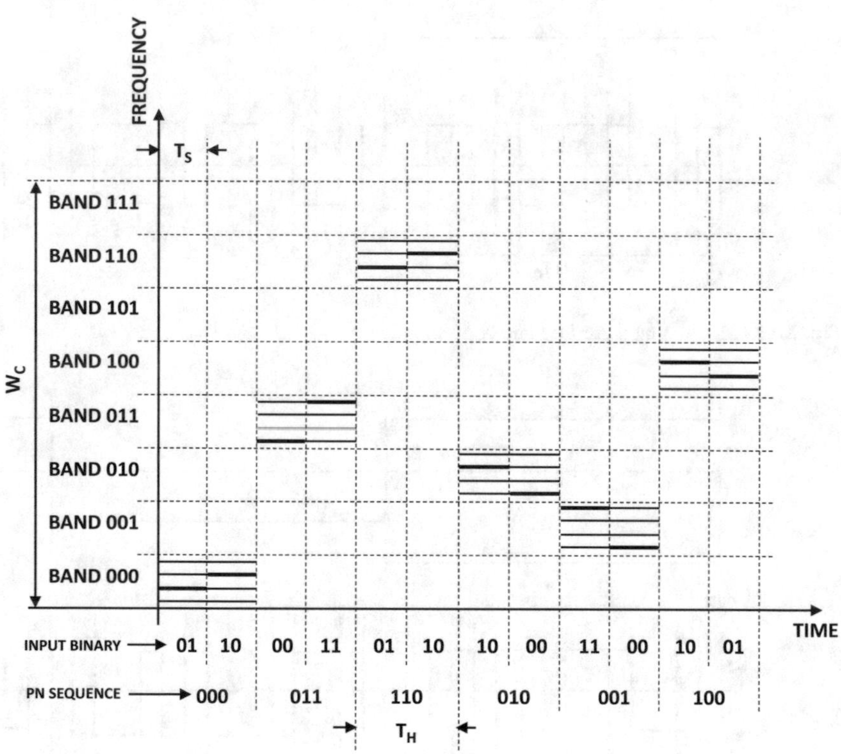

Fig. 3.22 Slow FHSS modulation example

pseudo-random or *pseudo-noise* (PN) sequence, FHSS relies on translating the narrowband modulating wave across different predefined subbands associated with carrier locations in the frequency domain. The binary stream is first modulated by means of M-FSK and then transmitted through carrier modulation in a specific frequency subband determined by the pseudo-random code. Because M-FSK is involved, this type of modulation is also known as FHSS/MFSK. Note that under FHSS, the spreading factor is given by the ratio between spread signal and the narrowband transmitted wave (i.e. SF $= \frac{W_c}{W_s}$).

It is possible to define to two rates, the symbol rate $R = \frac{Rb}{n} = \frac{1}{nT_b}$ where $n = \log_2 M$ and the hopping rate R_h that indicates how fast a different subband is used. Depending on how fast frequency hopping is carried out, there are two possibilities; if in a specific subband multiple symbols are transmitted, that is, $R > R_h$, the modulation is called *slow FHSS*, and, otherwise, if a single symbol is transmitted over multiple subbands, that is, $R \leq R_h$, the modulation is called *fast FHSS*.

Figure 3.22 shows tones generated when the binary sequence *0110001101101000 11001001* is modulated with slow FHSS. Specifically, $k = 3$ bits from a pseudo-random code *000011110010001100* are taken to generate up to $2^k = 8$ possible carrier frequencies associated with eight subbands. Within each subband, $n = 2$ bits of the input binary sequence are modulated with M-FSK where $M = 2^n = 4$. The total number of possible modulated FHSS tones is, therefore, $2^k M = 8 \times 4 = 32$. Since it is a slow FHSS example, multiple symbols or tones are transmitted in any single subband. For example, the first two binary symbols *01* and *10* are modulated over the same subband associated with the PN sequence *000*. When dehopped, the modulated wave exhibits just four tones that can be demodulated by means of conventional M-FSK mechanisms.

Figure 3.23 shows tones generated when the binary sequence *01100011* is modulated with fast FHSS. As in the slow FHSS case, $k = 3$ bits from the same pseudo-random sequence *000011110010001100* are taken to generate up to $2^k = 8$ possible carrier frequencies associated with eight subbands. Within each subband, $n = 2$ bits of the input binary sequence are modulated with M-FSK where $M = 2^n = 4$. The total number of possible modulated FHSS tones is, therefore, $2^k M = 8 \times 4 = 32$. Since it is a fast FHSS example, a single symbol is transmitted over two subbands. For example, the first binary symbol *01* is modulated over two subbands associated with the PN sequences *000* and *011*. When dehopped and, as for the slow FHSS case, the modulated wave exhibits just four tones that can be demodulated by means of conventional M-FSK mechanisms. Note that spread spectrum techniques are typically associated with *Code Division Multiple Access* (CDMA) where communication is only possible if devices and applications know the associated code.

3.2 Wireline

In the context of IoT, wireline communications are provided, among others, by three main technologies: Ethernet, PLC, and MS/TP. Well-known Ethernet provides high transmission rates and low latency but requires dedicated infrastructure either through an electrical twisted pair or through fiber optics. Ethernet is a general-purpose networking physical and link layer protocol that can be used in IoT networks. On the other hand, because the main goal is to support communication technologies that lower deployment costs and times, PLC overlaps modulating waves over pre-existent electrical wiring infrastructure. The price to pay for minimal deployment and operation costs is low transmission rates as well as high latency and packet loss. There are two main PLC technologies: (1) ITU-T G.9903 standardized from a well-known industrial technology known as G3-PLC and (2) IEEE 1901.2. MS/TP is a wireline medium access control method for the RS-485 physical layer that is primarily used in building automation networks like *BACnet*. RS-485, also known as TIA/EIA-485, defines the electrical characteristics of drivers and devices for use in certain architectures that provide serial communication.

3.2.1 Ethernet

Ethernet was originally designed to provide link-layer connectivity through packet switching. It converts upper network layer datagrams into frames that are transmitted over wireline links. Ethernet is ruled by the IEEE 802.3 standard that supports nominal transmission rates as fast as 400 Gbps [10]. It comprises several signaling and wiring variants of the IETF/OSI physical and data link layer layers. Although Ethernet has been traditionally used with higher layer networking protocols like IP, UDP, and TCP, it provides a natural solution to high-end IoT communication.

3.2.1.1 Physical Layer

Ethernet relies on different media for transmission and propagation. Initially, Ethernet was designed to work with coax cables with twisted pair and fiber optics support added later. As these technologies evolved, transmission rates also improved. Initially, nominal maximum transmission rates were 10 and 100 Mbps, and they were later extended, with the support of fiber options, to 1 Gbps, 10 Gbps, 40 Gbps and 100 Gbps. In 2017 200 and 400 Gbps Ethernet transmission rates were standardized as IEEE 802.3bs-2017. As with most wireline technologies, the Ethernet physical layer relies on line codes that range from Manchester to multilevel PAM. Ethernet modulation and demodulation consist of the generation and detection of these line codes. Synchronization between endpoints results from a well-known pattern of bit transitions that are described in the following section.

3.2.1.2 Link Layer

Figure 3.24 shows an Ethernet frame that encapsulates a network layer datagram. The preamble is a 7-byte pattern of alternating zeros and ones used by endpoints to determine bit synchronization including synchronization information extraction as described in Sect. 3.1. The *Start Frame Delimiter* (SFD) is a *10101011* sequence that indicates the actual start of the frame and enables the receiver to locate the subsequent fields of the frame. The destination address (DA) field identifies the station or stations for which this frame is intended. It may be a unicast (one endpoint), a multicast (a group of endpoints), or a broadcast address (all endpoints). The source address (SA) identifies the station that is transmitting the frame. If the Ethernet frame is fully compliant with IEEE 802.3, the length field specifies the size of the data payload field in units of bytes. If the Ethernet frame, on the other hand, complies with the older Ethernet II specification, the length encodes an ethertype field that indicates the type of datagram transported as payload. Since the MTU size of Ethernet is 1526 bytes, a payload length equal or larger than 1536 (hexadecimal 0x800) signals an ethertype instead. For example, a 1536 ethertype value indicates that the payload is an upper layer IPv4 datagram; on the other hand, a 34,525 ethertype value (hexadecimal 0x86DD) indicates that the payload is

Fig. 3.23 Fast FHSS modulation example

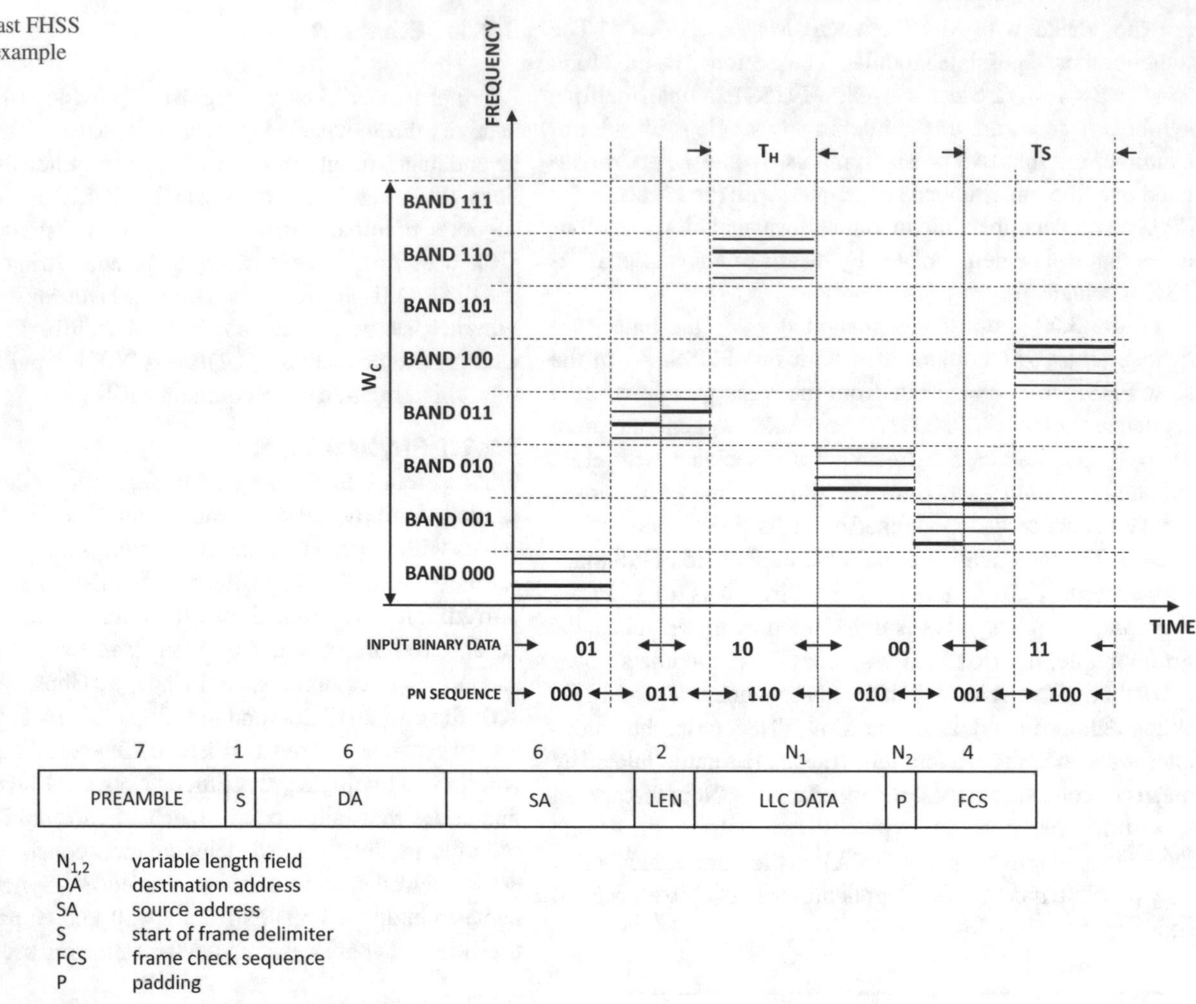

Fig. 3.24 Ethernet frame

an upper layer IPv6 datagram. For IEEE 802.3 compatible frames that do not include an ethertype, upper layer generated payloads are signaled through an additional LLC layer that is placed as Ethernet data. For all cases, the frames must be long enough to comply with collision detection requirements. If frames are not long enough, padding bits are typically added. The very last field of an Ethernet frame is the *frame checksum* (FCS) that through a 32-bit *cyclical redundancy checking* (CRC) block code provides basic channel encoding.

IEEE 802.2, the LLC protocol that is transported on top of standard IEEE 802.3, provides both unreliable connectionless and reliable connection-oriented delivery of frames. Connectionless delivery, in addition, supports optional acknowledgments. Figure 3.25 shows the *Service Access Point* (SAP) header that includes a *Destination Service Access Point* (DSAP), a *Source Service Access Point* (SSAP), and a control field. Both DSAP and SSAP identify the source and destination addresses of the upper layer protocols multiplexed under IEEE 802.2. The control field signals mode selection as well as the messaging mechanism to provide connection-oriented and connectionless communications. IEEE 802.2 can be extended by means of the *Subnetwork Access Protocol* (SNAP), also shown in Fig. 3.25, that supports the identification of upper layer datagrams by means of an organization identifier and an ethertype field.

Ethernet as other wireline physical layer technologies relies on full duplex communications where devices and applications can simultaneously transmit and listen for frames. This fact leads to *Carrier Sense Multiple Access with Collision Detection* (CSMA/CD) that is a generic mechanism for MAC under Ethernet.

The medium is shared so a device transmits frames if the channel is idle. If it is busy, a device continues to listen until the channel is idle, at this point it transmits immediately. If a collision is detected during transmission, the device transmits a brief jamming signal to assure that all stations know that there has been a collision, so they can cease transmitting. The jamming signal is a 32-bit to a 48-bit pattern of alternating ones and zeros. After transmitting the jamming signal, the device waits for a random amount of time, known as exponential backoff, and then it attempts to transmit the frame again. Essentially retransmissions are delayed by an amount of time derived from a fixed interval of time that depends

Fig. 3.25 IEEE 802.2

1	1	1/2	variable
DSAP	SSAP	CTRL	DATA

IEEE 802.2 (SAP)

1	1	1/2	3	2	variable
DSAP	SSAP	CTRL	OID	ETHERTYPE	DATA

IEEE 802.2 (SNAP)

DSAP destination service access point
SSAP source service access point
CTRL control field
OID organization id

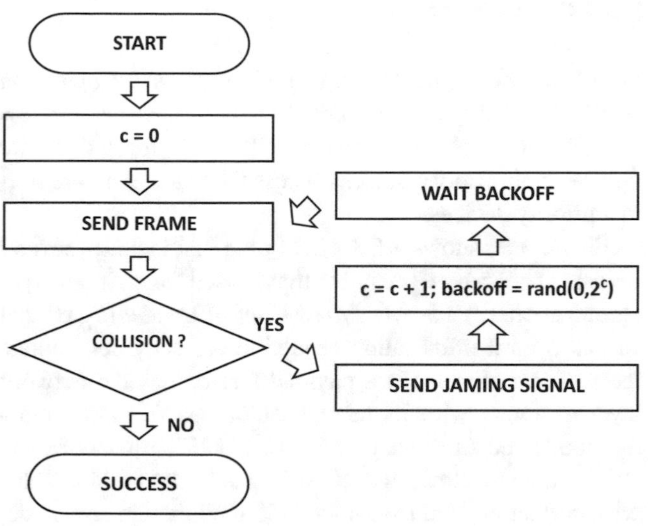

Fig. 3.26 CSMA/CD

on the transmission rate known as slot time and the number of retransmission attempts. After a c number of collisions, a random number of slot times between 0 and $2^c - 1$ is selected as exponential backoff. Figure 3.26 illustrates the flow diagram of the CSMA/CD algorithm.

3.2.2 ITU-T G.9903

ITU-T G.9903 provides physical and link layer mechanisms for low-rate data transmission over electrical wiring that are based on the protocols introduced by mainstream industrial G3-PLC [21]. The advantage of this technology is that deployment and maintenance costs are minimized since it relies on pre-existent wiring infrastructure. 3G-PLC has been mainly used for remote metering in smart grid scenarios, but its use has been extended to other applications ranging from home automation to plain Internet access. From an IoT perspective, 3G-PLC has been standardized as ITU-T G.9903 by the G3-PLC Alliance in order to provide large-scale connectivity over the electrical grid. IPv6 support, when ITU-T G.9903 is in place, results from the 6LoWPAN mechanism introduced in Sect. 4.3.

3.2.2.1 Physical Layer

Example 3.3 Consider a sensor that transmits 5-byte readouts every 20 ms and assume that the overhead due to network, transport, and session layer headers is 48 bytes. What is the efficiency of the communication system if the physical and link layers are 10 Gbps Ethernet? The efficiency η is, as described in Problem 2.7, given by $\eta = \frac{R}{C}$ where R is the transmission rate and C is the channel capacity.

Solution The transmission rate R is given by $R = \frac{L}{T} =$ 29,200 bps where $L = 8 \times (5 + 48 + H)$, $H = 20$ is the Ethernet header size as per information shown in Fig. 3.24 and $T = 20$ ms. The efficiency is given by $\eta = \frac{R}{C} = 2.92 \times 10^{-6}$ where $C = 10 \times 10^9$ bps. The efficiency is therefore around 0.0003%.

Electrical power lines are not good at transmitting pulses and line codes as they are intended for the transmission of AC signals over very long distances. Electrical power is usually propagated over wires as a high-voltage sinusoidal of fixed frequency (50 or 60 Hz depending on the country). This power sinusoidal is combined with device data streams modulated through low-voltage carriers. ITU-T G.9903, which standardizes 3G-PLC, establishes OFDM as the modulation mechanism that divides a 500 KHz channel into 128 subschannels. At any given time, only 36 subchannels are active for communications. Within each subchannel, the carrier is modulated by means of DPSK, DQPSK, D8-PSK, and 16-QAM that carry 1, 2, 3, and 4 bits per symbol, respectively. FEC is provided by means of a cascade between a convolutional encoder, a Reed-Salomon block encoder, and a scrambler that swaps bits of the output stream. Figure 3.27 shows the building blocks of the ITU-T G.9903 physical layer. Effective transmission rates of G.9903 range from 2.4 to 34 Kbps. This is fast enough to support many IoT applications.

3.2.2.2 Link Layer

Because a power line relies on a single wire, 3G-PLC devices can either transmit or receive packets but cannot do both tasks

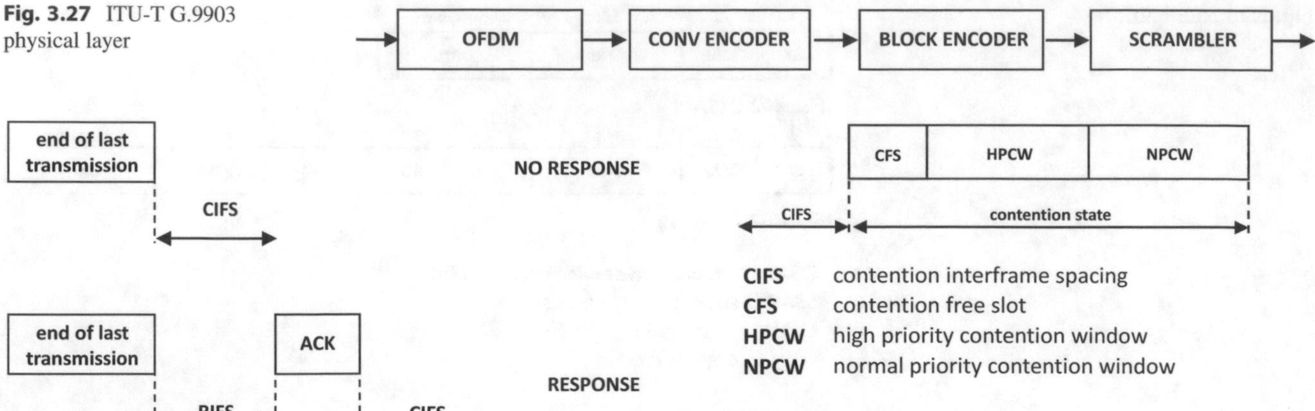

Fig. 3.27 ITU-T G.9903 physical layer

Fig. 3.28 End of transmission

Fig. 3.29 Priority resolution

at the same time. This limitation leads to a MAC mechanism that is performed by means of *carrier sense multiple access with collision avoidance* (CSMA/CA). This mechanism provides transmitters with access to the channel after a random backoff time that minimizes the collisions between contending devices. When a device wants to transmit frames, it waits for a random time period, after which if the channel is found idle, transmission starts. If, on the other hand, the channel is busy, the device waits for another random time period before attempting to retransmit.

As in the Ethernet case described in Sect. 3.2.1, the random exponential backoff mechanism tends to progressively reduce the probability of collisions with the number of retransmission attempts. The contention period starts after the last transmission is detected, but there are two possibilities depending on whether this transmission requires a response from its intended receiver. When a response is needed, it can be in the form of an acknowledgment (ACK) or a negative acknowledgment (NACK). In this case, after the transmission is finished, there is a fixed time period known as the *response interface spacing* (RISF) before the receiver transmits an ACK or a NACK. If the transmitter does not receive an ACK after a waiting time period, it assumes that the transmission was unsuccessful and retries the frame transmission. If an ACK is still not received after several retries, the transmitter can choose either to terminate the transaction or to start again. If no response is needed or if response is needed and it has been received, the contention period starts after a fixed time period known as *contention interframe spacing* (CIFS). This mechanism is illustrated in Fig. 3.28.

Under ITU-T G.9903 transmission supports two levels of priority; high and normal levels that map into two contention time windows shown in Fig. 3.29. The first contention slow is called *Contention Free Slot* (CFS) that is used for transmission of subsequent segments of a MAC frame without the backoff mechanism in order to prevent interruption from other devices. This happens when the first segment is sent either as a high or normal priority segment and all subsequent segments are sent in the CFS. The high and normal priority devices compete during the *High Priority Contention Window* (HPCW) and *Normal Priority Contention Window* (NPCW), respectively. Because HPCW is located before NPCW, high-priority devices access the channel before normal priority devices.

Figure 3.30 shows a G3-PLC frame that includes several fields; (1) a 9.5-byte preamble that is used for endpoint synchronization, (2) a 5-byte *Frame Control Header* (FCH) that carries control information needed to correctly demodulate the received signal, (3) a payload that carries the network layer datagram with its length determined from the transmission mode (normal or robust), (4) FEC that relies on Reed-Salomon block and convolutional codes and enables error correction, and (5) 2-byte FCS field that, in any mode, enables error detection. In robust mode, to support additional reliability, a repetition code is used after the convolutional encoder to repeat the bits at the output of the convolutional encoder four times.

3.2.3 IEEE 1901.2

IEEE 1901.2 is another PLC mechanism similar to 3G-PLC that supports physical and link layer mechanisms for low-rate data transmission over electrical wiring [5]. It supports slightly higher transmission rates than 3G-PLC, but as the latter, it relies on 6LoWPAN adaptation to support IPv6. Standardization started in 2009 triggered by interest of the automotive industry in providing electric vehicles to charging station communication. Additional applications include grid to utility metering and HANs.

3.2.3.1 Physical Layer

As G3-PLC, IEEE 1901.2 relies on OFDM for low-frequency modulation, under 500 KHz, in narrowband power line devices via AC and DC power lines. The standard supports indoor and outdoor transmissions over several types of lines: (1) low-voltage lines, less than 1000 V, like the lines between transformers and meters; (2) low-voltage to medium lines, between 1000 V and 72 KV, like the lines between transformers; and (3) long distance for multirange rural communications. Based on the geographical region, three frequency bands are available. As for any PLC technology, and because of the inherently low EMI, devices can transmit in frequency bands that relying on relatively small guardbands exhibit little disturbance. Depending on the transmission rate and reliability, the number of subchannels varies.

In a typical example, the 500 KHz band provides an OFDM scheme over 72 subchannels. Supported per subchannel modulations are DPSK, DQPSK, D8-PSK, BPSK, QPSK, 8-PSK, and 16-QAM. Typical transmission rates are around 500 Kbps. IEEE 1901.2 relies on *adaptive tone mapping* (ATM) that dynamically selects the usable tones and the optimum modulation and coding schemes that ensure reliable communication over the wireline channel.

3.2.3.2 Link Layer

Two types of frames are supported: (1) data frames and (2) ACK/NACK frames. Figure 3.31 shows the frame structure where each frame starts with a (1) preamble used for synchronization and detection as well as to signal *automatic gain control* (AGC). The preamble is followed by a FCH that specifies the number of symbols depending on the number of carriers used by the OFDM scheme. This header also contains information regarding modulation and the length of the current frame in symbols. The FCH also includes a FCS field that is used for error detection. The size of the FCS depends on the frequency band being utilized. From a topology perspective, IEEE 1901.2 networks are typically deployed following a full mesh structure. Because IEEE 1901.2 frames follow a similar structure to that of IEEE 802.15.4 frames presented in Sect. 3.3.3, they rely on 6LoWPAN adaptation to support IPv6.

3.2.4 MS/TP

The *Master-Slave/Token-Passing* (MS/TP) protocol relies on a RS-485 physical layer that enables devices to relay and exchange information. In addition, MS/TP and RS-485 are, respectively, the link and physical layers of the BACNet protocol stack [23]. MS/TP is a peer-to-peer, master/slave protocol that supports multiple masters. The mechanism is based on a token passing scheme.

3.2.4.1 Physical Layer

RS-485 enables communications in inexpensive LANs with multidrop links relying on differential signaling over twisted pairs. Differential signaling consists of transmitting data by means of two complementary electrical signals that, when subtracted at the receiver, minimize noise. Figure 3.32 illustrates an example of differential signaling when one bit is transmitted over two separate lines. The transmitter generates two pulses of different polarity that are affected by the same channel noise. On the receiver, the noisy pulses are subtracted, and the noise is canceled out resulting in a single large output pulse.

As shown in Fig. 3.33, MS/TP follows a bus topology with devices deployed in a *daisy chain* configuration where each device is connected point-to-point to one or two peers depending on its location in the network. Connectivity is

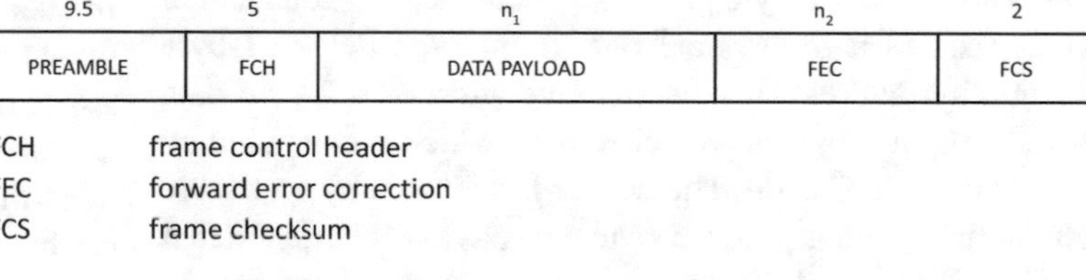

Fig. 3.30 G3-PLC frame

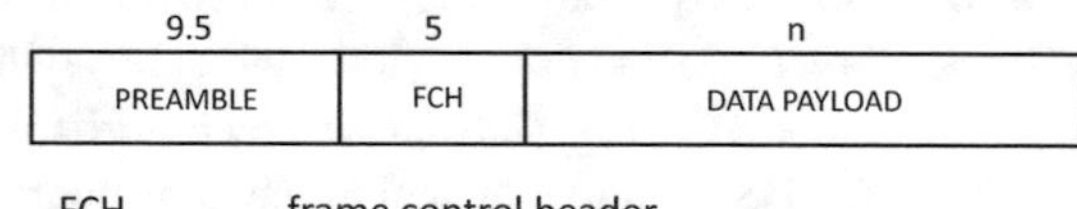

Fig. 3.31 IEEE 1901.2 frame

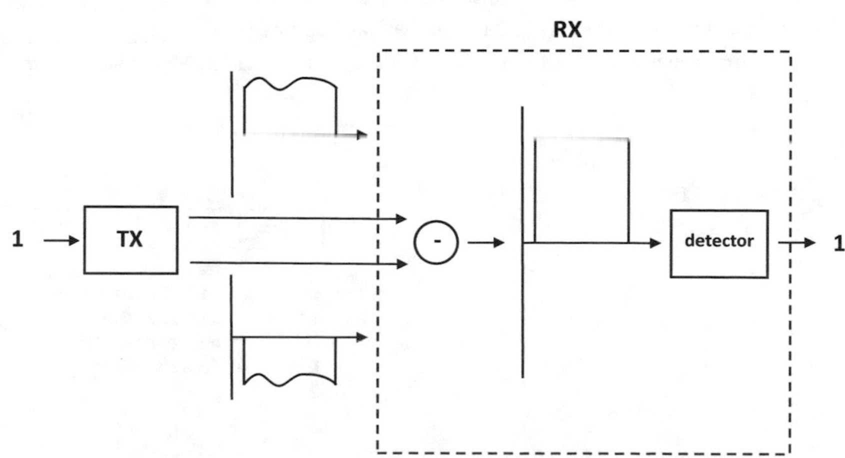

Fig. 3.32 Differential signaling

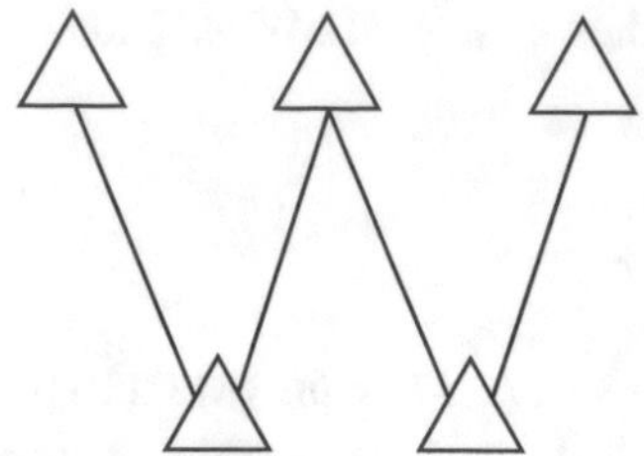

Fig. 3.33 Daisy chain configuration

provided by means of low capacitance shielded twisted pair cables. MS/TP networks support up to 64 devices, and the number of devices also determines the transmission range and data rate. The maximum bus length and the maximum transmission data rate are 1.2 km and 115.2 Kbps, respectively. In general, 115.2 Kbps is not achievable in a 1.2 km long network, and much lower transmission rates are typical. Addresses 0 through 127 are used to identify master devices, while the 255 destination address is used for broadcasting.

3.2.4.2 Link Layer

As its name indicates it, MS/TP relies on a master-slave scheme where transmissions are enabled by means of a token passing mechanism. In other words, the token is used to control device access on the bus. An MS/TP master endpoint can transmit a frame when it holds the token. When it is done transmitting, it passes the token to the next master device which is determined by its MAC address.

Figure 3.34 shows an MS/TP frame; it includes a header that starts with a 2-byte preamble encoded as a hexadecimal value 0x55ff, it follows a 1-byte frame type field, a 1-byte destination address, a 1-byte source address, a 2-byte data length, and a 1-byte header checksum which is computed by means of a CRC code. The header is followed by a variable length data payload and a 5-byte data payload checksum that is also CRC based. The frame finishes with an optional 1-byte padding. Note that data payload and data payload checksum are encoded using *Consistent Overhead Byte Stuffing* (COBS). COBS provides a technique that enables the removal of preamble sequences from these data fields. MS/TP IPv6 support is provided by means of partial support of IP header compression supported by 6LoWPAN and described in Sect. 4.

3.3 Wireless

As with any other IoT communication technology, and as indicated in Sect. 3.2, the main goal is to lower deployment costs and times. Because when comparing wireless and wireline technologies the former have considerable lower infrastructure cost, they are the preferred mechanisms for IoT scenarios. Under wireless communication, signal propagation is by means of antennas. Antennas provide the mechanisms for the conversion of electrical signals into electromagnetic waves and vice versa. There are two types of antennas; a transmitting antenna converts a wireline modulated signal into an electromagnetic wave that is radiated in a specific direction. Similarly, a receiving antenna converts the irradiated electromagnetic wave back into a wireline modulated signal. The electromagnetic wave power is measured in terms of *effective irradiated power* (EIRP) that assumes transmission as being originated by an isotropic point source. An isotropic point source is an ideal antenna that transmits power uniformly in all directions.

The wireless radio-frequency section of the electromagnetic spectrum lies between the frequencies of 9 kHz and 300 GHz such that different bands of the spectrum are used to deliver different services. The wavelength and frequency of electromagnetic radiation are related via the speed of light through the $\lambda = \frac{c}{f}$ expression where λ is the wavelength, c is the speed of light, and f is the frequency. Table 3.1 shows the subdivisions of the radio-frequency spectrum. IoT wireless solutions heavily rely on unlicensed *Instrument, Scientific, and Medical* (ISM) bands [25]. There are many frequencies that fall under the umbrella of ISM bands, but the three most important lie at 915 MHz (868 MHz in Europe), 2.4, and 5.8 GHz. Additional unlicensed ISM bands are also available for other applications like the 13.56 MHz ISM band that is intended for *near-field communication* (NFC) or the 433 MHz ISM band that is used for radiolocation. In the case of ISM bands, higher frequencies translate into smaller antennas that enable smaller hardware form factors. Higher frequencies, however, are typically associated with worse signal penetration and propagation. Therefore Sub-GHz bands exhibit better propagation properties than super-GHz bands. Super-GHz bands, however, are less restrictive and leverage from more relaxed regulations that, for example,

Fig. 3.34 MS/TP frame

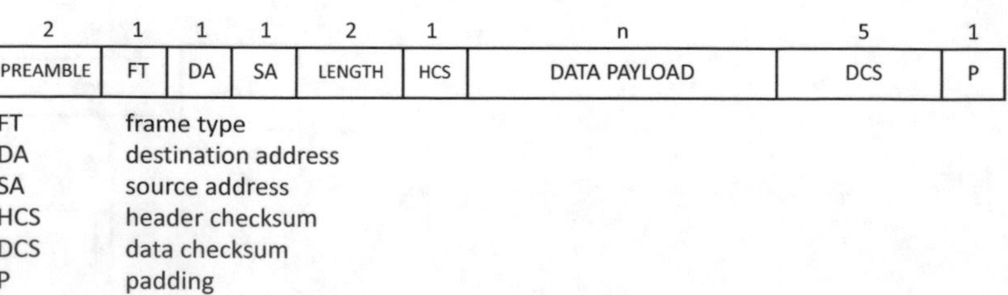

Table 3.1 Radio-frequency spectrum

Transmission type	Frequency	Wavelenth
Very low frequency (VLF)	9–30 kHz	33–10 km
Low frequency (LF)	30–300 kHz	10–1 km
Medium frequency (MF)	300–3000 kHz	1000–100 m
High frequency (HF)	3–30 MHz	100–10 m
Very high frequency (VHF)	30–300 MHz	10–1 m
Ultra high frequency (UHF)	300–3000 MHz	1000–100 mm
Super high frequency (SHF)	3–30 GHz	100–10 mm
Extremely high frequency (EHF)	30–300 GHz	10–1 mm

Table 3.2 Radio-frequency spectrum

RF band	Wireless network specification
915/868 MHz ISM	IEEE 802.15.11ah, IEEE 802.15.4, ITU-T G.9959, BLE, LoRa, SigFox
2.4 GHz ISM	IEEE 802.11ax, IEEE 802.15.4, BLE, LoRa
5.8 GHz	IEEE 802.11ac, IEEE 802.11ax

do not impose power duty cycle limits that lower transmission rates.

Table 3.2 shows the some wireless IoT technologies associated with the ISM bands. The table shows both WPAN and LPWAN technologies discussed in this chapter and in Chap. 8. In general, regulatory authorities in countries and regions control the use of the spectrum by means of frequency band allocation for both licensed and unlicensed services including the maximum allowable transmission power levels. These legislation differences make very difficult the full standardization of frequency band allocation and maximum power levels by ITU and other standardization bodies. Moreover, they introduce hardware and software design interoperability issues that affect deployment and maintenance costs. The use of unlicensed bands under ISM implies that anyone can freely use them if transmission power is low enough to minimize interference. However, if too many devices use the same unlicensed band, it can eventually become unusable. This is the case of the 2.4 GHz ISM band that is overcrowded with device communications ranging from cordless phones to IoT sensors.

Wireless, when compared to wireline communication, is affected by three well-known problems: (1) data link reliability issues due to the fading caused by multipath phenomena that results on bit errors that lead, in turn, to packet loss, (2) media access issues due to the contention of multiple devices attempting to simultaneously transmit data over the same wireless channel, and (3) security due to the fact that wireless transmissions by virtue of being open are more likely to be intercepted than those in a wireline scenario. In the case of IoT, reliability is further degraded because of the power limitations that result from the need of extended battery life. Similarly, media access contention is aggravated by the fact that collisions can be avoided but not detected because devices either transmit or receive frames at any given time but cannot do both things at the same time.

Media access is also affected by two situations indicated in Fig. 3.35: (1) the hidden station problem that results from stations A and C transmitting simultaneously to station B and not detecting each other since they are out of range and (2) the exposed station problem that results from station C transmitting to station D and preventing station B to transmit to station A. In general terms, the hidden station problem leads to excessive bit errors and therefore packet loss. The exposed station problem leads to excessive latency that, in turn, becomes packet loss from an application perspective. Essentially, packets that arrive too late are not useful and can, therefore, be considered lost by the application. Again, this depends on the nature of the application and QoS goals.

This section introduces several wireless physical and link layer technologies that are selected based on their native support of IPv6. This support is either direct or by means of an adaptation mechanism. This means that some well-known stacks like ANT+ are not included in this section because they do not have standard support of IPv6 adaptation.

3.3.1 IEEE 802.11

Wireless Fidelity (Wi-Fi), also known as IEEE 802.11, is an umbrella term for a number of technologies that are the wireless equivalent of Ethernet in the IoT world [8]. It includes many standard protocols that support a wide range of transmission rates and exhibit different levels of power consumption. A few of these protocols are optimized for IoT scenarios and have power efficiency as main goal. The IEEE 802.11 standard provides a physical and link layer mechanism for transmission in WLANs and, especially when considering IoT, WPANs.

IEEE 802.11 networks rely on three main elements: (1) stations that play the role of IoT access devices, (2) *access points* (APs) that play the role of IoT gateways interfacing between access and core networks, and the (3) distribution system that provides the backbone for connectivity to applications performing analytics. Stations are grouped in what is known as the *Basic Service Set* (BSS). APs, as most IoT gateways, include multiple communication stacks; IEEE 802.11 to interact with the access side and some other technology, like Ethernet, to interact with the core side. When considering both access and core networks, the whole infrastructure is known as the *Extended Service Set* (ESS).

With IEEE 802.11, a BSS responds to a WLANs or WPANs that is under the control of a single AP. The AP, in turn, enables communications by devices that use a common

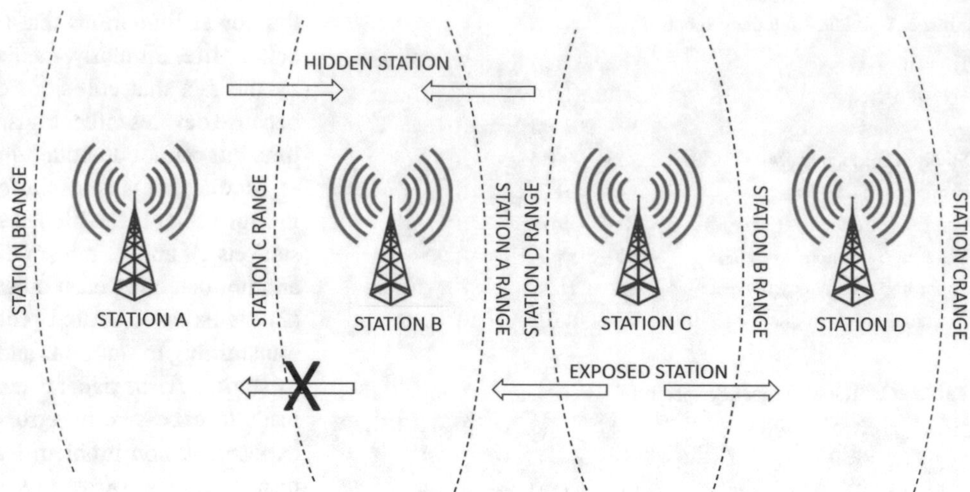

Fig. 3.35 Contention problems

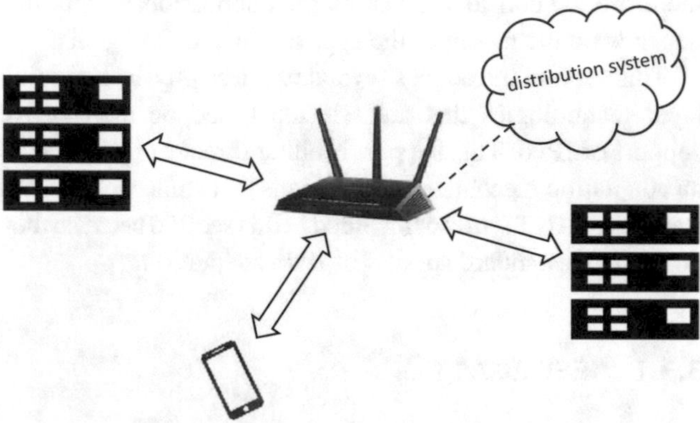

Fig. 3.36 Basic service set (BSS)

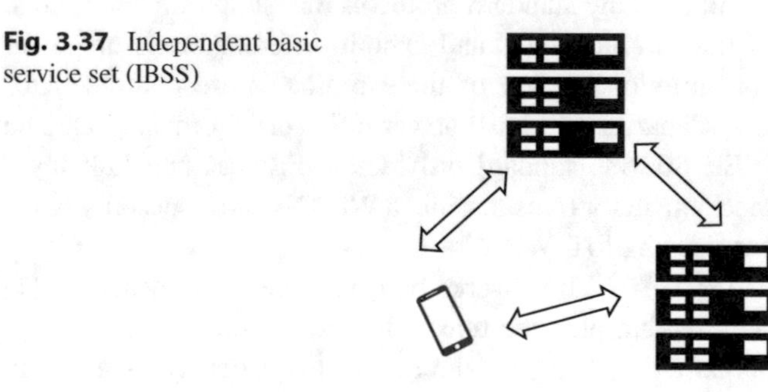

Fig. 3.37 Independent basic service set (IBSS)

BSS Identifier (BSSID) in order to share the wireless channel. The BSSID as well as other physical layer parameters like transmission rates and other modulation attributes are transmitted in beacon frames that are broadcasted at regular intervals by APs and devices.

Traditionally IEEE 802.11 defines two modes of operation of a single BSS; ad hoc and infrastructure modes that are representative of the topologies shown in Figs. 3.36 and 3.37, respectively. In an ad hoc IEEE 802.11 network, a BSS is known as *Independent BSS* (IBSS) since devices communicate directly with each other without the intervention of an AP. Moreover, the lack of an AP also implies that there is no core network and therefore no ESS. Note that in an IBSS, a randomly generated SSID and the other physically layer parameters are broadcasted by beacon frames transmitted by devices. Under infrastructure mode, on the other hand, all communication between devices on the same BSS is through an AP that acts as a gateway and provides connectivity to the core side. This superset of access devices, AP, and core network makes the ESS. Note that a single ESS can be composed of multiple BSS with several APs providing inter-BSS device connectivity.

The presence of an AP determines how communication between devices in the same BSS is. In infrastructure mode, packets sent from one device to another are transmitted through an AP, while in ad hoc mode packets go directly from source to destination. Infrastructure mode, therefore, exhibits higher latency, but it also improves reliability as the AP can buffer packets if the destination device is out of reach, sleeping, or disabled. This is particularly important in the context of IoT where connectivity is M2P between sensing and actuation devices and applications performing analytics. In this context, communication between devices is not common, and the use of IEEE 802.11 is almost exclusively restricted to infrastructure mode.

► **Infrastructure BSS and IoT** IoT architectures, that are based on devices communicating with applications by means of gateways, fit very well the infrastructure operation model of IEEE 802.11. In this context, IoT devices are IEEE 802.11 stations and IoT gateways are IEEE 802.11 APs. The distribution system is the IP core that carries IP traffic to and from the application as shown below.

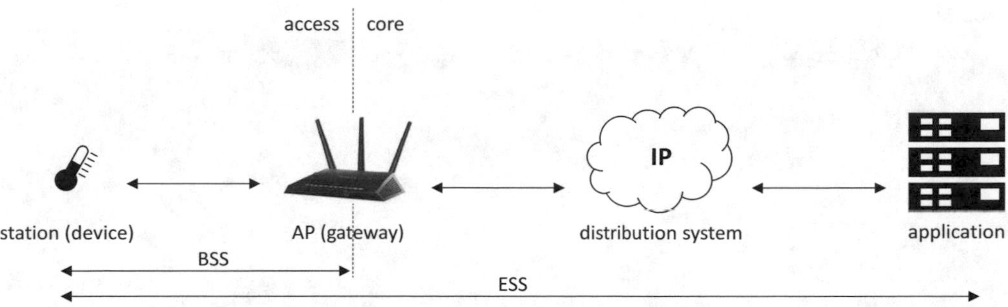

3.3.1.1 Physical Layer

The initial standardization of IEEE 802.11 presented two alternative physical layer mechanisms: FHSS and DSSS operating in the ISM 2.4 GHz band. The standard also introduced an *infrared* (IR) communication mechanism that was never fully implemented. Depending on negotiated parameters the nominal transmission rate for this basic IEEE 802.11 support is either 1 or 2 Mbps. Later amendments to the standard led by power and modulation scheme changes improve transmission rates, energy consumption, and signal range. Chronologically speaking, these amendments are IEEE 802.11a and IEEE 802.11b in 1999, IEEE 802.11g in 2003, IEEE 802.11n in 2009, IEEE 802.11ac in 2013, and IEEE 802.11ax in 2016. IEEE 802.11ad released in 2012 and IEEE 802.11aj released in 2018 introduce support of unlicensed bands in the 45–70 GHz spectral range available in certain countries. These two standards enable very high transmission rates in the Gbps range. From the IoT perspective (1) IEEE 802.11p was introduced in 2010 to support *Wireless Access in Vehicular Environment* (WAVE), (2) IEEE 802.11af in 2014 to provide low transmission rates by sending data over digital television guardbands, (3) IEEE 802.11ah for generic IoT support in 2016, and (4) IEEE 802.11ba for power efficient, by means of an aggressive *Wake-up Radio* (WUR) duty cycle, *narrow band* (NB) IoT access in 2020.

Many of the amendments to IEEE 802.11 evolve from the original IEEE 802.11a standard. IEEE 802.11a relies on OFDM operating in the ISM 5 GHz band and providing non-overlapping 20 MHz channels for modulation. Depending on different country regulations the number of channels and the maximum EIRP requirements may change. In many scenarios upper channels are reserved for outdoor use and therefore increase the EIRP limit. Each of these 20 MHz channels support 52 OFDM subchannels with carriers separated every 312.5 KHz. Out of the 52 OFDM subchannels, 48 are used for data transmission, while the remaining four provide frequency and phase signal reference. IEEE 802.11a allows a maximum of 15 9-μs backoff timeslots.

Table 3.3 shows the OFDM subchannel modulation schemes used under IEEE 802.11a to support maximum nominal transmission rates between 6 and 54 Mbps. The table includes the total number of code bits per symbol that results from aggregating all the code bits from the individual subchannels. The table also presents the FEC code rate that leads to the number of data bits per symbol. The number of data bits per symbol, in turn, is used to obtain the actual transmission rate. For example, the fifth row in the table is 16-QAM which includes 4 bits per subchannel and, at 48 subchannels per symbol, leads to 192 code bits per symbol. With a 1/2 code rate, this translates to $\frac{192\times 1}{2} = 96$ data bits per symbol that sent every 4 μs corresponds to a $\frac{96}{4\times 10^{-6}} = 24$ Mbps transmission rate.

IEEE 802.11b was released around the same time as IEEE 802.11a, but it was designed to operate on the mainstream and more crowded ISM 2.4 GHz band. This improves signal penetration at a cost of increased interference. For the same power consumption and due to its lower penetration, IEEE 802.11a typically requires a larger number of APs to support the same area coverage. IEEE 802.11b divides the ISM 2.4 GHz band into several overlapping 22 MHz channels. Depending on the world region, the total number of channels is anywhere between 11 and 14. IEEE 802.11b supports up to 31 20-μs timeslots.

IEEE 802.11b extends traditional IEEE 802.11 by introducing additional DSSS based transmission rates. Table 3.4 illustrates the different modulation schemes that are used under IEEE 802.11b. Note that the first two rows of the table present the two original IEEE 802.11 modulation schemes. Through 11-chip codes that are used to generate symbols at a rate of 10^6 symbols per second and relying on BPSK and QPSK modulations, the maximum nominal transmission rates are 1 and 2 Mbps, respectively. Higher transmission rates are obtained by changing the modulation method to a differential scheme that relies on D-QPSK and D8-PSK to accomplish 5.5 and 11 Mbps rates, respectively. Note that in this case, the code length is shorter at 8 chips, while the symbol rate is slightly higher at 1.375×10^6 symbols per second.

IEEE 802.11a and IEEE 802.11b rates are dynamically selected by means of *dynamic rate shifting* (DRS). This mechanism allows the transmission rate to be selected in order to compensate for multipath fading and other types of interference. If a link is not reliable, devices select lower transmission rates until the communication path is stable.

Table 3.3 IEEE 802.11a

Modulation	Code bits/subchannel	Code bits/symbol	Code rate	Data bits/symbol	Data rate (Mbps)
BPSK	1	48	1/2	24	6
BPSK	1	48	3/4	36	9
QPSK	2	96	1/2	48	12
QPSK	2	96	3/4	72	18
16-QAM	4	192	1/2	96	24
16-QAM	4	192	3/4	144	48
64-QAM	6	288	2/3	192	48
64-QAM	6	288	3/4	216	54

Table 3.4 IEEE 802.11 DSSS

Modulation	Code bits/subchannel	Code bits/symbol	Code rate	Data bits/symbol	Data rate (Mbps)
BPSK	11	1	1	1	
QPSK	11	1	2	2	
DQPSK	8	1.375	4	5.5	
D8-PSK	8	1.375	8	11	

Whenever possible if interference reduction is detected, a higher transmission rate is selected. This reduction is done automatically at the physical layer and is transparent to higher layers.

IEEE 802.11g uses the same OFDM modulation scheme as IEEE 802.11a but operates on the ISM 2.4 GHz band and supports transmissions rates between 6 and 54 Mbps. In addition, and to provide backward compatibility since they both operate on this band, IEEE 802.11g radios also support IEEE 802.11b transmission modes at a cost of lower throughput. Specifically, IEEE 802.11b devices can associate with IEEE 802.11g APs through a number of protection mechanisms that enable interoperation; a device sends a *Request to Send* (RTS) control frame to the AP and, in turn, the AP issues a *Clear to Send* (CTS) response to grant transmission. This interaction minimizes the collisions between IEEE 802.11b and IEEE 802.11g transmission attempts but increases latency thus lowering throughput. IEEE 802.11b devices can also transmit without the RTS/CTS frame exchange and just relying on the channel state to guarantee that it is clear before transmission. This is particularly useful in BSSs with very few devices since it can lead to lower latency and higher throughput. IEEE 802.11g backoff slot parameters follow those of IEEE 802.11a unless when it is operating under mixed mode where an IEEE 802.11g AP interacts with IEEE 802.11b devices. In this latter case, IEEE 802.11g backoff slot parameters are those of IEEE 802.11b, and the overall network throughput is lowered.

IEEE 802.11g introduces two mechanisms to improve throughput: (1) packet bursting that consists in buffering a number frames before transmitting them all together in order to lower average channel contention delay and (2) channel bonding that consists of transmitting frames of the same physical device over multiple non-overlapping subchannels. Packet bursting improves throughput but introduces additional network latency as frames are buffered before being transmitted. It also prevents other devices from transmitting as a source station accesses the channel for a comparatively longer amount of time. Similarly, channel bonding also improves throughput by increasing the transmission rate without affecting latency. The drawback of channel bonding, however, is more expensive and complex hardware. In opposition, packet bursting can be purely implemented in software.

MIMO is a technique that consists of using multiple outgoing and incoming antennas that, exploiting spatial diversity, lead to extra gain that can be used to defeat channel noise. MIMO mechanisms increase transmission rates. With spatial diversity the same signal travels over different paths and thus is more likely to overcome wireless channel fading. MIMO is not the same as channel bonding, since unlike channel bonding, MIMO achieves higher data rates without increasing the number of subchannels used. IEEE 802.11n introduces maximum nominal transmission rates of 600 Mbps by combining OFDM with MIMO. Moreover, IEEE 802.11n extends channel bandwidths by combining up to two 20 MHz channels in the ISM 2.4 GHz and ISM 5 GHz bands. IEEE 802.11n uses complex synchronization mechanisms that attempt to guarantee the highest throughput by dynamically adapting channel selection, antenna configuration, modulation schemes, and code rates.

Table 3.5 indicates the modulation schemes used by IEEE 802.11n to accomplish the highest transmission rates. For example, the last row in the table is 64-QAM which includes 6 bits per subchannel and, at 432 subchannels per symbol, leads to 2592 code bits per symbol. With a 5/6 code rate, this translates to $\frac{2592\times 5}{6} = 2160$ data bits per symbol that

Table 3.5 IEEE 802.11n

Modulation	Code bits/subchannel	Code bits/symbol	Code rate	Data bits/symbol	Data rate (Mbps)
BPSK	1	432	1/2	216	54
QPSK	2	864	1/2	432	108
QPSK	2	864	3/4	648	162
16-QAM	4	1728	1/2	864	216
16-QAM	4	1728	3/4	1296	324
64-QAM	4	2592	2/3	1728	432
64-QAM	6	2592	3/4	1944	486
64-QAM	6	2592	5/6	2160	540

Table 3.6 IEEE 802.11ac

Modulation	Code bits/subchannel	Code bits/symbol	Code rate	Data bits/symbol	Data rate (Mbps)
BPSK	1	1872	1/2	936	260
QPSK	2	3744	1/2	1872	520
QPSK	2	3744	3/4	2808	780
16-QAM	4	7488	1/2	3922	1040
16-QAM	4	7488	3/4	5883	1560
64-QAM	6	11,232	2/3	7488	2080
64-QAM	6	11,232	3/4	8424	2340
64-QAM	6	11,232	5/6	9360	2600
256-QAM	8	14,976	3/4	11,232	3120
256-QAM	8	14,976	5/6	12,480	3467

sent every 4 μs that corresponds to a $\frac{2160}{4\times10^{-6}} = 540$ Mbps transmission rate. By reducing by half the symbol guard period, a symbol period of 3.6 μs leads to transmission rate of $\frac{2160}{3.6\times10^{-6}} = 600$ Mbps.

IEEE 802.11ac, also known as Wi-Fi 5, extends some of the features of IEEE 802.11n to reach a transmissions rate of 3.4 Gbps. IEEE 802.11ac relies on 80 and 160 MHz channels operating only over the ISM 5 GHz band. Besides downlink MIMO transmissions, IEEE 802.11ac introduces the utilization of beamforming that is signal processing technique to support directional signal manipulation with the goal of minimizing interference.

Table 3.6 illustrates the modulation schemes used by IEEE 802.11ac to accomplish its highest transmission rates [24]. For example, the second to last row in the table is 256-QAM which includes 8 bits per subchannel and, at 1872 subchannels per symbol, leads to 14,976 code bits per symbol. With a 3/4 code rate, this translates to $\frac{14{,}976\times5}{6} = 11{,}232$ data bits per symbol that sent every 3.6 μs that corresponds to a $\frac{11{,}232}{3.6\times10^{-6}} = 3120$ Mbps transmission rate.

Wi-Fi 5 has led to the development of the next-generation Wi-Fi 6 that has been standardized as IEEE 802.11ax. One important difference with IEEE 802.11ac is that IEEE 802.11ax operates in the 2.4 GHz and the 5 GHz ISM bands. Additionally, IEEE 802.11ax improves MIMO by supporting both downlink and uplink transmissions. Efficiency and performance are also improved by means of other mechanisms introduced by IEEE 802.11ax like trigger-based random access that provides flexibility in the allocation of spectrum for uplink transmissions or dynamic fragmentation that is used to chunked frames into variable size fragments such that overall overhead is reduced. Both nominal transmission rates and latency are greatly improved.

Table 3.7 shows the modulation schemes used by IEEE 802.11ax to reach its highest transmission rates. For example, the second to last row in the table is 1024-QAM which includes 10 bits per subchannel and, at 4608 subchannels per symbol, leads to 46,080 code bits per symbol. With a 3/4 code rate, this becomes $\frac{46{,}080\times5}{6} = 34{,}560$ data bits per symbol that sent every 4 μs corresponds to a $\frac{34{,}560}{4\times10^{-6}} = 9600$ Mbps or 9.6 Gbps transmission rate.

IEEE 802.11p is based on IEEE 802.11a also operating in the ISM 5 GHz band and providing support for *vehicle-to-everything* (V2X) communication where transmissions are between vehicles (V2V) or between vehicles and roadside infrastructure (V2I). IEEE 802.11p relies on 10 MHz channels that double the transmission times and therefore reduce the data rates by half. Transmissions rates of 6, 9, 12, 18, 24, 36, 48, and 54 Mbps shown in Table 3.3 for IEEE 802.11a become 3, 4.5, 6, 9, 12, 18, 24, and 27 Mbps under IEEE 802.11p.

IEEE 802.11af, also known as Super Wi-Fi, relies on transmitting over unused frequency bands of the licensed TV spectrum between 54 and 790 MHz. The technology uses *cognitive radio* to dynamically select the portions of the spectrum that minimize interference. IEEE 802.11af uses the same OFDM modulation scheme as IEEE 802.11ac with 6, 7, and 8 MHz channels.

Table 3.7 IEEE 802.11ax

Modulation	Code bits/subchannel	Code bits/symbol	Code rate	Data bits/symbol	Data rate (Mbps)
BPSK	1	4608	1/2	2304	576
QPSK	2	9216	1/2	4608	1152
QPSK	2	9216	3/4	6912	1728
16-QAM	4	18,432	1/2	9216	2304
16-QAM	4	18,432	3/4	13,824	3456
64-QAM	6	27,648	2/3	18,432	4608
64-QAM	6	27,648	3/4	20,736	5184
64-QAM	6	27,648	5/6	23,040	5760
256-QAM	8	36,864	3/4	27,648	6912
256-QAM	8	36,864	5/6	30,720	7680
1024-QAM	10	46,080	3/4	34,560	8640
1024-QAM	10	46,080	5/6	38,400	9600

Table 3.8 IEEE 802.11af

Modulation	Code bits/subchannel	Code bits/symbol	Code rate	Data bits/symbol	Data rate (Mbps)
BPSK	1	2128	1/2	1064	43.2
QPSK	2	4256	1/2	2128	84.8
QPSK	2	4256	3/4	3192	128
16-QAM	4	8512	1/2	4256	171.2
16-QAM	4	8512	3/4	6384	256
64-QAM	6	12,768	2/3	8512	340.8
64-QAM	6	12,768	3/4	9576	384
64-QAM	6	12,768	5/6	10,640	427.2
256-QAM	8	17,024	3/4	12,768	512
256-QAM	8	17,024	5/6	14,168	569.6

Table 3.8 shows the modulation schemes used by IEEE 802.11af to accomplish the highest transmission rates. For example, the last row in the table is 256-QAM which includes 8 bits per subchannel and, at 2128 subchannels per symbol, leads to 17,024 code bits per symbol. With a 5/6 code rate, this translates to $\frac{17{,}024\times 5}{6} = 14{,}168$ data bits per symbol that sent every 24.75 μs corresponds to a $\frac{14{,}168}{24.75\times 10^{-6}} = 569.6$ Mbps transmission rate.

IEEE 802.11ah, also known as Wi-Fi HaLow, is a Wi-Fi standard intended exclusively for IoT use. As most other IEEE 802.11 deployments, IEEE 802.11ah networks are deployed as stars with the AP acting as IoT gateway. As opposed to conventional IEEE 802.11 standards, however, IEEE 802.11ah operates on the ISM 915/868 MHz band with 1, 2, 4, 8, and 16 MHz channels. As IEEE 802.11af, IEEE 802.11ah reuses the MIMO OFDM modulation scheme introduced by IEEE 802.11ac by changing symbol interspace and length to accomplish a wide range of transmission rates depending on power consumption. Also, as opposed to conventional IEEE 802.11 standards, IEEE 802.11ah supports a comparatively large number of devices in a single BSS in order to enable LPWAN support. IEEE 802.11ah provides a typical range of 100 m–1 km for direct connectivity that is supported by means of very low energy consumption that relies on power saving strategies.

Table 3.9 shows the modulation schemes used by IEEE 802.11ah to accomplish the lowest transmission rates. For example, the last row in the table is 256-QAM which includes 8 bits per subchannel and, at 24 subchannels per symbol, leads to 192 code bits per symbol. With a 5/6 code rate, this translates to $\frac{192\times 5}{6} = 160$ data bits per symbol that sent every 40 μs corresponds to a $\frac{160}{40\times 10^{-6}} = 4$ Mbps transmission rate.

3.3.1.2 Link Layer

IEEE 802.11 devices cannot detect collisions between their own transmissions and those of other devices because under wireless communications it is not possible to simultaneously transmit and receive frames. This results in a scenario where collisions can only be avoided. Because avoiding collisions takes longer than detecting them, latency is inherently higher in wireless systems than in wireline ones.

Example 3.4 Consider a sensor that transmits 200-byte packets every 20 ms. What is the communication system efficiency for both non-IoT IEEE 802.11ax and IoT IEEE 802.11ah scenarios? Assume the lowest possible transmission rates for both technologies.

(continued)

Table 3.9 IEEE 802.11ah

Modulation	Code bits/subchannel	Code bits/symbol	Code rate	Data bits/symbol	Data rate (Mbps)
BPSK	1	24	1/2	12	0.3
QPSK	2	48	1/2	24	0.6
QPSK	2	48	3/4	36	0.9
16-QAM	4	96	1/2	48	1.2
16-QAM	4	96	3/4	72	1.8
64-QAM	6	144	2/3	96	2.4
64-QAM	6	144	3/4	108	2.7
64-QAM	6	144	5/6	120	3.0
256-QAM	8	192	3/4	144	3.6
256-QAM	8	192	5/6	1608	4

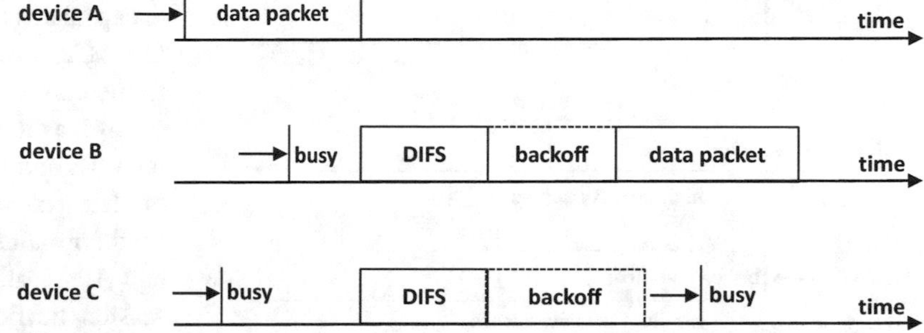

Fig. 3.38 IEEE 802.11 contention

(continued)
Solution The transmission rate R is given by $R = \frac{L}{T} = 80{,}000$ bps where $L = 8 \times 200$ bits and $T = 20$ ms. The efficiency under IEEE 802.11ax is given by $\eta_{\text{IEEE 802.11ax}} = \frac{R}{576{,}000{,}000 \text{ bps}} = 0.000138$ while efficiency under IEEE 802.11ah is given by $\eta_{\text{IEEE 802.11ah}} = \frac{R}{300{,}000 \text{ bps}} = 0.2667$. By virtue of being considerable slower, IoT technologies tend to occupy a comparative larger section of the available channel network bandwidth.

When many devices compete to simultaneously access a channel, MAC is by means of a contention-based mechanism known as *distributed coordination function* (DCF). DCF relies on the same CSMA/CA algorithm, described in Sect. 3.2.2, that 3G-PLC supports.

On the other hand, when the AP allocates specific timeslots for devices to individually access the channel, MAC is by means of a contention-free mechanism known as *point coordination function* (PCF). In the context of IoT where communications are M2P and traffic typically flows from devices to the AP but not between devices, PCF provides the most efficient mechanism for sensor and actuator data transmission. PCF is a master-slave architecture usually found in many WPAN solutions; however, its support is not mandatory under IEEE 802.11 so many implementations only rely on DCF for media access.

Figure 3.38 shows the time diagram of DCF mode where two devices, B and C, are trying to transmit data when a third device A is transmitting. First, both B and C wait for A to stop transmitting, and then, once the channel is free, they wait for a fixed amount of time known as *distributed interframe spacing* (DIFS). If at this point the channel is free, each device calculates a backoff time and starts transmitting if the channel is still free after this interval has elapsed. The backoff time C_w is randomly selected in the interval $(C_{w\min}, C_{w\max})$ in order to minimize the chances of repeated collision when devices are allowed to retransmit. Note that the C_w is computed as a multiple of the standard IEEE 802.11 slot time.

Besides DIFS, some devices access the channel by means of a shorter interframe spacing time known as *short interframe spacing* (SIFS). SIFS is typically used for immediate transmission of certain control frames and frame fragments. Figure 3.39 illustrates the time diagram when device A sends a frame to device B, while device C is also attempting to transmit a frame. Due to the shorter backoff time, device A transmits the frame first. This frame is marked to be acknowledged by the destination, so device B sends a special control frame that serves as acknowledgment. Because of the importance of this frame, the SIFS guarantees that the acknowledgment frame is transmitted right away, thus preventing other devices, including C, from sending any frames. Note that both SIFS and DIFS respectively provide high and low priorities for devices to access the channel. By

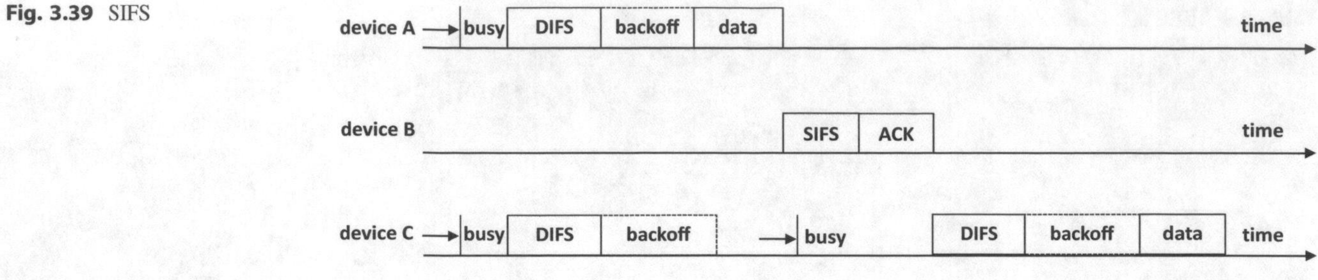

Fig. 3.39 SIFS

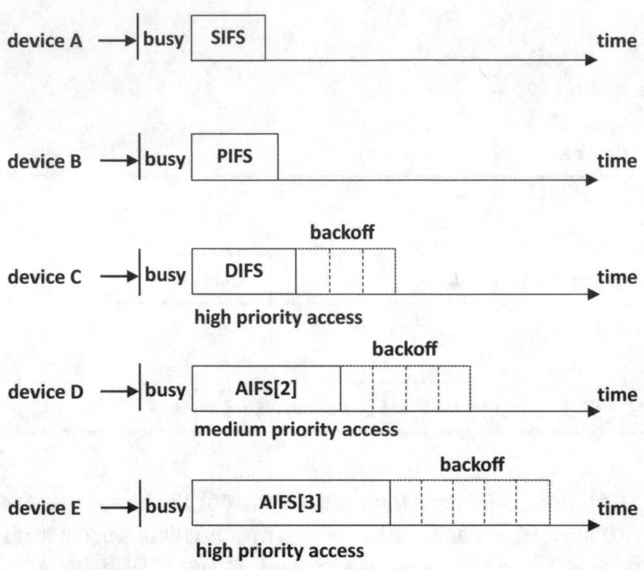

Fig. 3.40 SIFS

default, there are no other MAC priorities and no way to guarantee a specific QoS level. In general, given the nominal transmission rate of each of the standards that fall under the IEEE 802.11 umbrella, the actual transmission rates are much lower due to the latency introduced by CSMA/CA contention. The less severe the contention is, the closer to the nominal transmission rate the actual transmission rate becomes.

Devices that need predictability and guaranteed QoS can rely on PCF. The device that serves as point coordinator accesses the medium through an interframe spacing known as *PCF IFS* (PIFS) that lies somewhere in between the SIFS and DIFS spacings. Once the coordinator has control, it tells all other devices the duration of the contention free period. This enables these devices to access the medium, avoiding collisions, in a controlled fashion. Specifically, the point coordinator sequentially polls stations to check and verify whether they have frames to transmit. The PCF functionality, although part of the original IEEE 802.11 specifications, is typically implemented as part of enhancements introduced by IEEE 802.11e. IEEE 802.11e provides an extension to the basic DCF called *Enhanced DCF* (EDCF) that relies on an *Arbitrary Interframe Spacing* (AIFS). Essentially, AIFSs are mapped to three access priorities: high that supports a spacing like that of DIFS, medium, and low. Figure 3.40 shows the different AIFSs as well as SIFS and DIFS spacings.

Figure 3.41 shows a regular IEEE 802.11 frame that encapsulates a network layer datagram. Each frame consists of a MAC header, payload data, and a frame checksum. The frame control is a 2-byte field that indicates whether it is a control, a data, or a management frame. Control frames, like the CTS and RTS frames, are used to facilitate the exchange of data frames between devices. Management frames are used for maintenance of the communication between devices and APs. Data frames are used to carry real application and session traffic. The frame control field also indicates the encryption mechanism; *Wired Equivalent Privacy* (WEP), *Wi-Fi Protected Access* (WPA), or *Wi-Fi Protected Access II* (WPA2). This field, in addition, provides fragmentation as well as retransmission information. The duration id field has multiple uses; when included in data frames, it indicates the amount of air time the sending radio is reserving for the pending acknowledgment frame, while when included in management frames, it is used to indicate the AP association id. The addresses 1 and 2 are the receiver and sender MAC addresses, respectively. The addresses 3 and 4 are used for mesh routing and to provide connectivity between APs in an ESS. The sequence control field identifies the frame where the first 4 bits specify the fragmentation number, while the last 12 bits indicate the sequence number itself. Data frames are typically encapsulated following the IEEE 802.2 standard described in Sect. 3.2.1.

Once a device activates its IEEE 802.11 interface, it scans for other devices that are within range and ready for association. Scanning can be either passive or active. Under passive scanning, devices sequentially listen for beacon frames transmitted by other devices over IEEE 802.11 channels. These beacon frames provide relevant information related to synchronization and modulation as well as other physical layer parameters. Under active scanning, a device can send a probe frame indicating the SSID of the BSS it wants to be associated with. When the AP receives the probe, it replies with a probe response. Alternatively, active scanning can also involve a device sending a broadcast probe to all available devices. APs reply with response probes that are used by the devices to construct lists.

Fig. 3.41 IEEE 802.11 frame

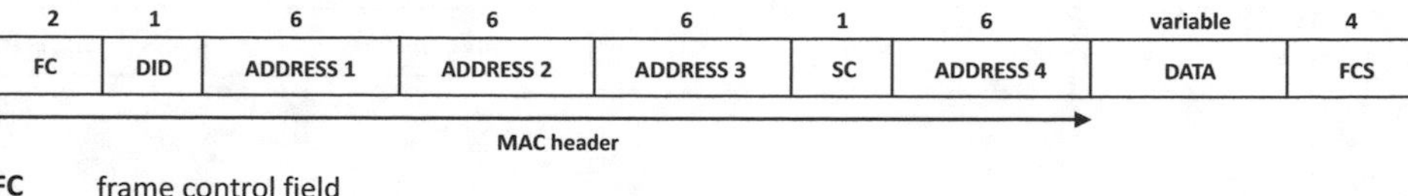

FC frame control field
DID duration id
SC sequence control field
FCS frame checksum

Devices that implement the IEEE 802.11 standard support two types of services: *station services* and *distribution services*. Station services deal with authentication and privacy between devices. Authentication is performed by devices before association. APs can be configured with open or pre-shared key authentication. Under open authentication, all devices are successfully authenticated providing very little security and validation. Shared key authentication relies on devices sharing a password that is pre-configured by means of an alternative channel like user configuration. Similarly, to authentication that occurs before association, deauthentication occurs before a device disassociates from a given device. Authentication and deauthentication are carried out through special management frames. Privacy between devices results from data frame encryption based on WEP, WPA, and WPA2. Distribution services deal with station services that span beyond communication between devices in each BSS. Association enables devices to logically connect with APs. APs cannot deliver any traffic from a device until it is associated. Disassociation is performed when a device leaves a network if, for example, it disables its IEEE 802.11 network interface. Reassociation allows a device to dissociate from an AP and associate to another one due to a stronger beacon being detected. In general, the distribution service provides the mechanisms for devices to send frames within BSSs and ESSs. Part of distribution services is integration that gives APs the capability to interface with networks that are not IEEE 802.11 based.

3.3.2 IEEE 802.15.3

IEEE 802.15.3 provides a physical and link layer specification for high-rate WPANs [15]. These WPANs are deployed as *piconets* with each piconet managed by a *Piconet Coordinator* (PNC) that supports a network of devices (DEVs) synchronized by means of beacons as illustrated in Fig. 3.42. The IEEE 802.15.3 nominal transmission rate is up to 55 Mbps with signals that are modulated as 64-QAM over the 2.4 GHz ISM band.

The isotropic coverage of a piconet is 10 m around the PNC that, in turn, can be fixed or mobile. In total, up to 245 devices can be part of a single piconet, with only one of them assuming the role of PNC. As mentioned above, the PNC periodically transmits beacon frames that are used by device

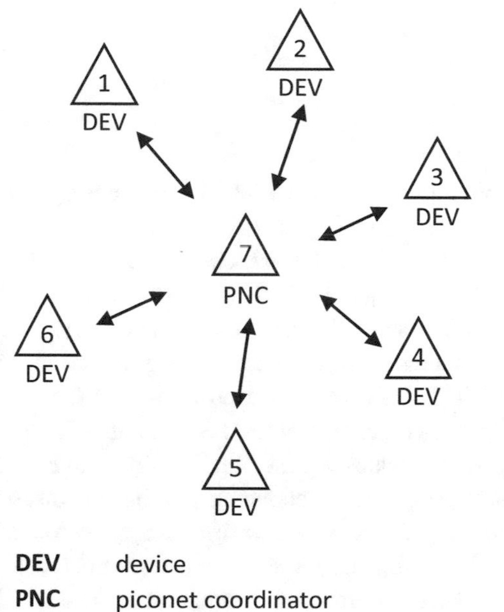

Fig. 3.42 IEEE 802.15.3 piconet

members for basic timing and synchronization. Moreover, the PNC also manages, among other things, power cycles, access control, and QoS requirements. In general, all traffic in the piconet is P2P single hop with the PNC always being the source or destination of frames.

Figure 3.43 shows the superframe structure IEEE 802.15.3 relies upon to provide timing. A superframe starts with a beacon; it continues with an optional *contention access period* (CAP) and ends with a *channel time allocation period* (CTAP). Management information as well as timing allocations are transmitted by the beacon. The CAP is provided by means of traditional CSMA/CA that supports the transmission of commands and asynchronous data between devices and the PNC. On the other hand, the CTAP is based on *time-division multiple access* (TDMA) that provides regular *Channel Time Allocations* (CTAs) and *Management Channel Time Allocations* (MCTAs). CTAs enable devices, including the PNC, to transmit commands and periodic data at fixed rates. This is useful in RTC scenarios where a device requests from the PNC a timeslot for transmission of real-time traffic. Transmission of asynchronous data is supported by means of the CAP, where devices attempt to transmit small chunks of data after requesting channel time for the total amount of time needed to transmit frames.

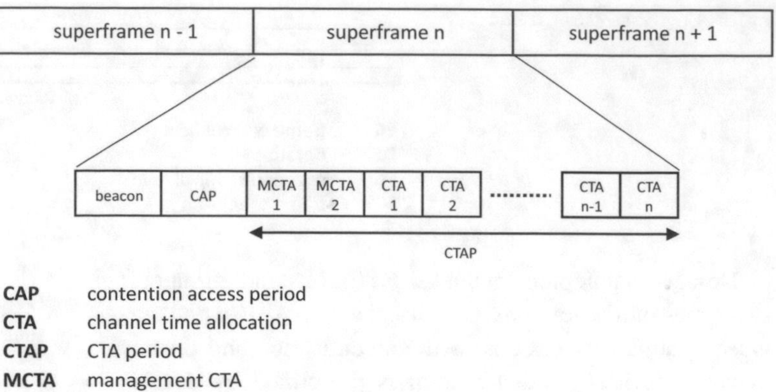

Fig. 3.43 IEEE 802.15.3 superframe

Devices join an IEEE 802.15.3 piconet by sending an association request to the PNC. The PNC processes the request and transmits an association response that assigns the device a unique 8-bit *Device Identifier* (DEVID) that identifies the device in the piconet. The device then sends a new association request now using the newly assigned DEVID. This handshake is used by the PNC to make sure that the device sends frames fast enough. In order to do so, the PNC creates a timer that expires after *Association Timeout Period* (ATP) seconds triggering the disassociation of the device. The device identifier plays the role of a short address by saving frame space when compared to the use of 64-bit MAC source and destination addresses. The PNC periodically checks for devices with expired ATPs to send a disassociation request to the devices. Similarly, if a device wants to leave the network, it can transmit a disassociation request to the PNC.

Acknowledgments are requested upon transmission of frames and four possible types may be requested: (1) no acknowledgment, (2) immediate acknowledgment that is transmitted upon the reception of a single frame, (3) delayed acknowledgment that is transmitted upon the reception of multiple frames, and (4) implied acknowledgment that is implied when the receiver transmits a frame back to the originator. This latter option was introduced in the IEEE 802.15.3b amendment of the original standard.

IEEE 802.15.3 also introduces resource sharing by providing mechanisms that enable devices to know what services are available in the piconet. Moreover, devices can also advertise their own services. Service information is carried by means of *Information Elements* (IEs) that are optionally exchanged during the piconet association process. Several commands provide the discovery of services by devices and the PNC including information request, probe request, and announce commands.

Piconets are identified through two parameters: the *Piconet Identifier* (PNID) and the *Beacon Source Identifier* (BSID). In order to start a new piconet, a device can act as a PNC that scans for the available channels in order to find one that is not being used. An open scan is typically

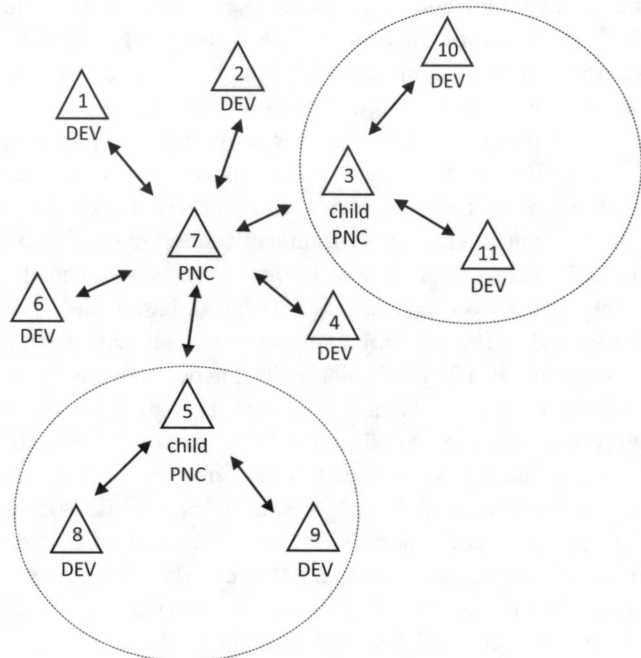

Fig. 3.44 Child piconets

initiated when the device looks for beacon frames to identify clear channels. If a clear channel is found, after waiting for a specified time period, the device becomes a PNC and starts sending beacons with specific PNID and BSID parameters. Potentially, if no free channels are found and a piconet is already present, the device can associate with it in order to interact with other devices. Otherwise the device can start a dependent piconet.

A piconet can be fully independent and have no child piconets, or it can be dependent and be a child or a neighbor of a parent piconet. A parent piconet allocates private CTAs for its child piconets to create their own networks. The child PNC coordinates its own devices while being a device of the parent piconet. Comparatively, a neighbor piconet is an autonomous network that is not a member of the parent piconet but that it relies instead on CTAs assigned by the parent PNC.

Figure 3.44 shows an IEEE 802.15.3 topology with a parent PNC and two child PNCs. The idea behind the child piconet support is double: (1) to extend the coverage area of the piconet and to (2) distribute computational and memory resources from the parent PNC to child PNCs. A parent piconet can have multiple child and neighbor piconets. Dependent piconets share the frequency channels of different piconets when no spectrum is available. The trade-off of physically extending a piconet is additional latency. Dependent piconets are functionally autonomous from their parent piconet and have different PNIDs, but they rely on CTAs in the parent superframe for transmission. A child PNC is a member of the parent piconet, and it can interact with any device in the parent piconet. Moreover, a child PNC is also a member of the child piconet and can exchange frames with any device in the child network. A neighboring PNC, however, is not a member of the parent piconet, and therefore it cannot exchange frames with devices in the parent piconet.

One issue with the IEEE 802.15.3 topology is that it does not allow, like many other WPAN technologies, support of a full mesh where network devices are directly connected to each other. Specifically, a child PNC can not only communicate with the devices in its child piconet, but it can also interact with its parent PNC and all devices in the parent piconet. Unfortunately, and regardless of the physical coverage, devices in the child piconet cannot communicate with the parent PNC or the devices in the parent piconet. This model of operation also introduces dependencies and single points of failure where if the parent PNC crashes all child piconets also stop working until the ATP expires. Similarly, if a parent PNC shuts down, it must select one and only one child PNC to take over the channel.

3.3.3 IEEE 802.15.4

The IEEE 802.15.4 specification introduces a set of physical and link layer technologies intended for use in WPANs [11]. As such it is one of the preferred mechanisms to support ultra low power consumption, and therefore long battery life, in the context of LLNs. IEEE 802.15.4 also serves as physical and link layer mechanisms for stand-alone standards like ZigBee [6], ISA 100.11a [3], and WirelessHART [7]. These technologies are well-known protocol stacks that are an integral part of many M2M and CPS solutions. While ZigBee relies on profiles that target home automation and smart energy scenarios, WirelessHART and ISA 100.11a target industrial automation and control. These standards use IEEE 802.15.4 in combination with upper proprietary layers that do not usually enable native IP connectivity.

In most modern IoT scenarios however, as shown in Fig. 3.45, IEEE 802.15.4 is used in combination with IETF protocols to provide efficient end-to-end IPv6 connectivity.

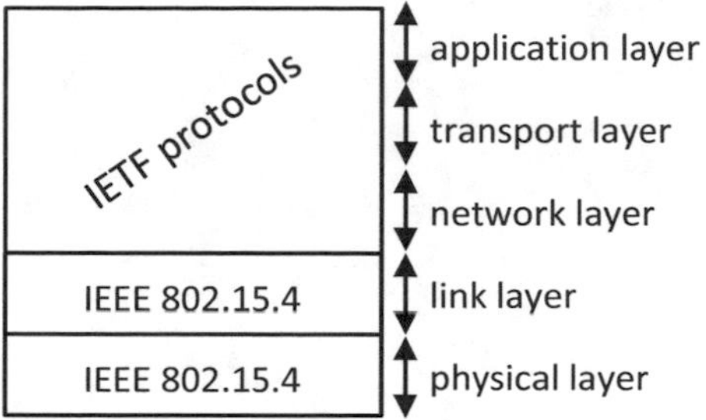

Fig. 3.45 IEEE 802.15.4 and IETF protocols

In all cases ultra low power consumption results from limiting both signal coverage and the amount of data being transmitted, thus decreasing transmission rates. This is additionally improved by managing duty cycles where power down and sleep modes are controlled.

The original IEEE 802.15.4 standard was released in 2003 and amended in 2006 to increase transmission rates by incorporating additional modulation schemes. This release led to IEEE 802.15.4a in 2007 that was consolidated as IEEE 802.15.4 in 2011 by further improving the offer of frequency bands and modulation schemes. In 2012 IEEE 802.15.4e introduced a channel hopping mechanism to improve resilience against channel interference by means of time synchronized multi-hop communications. IEEE 802.15.4e is specially designed to support industrial applications that are standardized in the context of IIoT. Note that IEEE 802.15.4e has been adopted as a link layer mechanism for both WirelessHART and ISA 100.11a. Other amendments to IEEE 802.15.4 include IEEE 802.15.4c and IEEE 802.15.4d that add support for specific frequency bands in China and Japan, respectively.

3.3.3.1 Physical Layer

The original IEEE 802.15.4 standard relies on DSSS modulated by means of OQPSK with 16 non-overlapping channels operating in the ISM 2.4 GHz band. The chip rate is 2×10^6 chips per second, and since each symbol is 32-chip long, it leads to a symbol rate of $\frac{2\times10^6}{32} = 62{,}500$ symbols per second. Because OQPSK implies 4-bit symbols, the overall transmission rate is $62{,}500 \times 4 = 250$ Kbps. The legacy IEEE 802.15.4 standard, also relying on DSSS but by means of BPSK, supports in the United States a transmission rate of 40 Kbps over ten ISM 915 MHz channels and in Europe a rate of 20 Kbps over a single ISM 868 MHz channel. In the current IEEE 802.15.4 standard, this support is extended to 100 and 250 Kbps in the ISM 868 MHz and 915 MHz bands, respectively.

IEEE 802.15.4a introduced two additional modulation schemes later integrated into IEEE 802.15.4: (1) one based on *direct-sequence ultra wideband* (DS-UWB) and (2) another based on *chirp spread spectrum* (CSS). DS-UWB supports precision ranging, and it is both efficient and robust for

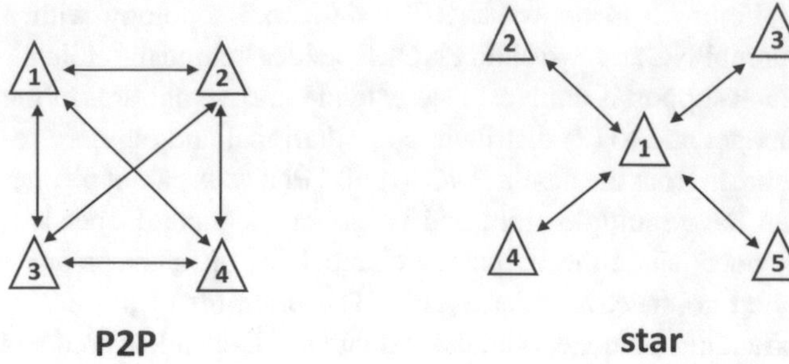

Fig. 3.46 RFD topologies

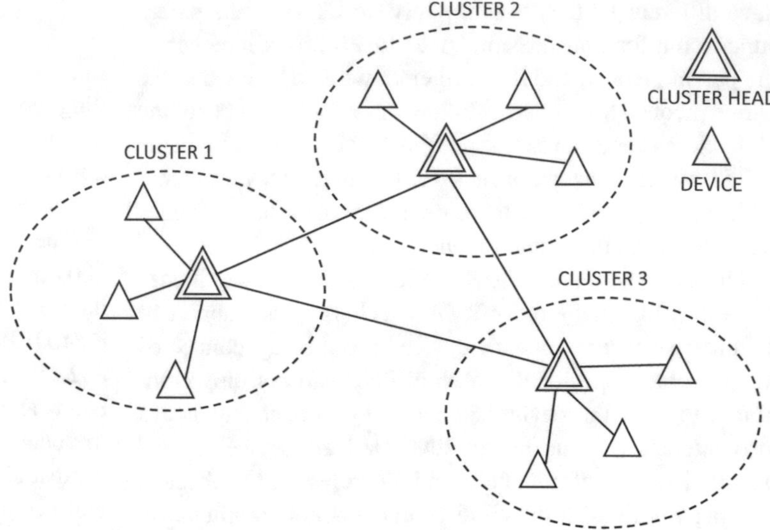

Fig. 3.47 Cluster networks

low transmission power. CSS is a particular type of spread spectrum that does not rely, as opposed to traditional DSSS and FHSS, on a pseudo-random sequence derived from a code to spread the spectrum. It spreads the spectrum, instead, by linearly varying the frequency of the sinusoidal carrier based on a spreading factor that results from the transmission of chirps. In either case, these changes in modulation attempt to improve the trade-off between power density and signal bandwidth in order to improve the SNR at the receiver.

Like the link layer of other technologies, the one of IEEE 802.15.4 provides many functions including the transmission of beacon frames, frame validation, node association, and security. IEEE 802.15.4 defines two device classes: (1) *Full Function Devices* (FFD) that can serve both as PAN coordinators and as regular devices in any topology and (2) *Reduced Function Devices* (RFD) that only provide node functionality in simple topologies like star and P2P. Figure 3.46 illustrates these two topologies supported by RFDs communicating with one or more FFDs. Summarizing, an FFD can coordinate a network of devices, while an RFD is only able to communicate with other devices. FFDs, besides supporting all topologies associated with RFDs, also support the cluster network topology shown in Fig. 3.47.

Figure 3.48 shows a scenario where multiple FFD and RFD topologies are combined. In fact, basic P2P can be extended into a mesh topology that supports up to 64,000 devices. In this situation efficient routing is many times accomplished by means of reactive mechanisms that, relying on requests and responses, outperform proactive table-driven routing.

Example 3.5 Consider a sensor that transmits 5-byte readouts every 20 ms and assume that the overhead due to network, transport, and session layer headers is 32 bytes. What is the efficiency of the communication system if the physical and link layers are IEEE 802.15.4? Assume that the IEEE 802.15.4 header size is minimum.

Solution The transmission rate R is given by $R = \frac{L}{T} =$ 18,800 bps where $L = 8 \times (5 + 32 + H)$, $T = 20$ ms and $H = 10$ is the minimum IEEE 802.15.4 header size as per information shown in Fig. 3.50. Note that the IEEE 802.15.4 header size in this case is 10 bytes long because it includes a minimum size 4-byte address field with no security. The efficiency is given by $\eta = \frac{R}{C} =$ 0.0752 where $C = 250{,}000$ bps, and it is therefore around 7.52%.

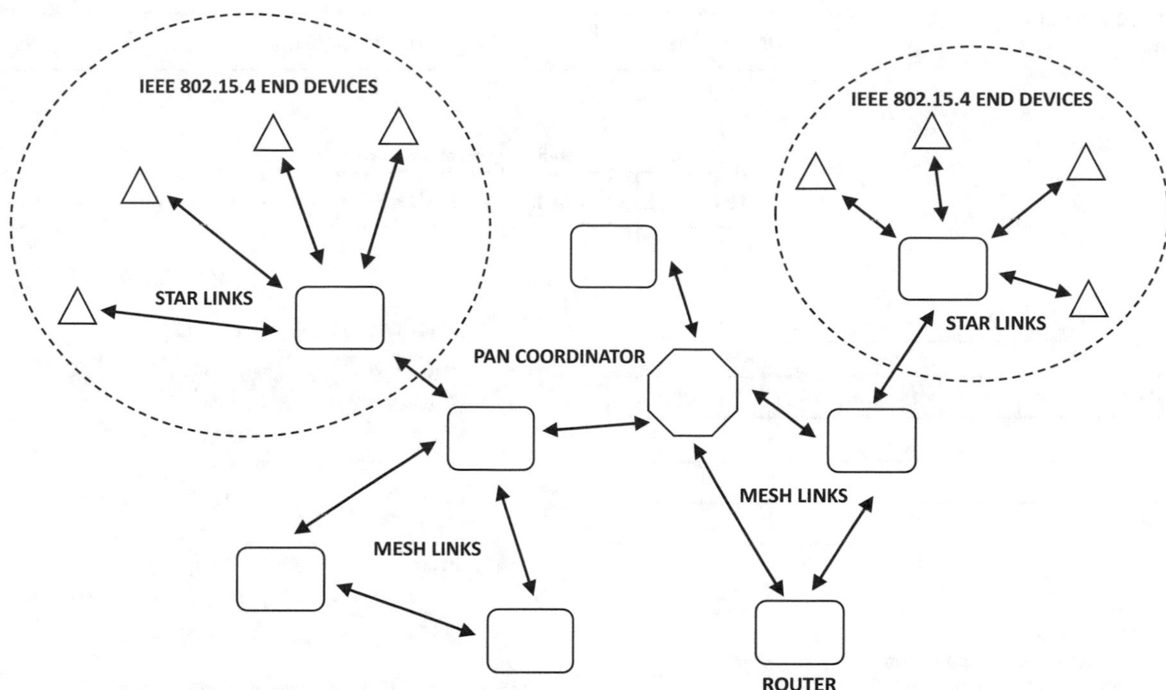

Fig. 3.48 FFD and RFD topologies

3.3.3.2 Link Layer

Under IEEE 802.15.4 media access is carried out by a combination of contention and contention free periods accessible via timeslots supported by both CSMA/CA and TDMA.

Figure 3.49 shows the structure of an IEEE 802.15.4 superframe that consists of a beacon transmission followed by 16 timeslots. The PAN coordinator transmits beacons that are used for device synchronization, to transmit configuration information and for identification. Beacons are sent at a fixed predefined interval between 15 ms and 252 s. Because beacon transmissions are quite infrequent and are not likely to suffer from collisions, they do not rely on CSMA/CA for media access. Contention-based access, that occurs right after beacon transmission, takes a section of the available 16 timeslots for devices to transmit relying on CSMA/CA. The remaining section of timeslots provides contention-free access where the PAN coordinator assigns, by means of TDMA, guaranteed access timeslots to certain devices. This is typically associated with scenarios where the application has predefined bandwidth requirements that require them to reserve exclusive timeslots for transmissions. The beacon carries a permit joining flag that is used to indicate devices that they can join the network. Under standard IEEE 802.15.4, network joining is initiated at the device, for example, by pushing a physical button. This contrasts with the *Thread* architecture presented in Chap. 9 that also relies on IEEE 802.15.4 but uses a different mechanism for devices to join a network.

The use of predefined intervals for the transmission of beacons combined with the presence of contention-free timeslots provide devices with a way to predict channel access and estimate duty cycles that are essential to minimize power consumption. During contention-based access, the CSMS/CA mechanism is like that of IEEE 802.11 described in Sect. 3.3.1.1. Devices sense the channel before transmission and backoff for a random time period before transmitting. Backoff is exponential as it doubles whenever a collision is detected. IEEE 802.15.4 supports basic reliability allowing devices to request acknowledgments of transmitted frames. If the far end device fails to send the acknowledgment due to network loss, the transmitter resends the frame. Only a fixed number of retransmissions are allowed before the frame is declared lost. An IEEE 802.15.4 network can also be configured in non-beacon mode when device transmissions are infrequent enough that contention-based access results in a lack of collisions. This scenario comparatively lowers topology and device complexity requirements.

Support of single channel communication, as enabled by the original IEEE 802.15.4 specification, results in unpredictable reliability especially in the context of multi-hop deployments where applications are time restricted. This is particularly true for certain IIoT applications that rely on media transmission. IEEE 802.15.4e addresses this problem by introducing the *Time Synchronized Mesh Protocol* (TMSP). TMSP provides a *Time Slotted Channel Hopping* (TSCH) MAC layer that relies on time synchronized channel hopping in order to overcome multipath fading and interference. Essentially, devices link themselves to groups of slots repeating over time where each slot is associated with

Fig. 3.49 IEEE 802.15.4 superframe

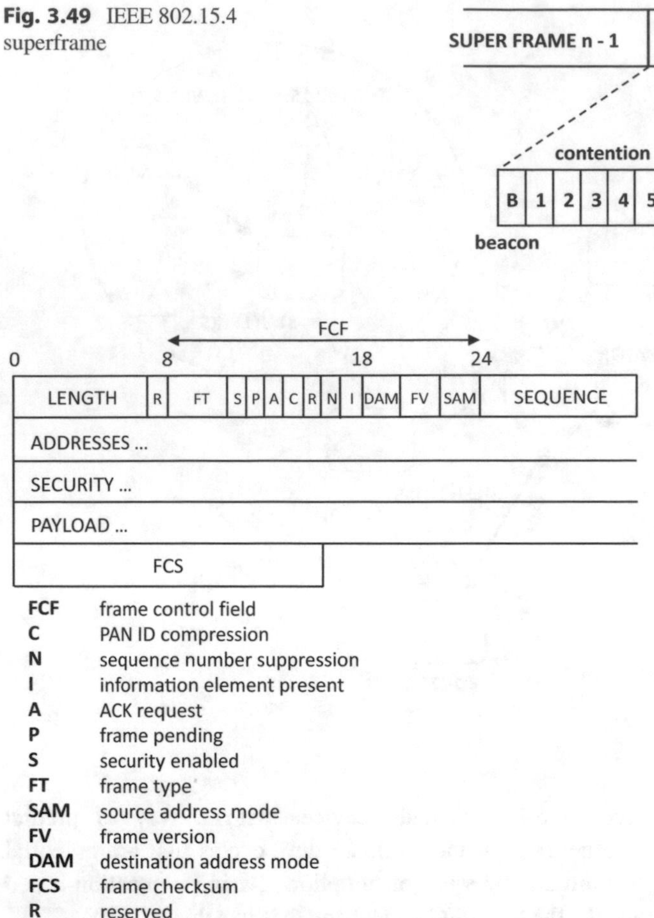

Fig. 3.50 IEEE 802.15.4 data packet

Table 3.10 Frame type encoding

Frame type value	Meaning
000	Beacon
001	Data
010	Acknowledgment
011	MAC command
100	Reserved
101	Multipurpose
110	Fragment
111	Extended

Table 3.11 SAM and DAM encoding

SAM/DAM	Meaning
00	Neither PAN ID nor the address field is given
01	Reserved
10	Address field contains a 16-bit short address
11	Address field contains a 64-bit extended address

a schedule that specifies what devices communicate with each other. In this context, synchronization of devices is key, two mechanisms exist: (1) acknowledgment based where the receiver computes the difference between the expected and actual arrival times of a frame and provides this information as feedback to the sender so it can synchronize itself against the receiver and (2) frame based where the receiver also computes the difference between the expected and actual arrival times of a frame but instead it adjusts its own clock to be synchronized with the sender.

IEEE 802.15.4 defines four frame types: (1) data frames that are used for transmission of regular frames; (2) acknowledgment frames that are transmitted by the receiver to confirm reception whenever requested by the sender; (3) MAC command frames that are typically used in beacon mode to enable MAC services like network association, dis-association, and management of synchronized transmission by the PAN coordinator; and (4) beacon frames that are used by the PAN coordinator to signal physical layer parameters and trigger communication with associated devices.

Figure 3.50 shows an IEEE 802.15.4 frame. The frame starts with a 7-bit frame length field and a reserved bit used for future extensions that are followed by a 16-bit *Frame Control Field* (FSF). The length value accounts for the entire frame size including training checksum bytes. It thus limits the maximum frame size to 127 bytes. The FSF includes several fields (1) *PAN ID compression* that is used to indicate whether the 16-bit *PAN Identifier* (PAN ID) is included as part of the MAC addresses, (2) *ACK request* to ask for the transmission of an acknowledgment frame when this frame is received, (3) *frame pending* that enables a sender to tell the receiver that it has more frames to send in order to maximize channel utilization, (4) *security enabled* that indicates that the frame includes security parameters, (5) *frame type* field to signal the nature of the frame as encoded in Table 3.10, (6) *frame version* that specifies the IEEE 802.15.4 revision that applies to this particular frame, and (7) *Source Address Mode* (SAM) and *Destination Address Mode* (DAM) fields that indicate how source and destination MAC addresses are encoded. SAM and DAM encoding is shown in Table 3.11, (8) *sequence number suppression* and (9) *IE present* bits. IEEE 802.15.4 enables the use of the FSF to indicate additional members that are used to send information between devices. Specifically, information between neighbors is carried by means of IEs that are included in the frame when the IE present bit in the FSF is set. IEs are not encrypted but they are authenticated.

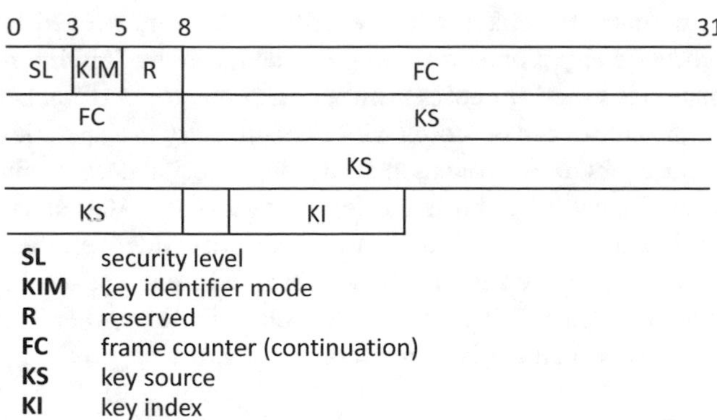

Fig. 3.51 Security header

Following the control field, the frame includes an 8-bit sequence number used for tracking of fragments and acknowledgments. The variable length source and destination addresses, as encoded by the SAM and DAM fields, follow. IEEE 802.15.4 relies on unique 64-bit long addresses that are hardcoded in all radio chips and 16-bit *short* addresses configured by the PAN coordinator. The 64-bit addresses are formed based on an IEEE 64-bit *Extended Unique Identifier* (EUI-64). Short addresses simplify addressing in restricted environments, while long addresses provide global reachability. If either 16-bit or 64-bit MAC addresses are transmitted, then 16-bit PAN ID values must also be included unless the PAN ID compression field is set. When set, this means that the PAN ID for both addresses is the same, and therefore it can be removed from the source address. This leads to an address field that can be anywhere between 4 and 20 bytes long.

Although IEEE 802.15.4 security, when enabled, is present at the link layer, it also protects the upper layers. In fact, IEEE 802.15.4 security is typically provided through hardware and firmware implementations that are highly efficient. This contrasts with the security provided by upper layer protocols that is implemented as user space software libraries. Of course, security services are terminated and reestablished between devices and entities at the end of the link. This can be considered a violation of the well-known *end-to-end principle* of network topologies. This principle states that, whenever possible, certain functions like security must be deployed on an end-to-end basis typically at the application layer. This principle, when carried out, guarantees maximum efficiency improving overall latency, throughput, and security.

If security is enabled in the IEEE 802.15.4 FSF, the auxiliary security header, shown in Fig. 3.51, follows the address field. The frame ends with a 16-bit CRC block code that is used as FCS. This header includes an 8-bit security control field that specifies the 3-bit security level described in Table 3.12. This level indicates whether the payload is

Table 3.12 IEEE 802.15.4 security modes

Security mode	Security provided
No security	Data is not encrypted Data authenticity is not validated
AES-CBC-MAC-32	Data is not encrypted Data authenticity is 32-bit MIC
AES-CBC-MAC-64	Data is not encrypted Data authenticity is 64-bit MIC
AES-CBC-MAC-128	Data is not encrypted Data authenticity is 128-bit MIC
AES-CTR	Data is encrypted Data authenticity is not validated
AES-CCM-32	Data is encrypted Data authenticity is 32-bit MIC
AES-CCM-64	Data is encrypted Data authenticity is 64-bit MIC
AES-CCM-128	Data is encrypted Data authenticity is 128-bit MIC

Table 3.13 KIM encoding

KIM	Meaning
00	Key identified by source and destination address
01	Key identified by macDefaultKeySource + 1-byte key index
10	Key identified by 4-byte key source + 1-byte key index
11	Key identified by 8-byte key source + 1-byte key index

protected by encryption and/or message authentication. The 32-bit frame counter is used to provide replay protection and to keep track of the sequence of frames in order to apply chain encryption mechanisms. Additionally, the security control field also includes a *Key Identifier Mode* (KIM) shown in Table 3.13 that indicates how the keys required to process security are to be determined by the devices. IEEE 802.15.4 relies on 128-bit keys that may be implicitly known by transmitter and receiver, or they may be derived from information transported in the 64-bit *Key Source* (KS) and 8-bit *Key Index* (KI) fields. In general, the KS field indicates the group key originator, while the KI field is used to identify a key from a specific source. KS and KI together form the *Key Control* (KC) field.

The security modes shown in Table 3.12 carry information related to security in the different configurations shown in Fig. 3.52. For scenarios where only confidentiality is needed, the payload is encrypted by means of AES-CTR that relies on *Advanced Encryption Standard* (AES) encryption in *counter* (CTR) mode. For scenarios where message authentication is needed, a *Message Integrity Code* (MIC) also known as *Message Authentication Code* are calculated over the IEEE 802.15.4 header and payload by means of AES-CBC that relies on AES encryption in *Cipher Block Chaining* (CBC) mode. This leads to AES-CBC-MAC-32, AES-CBC-MAC-64, and AES-CBC-MAC-128 based on whether the MIC is, respectively, 32, 64, or 128 bits long. The MIC is then appended to the unencrypted payload. For scenarios where both confidentiality and message authentication are needed, CTR and CBC modes are combined into a counter combined mode that leads to security modes AES-CCM-32, AES-CCM-64, and AES-CCM-128 for MIC sizes of 32, 64, and 128 bits, respectively. In this case, encryption is applied after message authentication is performed. Note that the frame counter and the KC fields are set by the sender whenever encryption is in place. The frame counter is typically incremented each time a frame is hardware encrypted, while the KC is set by the application and contains the KI that identifies the key in use.

The original datagram is chunked into several 16-byte blocks that are identified through a block counter. Each block is encrypted using a different *initialization vector* (IV) that is computed on-the-fly for each frame as illustrated in Fig. 3.53. Essentially the IV consists of a 1-byte flag, the 64-bit long source address, the 32-bit frame counter, the 8-bit frame control, and the 16-bit block counter. Note that the block counter is not sent so the receiver typically estimates its value based on the order of block within the frame.

IEEE 802.15.4 introduces support of access control by allowing devices to use the source and destination addresses

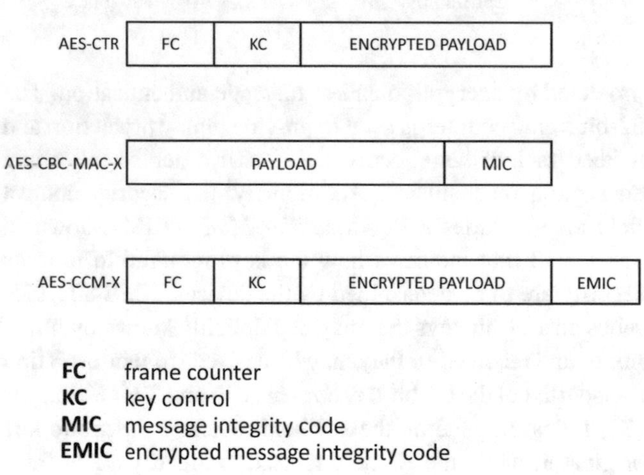

Fig. 3.52 Payload data formats with IEEE 802.15.4 security

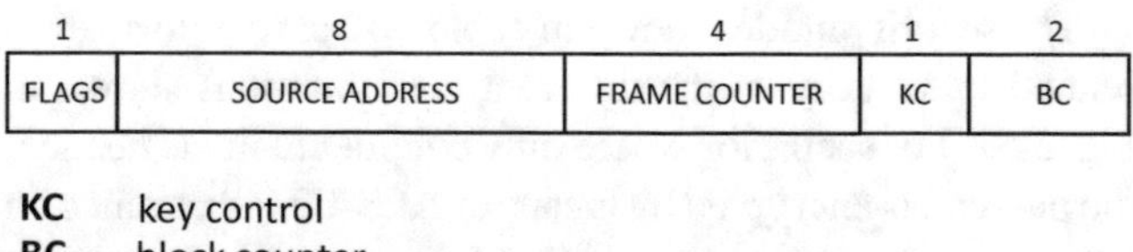

Fig. 3.53 Format of the IV

ADDRESSES	SECURITY SUITE	KEY	LIV	RC

LIV last IV
RC replay counter

Fig. 3.54 IEEE 802.15.4 ACL entry

of a frame to look for the security parameters needed to process encryption and message authentication. This information is stored as entries in an *access control list* (ACL). Up to 255 entries can be stored with a default entry that specifies security for those frames that are not associated with any entry. Figure 3.54 shows the format of a single ACL entry. Each entry includes the addresses, a security suite identifier, the cryptographic key, and, for scenarios with encryption, the last available IV. If replay protection is in place, then the latest frame counter is stored too.

3.3.3.3 TSCH

TSCH, introduced in IEEE 802.15.4e, is the MAC layer of the TMSP. It provides a way to map timeslots to user channels based on a preassigned hopping sequence. The main goal is to make sure that no frames collide on the network and, therefore, frames are scheduled and synchronized for energy efficiency with no need for an extra preamble or guardbands. Collision-free communication is accomplished by means of the aforementioned schedule that allows multiple devices to communicate at the same time over different frequencies. As a collision-free mechanism, synchronization is critical under TSCH. Specifically, synchronization is carried out by means of frames exchanged between neighbors. If devices do not receive any traffic for a predefined time period, they can transmit dummy frames to trigger synchronization.

Under TSCH all devices are fully synchronized with timeslots that are long enough to fit a full size frame and its acknowledgment. Timeslot durations of 10 ms are typical, providing enough time for transmission, propagation, and processing delays. Timeslots are grouped into slotframes that are periodically transmitted. Depending on the application, a single slotframe can hold anywhere between tens and thousands of timeslots. Shorter slotframes are associated with higher transmission rates but also higher energy consumption. TSCH allows a single schedule to rely on multiple simultaneous slotframes. At some point in time, a device can perform two simultaneous tasks on two different slotframes like receiving frames from device 2 in slotframe 1 and transmitting frames to device 3 in slotframe 2. Two rules specify how devices behave in a multi-slotframe scenario: (1) transmissions have higher priority than receptions, and (2) lower slotframes have higher priority than higher slotframes.

The TSCH schedule specifies what a given device does in a timeslot. Specifically, it indicates the channel offset associated with hopping and the address of the neighbor with which to communicate. Possible actions in the schedule include receiving or transmitting data as well as sleeping. For each transmitting timeslot, the device checks whether there are pending frames in the outgoing queue for the destination. If there are no frames, the device keeps its radio turned off in order to save energy; otherwise it transmits the pending frame requesting an acknowledgment whenever needed. Similarly, for each receiving timeslot, the device listens for incoming frames; if no frames are received for a certain amount time, the device shuts off its radio; otherwise if it receives a frame, the device can send an acknowledgment if needed.

The combination of the timeslot and channel is called a cell. A cell represents an atomic unit of scheduling. A single device can communicate with another one through one or more cells. If multiple cells carry traffic between two devices, then the scheme is called a bundle. The larger the bundle is, the higher the transmission rate from one device to the other. TSCH also provides a mechanism that enables multiple devices to simultaneously transmit over the same cell. In this case of shared cells, contention is prevented by means of a backoff algorithm.

Each cell is determined by both parameters: the channel offset and the timeslot offset that indicate the spectral and temporal position of the cell. TSCH introduces a global timeslot counter called *Absolute Slot Number* (ASN) that specifies the number of timeslots that have been scheduled since the start of the network. In general, $\text{ASN} = k \times S + t$ where k and S are the slotframe cycle and size, respectively, and t is the timeslot offset. The ASN value is broadcasted throughout the network for devices to synchronize when they join the network. The ASN is 40 bits long in order to support hundreds of years without overflowing.

Although the channel offset is constant for any given transmission from one device to another, this does not mean the transmission frequency is constant. In fact, channel hopping is supported by a transmission frequency calculated as $\text{frequency} = F((\text{ASN} + c) \mod N)$ where c is the channel offset, N is the maximum number of frequencies, and $F(\ldots)$ is a mapping to one of the possible N frequencies. Because ASN is an incrementing counter, for any fixed cell, the transmission frequency changes in each slotframe. This enables channel hopping even when keeping the schedule simple by relying on fixed cells associated with fixed channel and timeslot offset. Channel hopping is essential in order to prevent multipath fading and interference as these impairments are frequency dependent.

The TSCH network schedule is linked to the application needs; a sparse schedule implies a solution where devices consume very little energy, but throughput is very low too, while a dense schedule implies a solution where devices generate a lot of traffic and consume a lot of energy. TSCH introduces several IEs that can be used for communication in the context of this mechanism.

IEEE 802.15.4e [4] adapts security to support TMSP; it defines the possibility of using optional 40-bit frame counters that are set to the ASN of the network. The use of the ASN provides time-dependent security as well as replay protection. In order to support the use of the 40-bit frame counter, IEEE 802.15.4e modifies the security control field to use two of the originally reserved bits to (1) enable suppression of the frame counter field and to (2) indicate whether the frame counter field is 32 or 40 bits long. When ASN-based encryption is in use, the 40-bit frame counter is followed by the key control field as indicated by Fig. 3.52.

3.3.3.4 Limitations

IEEE 802.15.4 exhibits some limitations that affect its reliability in the context of LLNs. Specifically, one of the most important limitations affecting not only IEEE 802.15.4 but also many other IoT technologies is due to the interference associated with other transmissions over the ISM bands. This is particularly critical when considering the overcrowded 2.4 GHz band used by different versions of IEEE 802.11 and BLE as well as other devices ranging from remote control toys to cordless phones. Additionally, radio propagation is mainly by the way of scattering over surfaces and diffraction over and around them in a situation typical of a multipath environment. Essentially, in this scenario, multipath fading results from signals taking different paths and arriving at the receiver with different phases. If all signals interact positively (i.e. they have same phase), then there is constructive interference or upfading, while if, on the other hand, signals interact negatively (i.e. they have opposite phase), then there is destructive interference or simply fading. As opposed to IEEE 802.15.4 that is greatly affected by fading, IEEE 802.15.4e attempts to address some of these problems through TMSP.

IEEE 802.15.4 also limits the communication range. Specifically, the coverage is around 200 m in outdoor environments. Longer coverage is possible; however, it requires the use of multi-hop transmissions in the context of mesh forwarding. Devices that provide message relaying introduce three problems: (1) additional deployment costs, (2) increased latency, and (3) reduction of communication reliability. This latter point is associated with the fact that multi-hop communication relies on systems that are "in series" and as such they have multiple points of failure when compared to one-hop communication schemes. One way to improve reliability is by increasing transmission power; however, the nature of IoT solutions typically prevents this.

3.3.4 Bluetooth Low Energy

Bluetooth is a full protocol stack that supports small coverage while providing low to medium transmission rates under wireless communications [13]. Initially released in 1998 as Bluetooth 1.0 and 1.0B by the Bluetooth Special Interest Group, it has been updated several times since then. In 2002 it was amended as Bluetooth 1.1, also released as IEEE 802.15.1-2002, to address several deficiencies of the original specification and to support *Received Signal Strength Indicator* (RSSI) as a mechanism for power control. This standard led to Bluetooth 1.2, also released as IEEE 802.15.1-

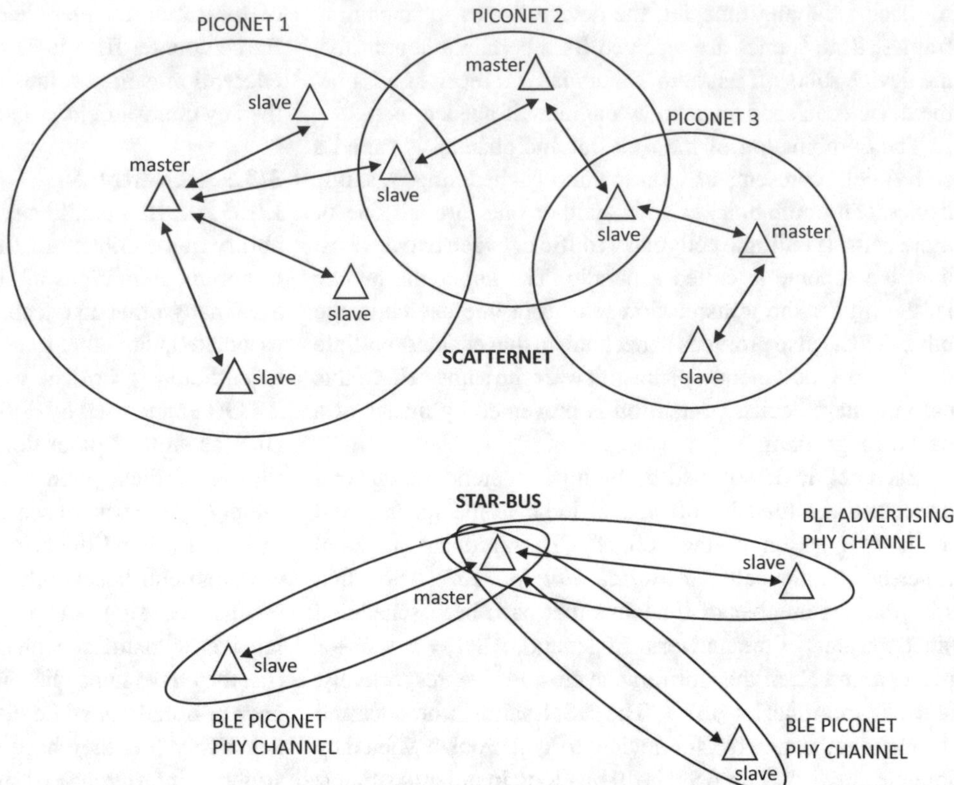

Fig. 3.55 BLE and Bluetooth topologies

2005, to improve modulation and transmissions rates among other things. *Enhanced Data Rate* (EDR), introduced in 2004 as Bluetooth 2.0, further improves transmission rates and energy consumption by incorporating power duty cycles. Bluetooth 2.1 was released in 2007 to simplify the device pairing process. In 2009, Bluetooth 3.0 introduced *High Speed* (HS) support by means of co-located IEEE 802.11 links. BLE, also known as Bluetooth Smart and standardized as Bluetooth 4.0, was released in 2010 as a brand new mechanism to support low power consumption [18]. As such it has been an excellent candidate for many M2M, CPS, and hybrid IoT solutions. BLE is not backwards compatible, so classic Bluetooth devices cannot directly interact with pure BLE-based topologies. Bluetooth 4.0 relies on devices supporting both mechanisms independently. Bluetooth 4.1 and Bluetooth 4.2 were, respectively, released in 2013 and 2014 to address certain requirements in order to enhance accessibility of connected devices. Bluetooth 5.0, released in 2016, boosted BLE modulation and transmission rates. In 2019 Bluetooth 5.1 added *angle of arrival* (AoA) and *Angle of Departure* (AoD) to improve asset tracking complementing the use RSSI [14].

It is important to emphasize that IEEE 802.15.1 [1] is the standardization of Bluetooth 1.1 and Bluetooth 1.2 and it does not refer to any other version of Bluetooth including BLE. The physical and link layers of BLE, in a similar way to ZigBee, are mechanisms that by means of adaptation can be used to transport IPv6 traffic that is essential to IoT.

Under classic Bluetooth, a master device controls up to seven slaves on a single piconet. In the context of Bluetooth, a piconet is an ad hoc network that links a wireless user group of devices. Slaves communicate with the master, but they do not communicate with each other. A slave can also belong to more than one piconet. An example of a classic Bluetooth topology is shown in Fig. 3.55 with multiple piconets deployed in a scatternet configuration. Under BLE, however, slaves talk to the master on a separate channel. As opposed to classic Bluetooth piconets, where slaves listen for incoming connections from the master, a BLE slave initiates connections, and it is therefore in control of power consumption. A BLE master, which is assumed not to be as power constrained as a slave, listens for advertisements and initiates connections as a result of received advertisement frames. An example BLE topology is also shown in Fig. 3.55 deployed in a star-bus configuration.

3.3.4.1 Physical Layer

Classic Bluetooth relies on FHSS at a rate of 1600 hops per second with each code spanning over 79 channels in the 2.4 GHz ISM band. The code that controls the hopping pattern is derived from the 48-bit MAC address of the master device. BLE, on the other hand, uses a different FHSS scheme as it relies on 40 2 MHz channels that enable higher reliability over long-range coverage. Classic Bluetooth nominal transmission rates are 1, 2, and 3 Mbps, while the BLE nominal transmission rate is 1 Mbps with a typical throughput of 260 Kbps.

The individual channel modulation schemes used by FHSS to accomplish those transmission rates are *Gaussian FSK* (GFSK) for 1 Mbps, DQPSK for 2 Mbps, and D8-PSK for 3 Mbps. GFSK is a version of FSK where a type of digital filter known as Gaussian is used to smooth the transitions between symbols and therefore minimize the noise. One problem with FSK-based modulations is that transmitting an all-ones sequence like *1111111111...111* results in modulating the same symbol and therefore transmitting a one-tone signal for a long period of time. This signal confuses the detector that dynamically adjusts to small frequency changes and causes it to fail to demodulate the next zero. One way to prevent this situation is by introducing a whitening mechanism that randomizes in a controlled fashion by means of a scrambler the sequence of transmitted bits. Figure 3.56 shows the scrambler; it relies on a feedback shift register that provides a pseudo-random bitstream that is bitwise added to the input sequence. The resulting sequence serves as input to the modulator.

BLE power consumption is somewhere between 1 and 50% of that of classic Bluetooth. Power consumption also dictates the transmission rate as shown in Table 3.14. Note that classic Bluetooth, as opposed to BLE, defines three classes that link power consumption to distance. Class 1 and BLE radios implement *transmit power control* (TPC) that forces transmitting devices to adjust power in order to minimize interference and extend battery life. Under TPC, the RSSI is used to determine whether signals are received within an acceptable range in order to vary power accordingly. Note that TPC is optional in class 2 and 3 radios.

Fig. 3.56 Scrambler

Table 3.14 Power vs. range

Class	EIRP	Range (m)
BLE	10 μW < 10 mW	<50
1	2.5 < 100 mW	<100
2	1 mW < 2.5 mW	<10
3	<1 mW	0.1 < 1

3.3.4.2 Link Layer

The BLE link layer is based on the state machine shown in Fig. 3.57. The machine defines five states: (1) advertising, (2) init, (3) standby, (4) scan, and (5) connection. The link layer is in standby state when it first starts. In this state the link layer is inactive, and it can transition to and from the scan, advertising or init states. In addition, in the *connection* state, it can also transition back to standby. In advertising state, a device sends advertising frames and responds to incoming scan requests from other devices. A device in advertising state is discoverable and can transition, as a slave, to connection state when it receives a connection request from a master device in init state.

In scan state, a device listens for advertising frames in the wireless channel. Two types of scanning are possible: passive where the device listens for frames and active where the device sends scan requests to induce other devices to send advertising frames. If a device stops scanning, it transitions back to the standby state. When a device receives an advertising frame, it can attempt to connect to the advertising device by sending a connect request that triggers a transition, as a master, to the connection state. When the advertising device receives the connection request, it also transitions to the connection state as a slave. In connection state, master and slave devices send and receive data frames. A master sends periodic data frames to a slave in order to give it an opportunity to reply and send its own data frames. Specifically, when a slave receives an individual data frame, it can send back another data frame. If the slave wants to send more frames, it must wait for the master to send more of these periodic frames. A slave can go to sleep, by ignoring data frames from the

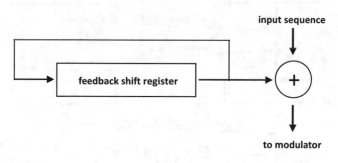

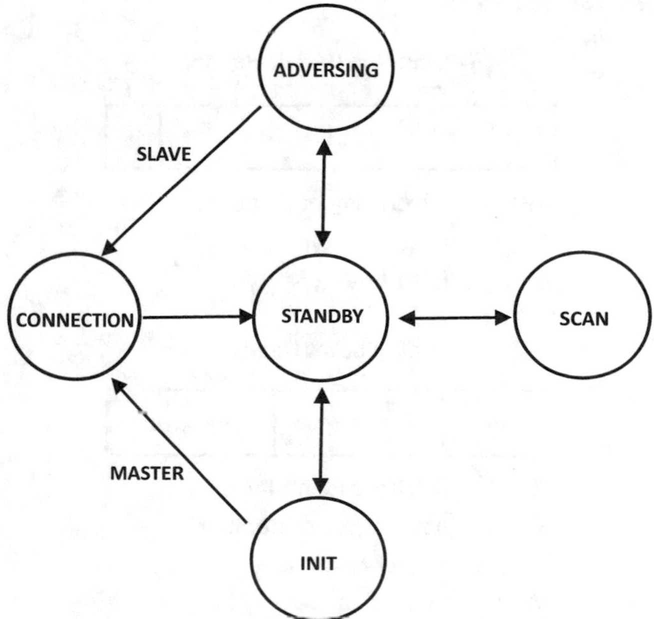

Fig. 3.57 Link layer state machine

master, and reduce power consumption. Advertising frames are used to set up connections between devices, while data frames are used for communication traffic between devices. Advertising frames are sent over three advertising channels, while data frames are sent over 37 different data channels. The upper sublayer of the BLE link layer is called the *Logical Link Control and Adaptation Protocol* (L2CAP) that enables the multiplexing of data with higher layer protocols and supports segmentations among other things.

Figure 3.58 shows a generic BLE frame. It starts with a preamble that is followed by an access address field that is used by the receiver to filter out this frame from channel noise. This field is in turn followed by a packet header field that is used to indicate the nature of the frame. The length field signals the size of the data payload. The last field of the frame is a FCS that by means of a CRC block code provides basic channel encoding. The preamble is an 8-bit sequence of alternating ones and zeros that are used by the receiver to detect the modulation frequencies and perform gain control. The access address is a 32-bit field that can be an advertising access address or a data access address. The advertising access address is always *10001110100010011011111011010110* and is used for data broadcasting as well as advertising and scanning frames.

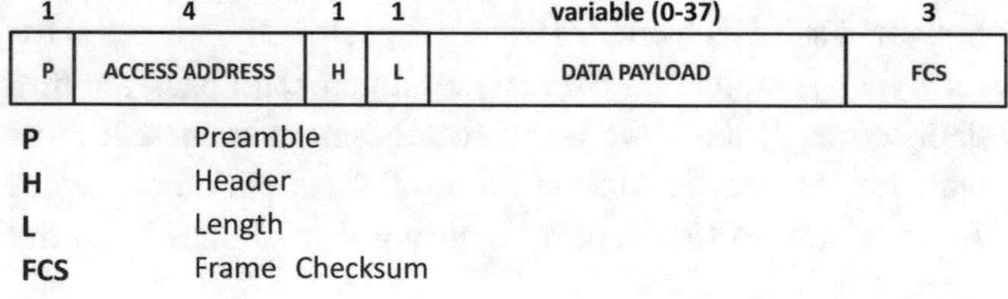

Fig. 3.58 BLE frame

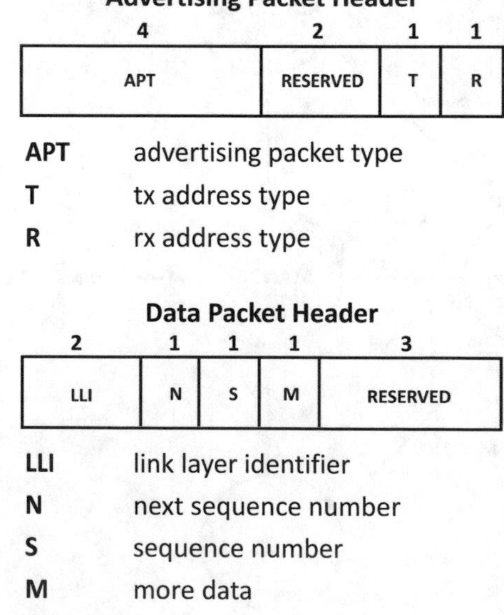

Fig. 3.59 Packet header

The data access address is a random 32-bit number that identifies a connection between two devices. The 8-bit packet header field is encoded as indicated in Fig. 3.59 depending on whether it applies to advertising or data frames. For advertising frames, the header includes an advertising frame type and two bits to indicate whether transmission and reception addresses are public or random. The possible advertising frame types are (1) advertising indication, (2) connection indication, (3) non-connectable indication, (4) scannable indication, (5) active scanning indication, (6) active scanning response, and (7) connection request. For data frames, the header includes a logical link identifier that indicates whether the frame is a segment of higher layer datagram or whether it is used for connection management. This header also includes a sequence number that identifies this data frame within the transmission stream and a next sequence number that tells the peer device what the expected receiving sequence number is. Finally, the *more* data field in this header tells the receiver that there are more frames to be sent and therefore the connection should stay active and power management in effect.

Figure 3.60 shows how first a device, an advertiser, sends an advertising frame that triggers another device, an initiator, to transmit a connection request. Once the connection is established, the initiator becomes master and the advertiser becomes slave. From this point on, data frames are sent by both devices. Whenever a data frame is sent, the sending device specifies the sequence number of the transmitted frame (S) and expected sequence number of next incoming frame (N) as well as whether more frames are to follow (M). If a

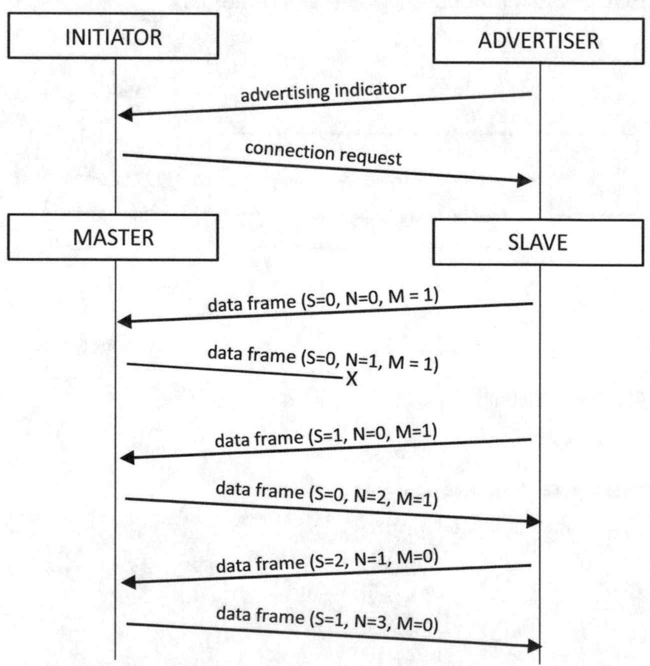

Fig. 3.60 Connection creation and data transfer

frame is lost and, therefore, it does not arrive at the endpoint, the transmitter knows it has to retransmit it when it analyzes the expected sequence number of the incoming frame. In the exchange of frames, each frame is further identified by device addresses that follow the length field as part of the data payload. Specifically, 48-bit initiator and advertising MAC addresses are used to identify the transmitting devices. Each MAC address can be a public device address that is assigned by the manufacturer or a random device address that can be either static if it is reset on power cycle or private if it periodically changed to prevent device tracking.

3.3.5 ITU-T G.9959

ITU-T G.9959 provides physical and link layer wireless mechanisms for low-rate data transmission that are derived from the well-known consumer-oriented Z-Wave technology [20]. Z-Wave focuses mainly on sensing and actuation functions in home and small business facilities. Some common examples of Z-Wave use include physical security, lighting and climate control, smoke detectors, appliances, remote controls, as well as smart meters. In a similar way to ZigBee and IEEE 802.15.4, Z-Wave and ITU-T G.9959 support wireless mesh topologies that enable devices to communicate with other devices even if they are not within range. This is done by means of traffic transmission over intermediate nodes. The architecture also supports master controllers that provide wider access to devices. An individual ITU-T G.9959 WPAN can integrate up to 232 devices, and multiple master controllers can be used to partition the network based on different functions.

3.3.5.1 Physical Layer

ITU-T G.9959 operates on the 915/868 MHz ISM band but relies on other bands in many other countries based on licensing and regulations. Modulation is based on plain FSK and GFSK in a similar way to BLE but without FHSS. Three different transmission rates are supported by this technology through physical layer configuration changes: R1 at 9.6 Kbps, R2 at 40 Kbps, and R3 at 100 Kbps. The signal coverage is up to 30 m depending on power and obstruction conditions. The electrical transmission power is 1 mW.

Figure 3.61 shows the building blocks of the physical layer. The data stream is first modulated by FSK or GFSK followed by Manchester or Polar NRZ line coding depending on the rate as shown in Table 3.15. Line coding is introduced to minimize the DC component of the transmitted signal and improve range as well as penetration.

3.3.5.2 Link Layer

ITU-T G.9959 relies on a traditional CSMA/CA MAC as most wireless technologies discussed in this section. Collision avoidance is active for all devices in the network if their radios are active. It is completely independent of the transmission rate mode R1, R2, or R3. ITU-T G.9959 also introduces both power efficiency by means of predictable duty cycles and reliability by means of retransmissions.

ITU-T G.9959 supports three channel configurations. Configuration one supports one channel named channel A. Configuration two supports two channels named channel A and channel B. Configuration three supports three channels named channel A, channel B, and channel C. Figures 3.62 and 3.63, respectively, show IEEE G.9959 unicast and multicast MAC frames for channel configuration one and two. The frame format for channel configuration three is identical, but it also includes a sequence number field. Each frame starts with a preamble that is used for device synchronization, follows with a *Start of Frame* (SoF) field and a *MAC Protocol Data Unit* (MPDU), and ends with an *End of Frame* (EoF) field. From a topology perspective, ITU-T G.9959 networks are typically deployed in mesh configurations.

The MPDU includes a *MAC Header* (MHR), a *MAC Service Data Unit* (MSDU), and a *MAC Footer* (MFR). The MHR starts with a 4-byte *HomeID* field that indicates the ITU-T G.9959 network the device is associated with. All devices in the network share the same HomeID. The MHR follows with a 1-byte *source node ID* that together with

Table 3.15 Z-wave modulation

Data rate	Modulation	Line coding	Rate (Kbps)
R_1	FSK	Manchester	9.6
R_2	FSK	NRZ	40
R_3	GFSK	NRZ	100

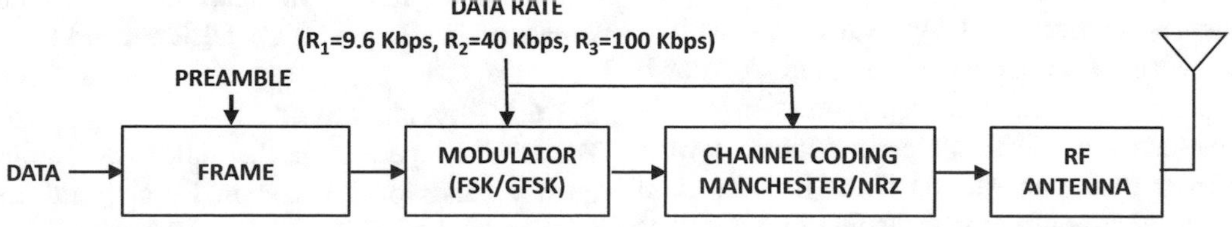

Fig. 3.61 Z-wave physical layer

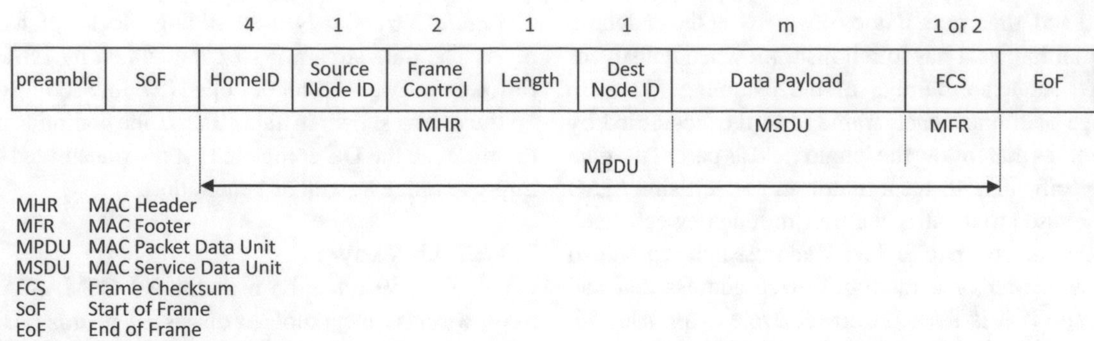

Fig. 3.62 ITU-T G.9959 unicast MAC frame

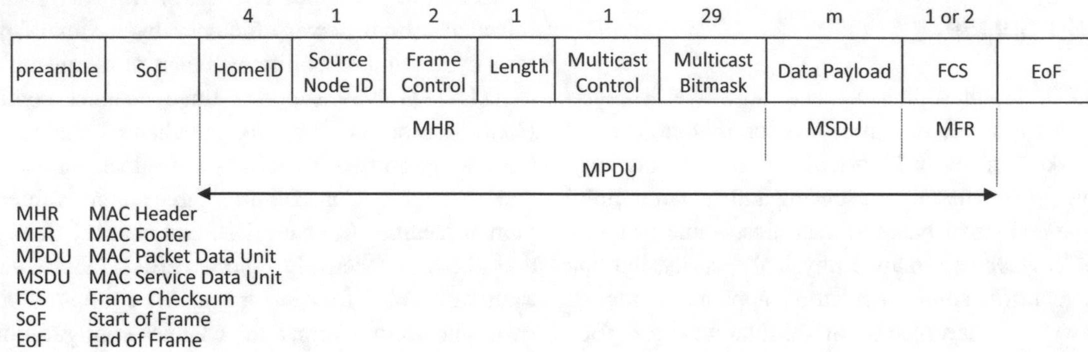

Fig. 3.63 ITU-T G.9959 multicast MAC frame

Table 3.16 Destination NodeID

Value	Meaning
00000000	Uninitialized
00000001 – 11101000	NodeID
11101001 – 11111110	Reserved
11111111	Broadcast NodeID

the HomeID identify the device that generated the frame. The MHR continues with a 2-byte frame control field that defines the frame type as well as other control flags. Possible frame types, among others, are unicast, multicast, and acknowledgment frames. The following fields are the 1-byte MPDU length and a 1-byte *destination node ID* that identifies the device destination. Possible values of the destination NodeID are shown in Table 3.16. Multicast frames, shown in Fig. 3.63, include instead a 1-byte multicast control field and 29-byte multicast bitmask that identify the destinations of the frame. For frames associated with channel configuration three, the sequence number is a 1-byte value. The MSDU field includes the actual upper layer datagram. Acknowledgment frames that are used when retransmissions are in place do not include the MSDU field. In general devices can only receive frames with MSDU fields that are short enough to comply with configured transmission rates. The MFR includes a 1-byte FCS. Because ITU-T G.9959 frames follow a similar structure to that of IEEE 802.15.4 frames presented in Sect. 3.3.3, they can support IPv6 with the right adaptation mechanism.

3.3.6 DECT ULE

Digital enhanced cordless telecommunications (DECT) is a cordless telephone technology that has been extended, as *DECT Ultra Low Energy* (DECT ULE), for use in wireless sensor and actuator networks in the context of smart home applications [9]. The protocol is managed by the ULE Alliance and intends to provide long-range and middle-rate transmission rates with very little power consumption and low latency. Low power consumption combined with low duty cycles leads to batteries lasting over 5 years. Typical coverage is up to 60 and 500 m for indoor and outdoor scenarios, respectively. The DECT ULE topology is a star with one-hop links between devices and the base station that plays the role of IoT gateway. Up to 400 devices are supported in a single DECT ULE network.

3.3.6.1 Physical Layer

DECT ULE operates on the unlicensed cordless phone frequency bands of 1.8 GHz in Europe and 1.9 GHz in US with channel spacing of 1.728 MHz supporting a maximum nominal transmission rate of 1 Mbps. The modulation relies on a GFSK scheme.

3.3.6.2 Link Layer

DECT ULE uses frames of 24 timeslots transmitted every 10 ms divided in two groups of 12 timeslots configured for downlink and uplink communications, respectively. Frames are transmitted on channels at different frequencies, thus supporting *frequency division multiple access* (FDMA). Each frame, in turn, supports full duplex timeslot groups relying on *time division duplex* (TDD). TDD is a mechanism by which uplink and downlink communications share the same carrier frequency, but they have different allocated timeslots. Within each TDD group, multiple timeslots support TDMA.

Figure 3.64 shows a single timeslot that consists of a 4-byte preamble that is used for device synchronization, a 1-byte network data header, a 5-byte network data payload, and a 2-byte network data checksum. The frame includes a variable length user data payload that, depending on the configuration, includes an embedded user data checksum. The timeslot also includes a 4-bit quality field that provides QoS support and a 60-bit (7½-byte) guard to prevent access collisions. This frame format leads to a packet size that is anywhere between 32 and 256 bytes. DECT ULE also supports security by means of pre-shared key-based encryption.

Figure 3.65 shows the DECT ULE protocol stacks; two planes are supported: a control plane (C-Plane) and a user plane (U-Plane). C-Plane supports reliable transmission of control traffic, while U-Plane supports speech and data transmission in circuit switched (CS) and packet switched (PS) modes, respectively. The link layer incorporates a *Data Link Control* (DLC) layer that is responsible for routing of C-Plane and U-Plane information. The network layer operates between peers with exchange of messages used for establishment as well as maintenance and release of sessions. In order to enable IPv6, 6LoWPAN datagrams are transmitted directly over DECT ULE DLC in the U-Plane.

3.3.7 NFC

Near-field communications (NFC) is a technology that provides ways for devices to communicate with each other by means of contactless transactions. NFC also enables efficient, convenient, and easy access to digital content. NFC relies on several contactless card mechanisms, and therefore it is backward compatible with existing infrastructure. NFC devices supports three modes of operation: (1) NFC card emulation that enables devices to act like smart cards so they can perform transactions like payments, (2) NFC reader/writer that enables devices to read information stored in NFC tags, and (3) NFC peer-to-peer that enables devices to communicate with each other. From an IoT perspective, NFC peer-to-peer mode is the mechanism to provide IPv6 connectivity.

3.3.7.1 Physical Layer

NFC relies on the ISO/IEC 18000-3 standard that enables passive *radio-frequency identification* (RFID) element identification by means of electromagnetic induction between two antennas located within their near fields [2]. This forms an air-core transformer that operates within the 13.56 MHz

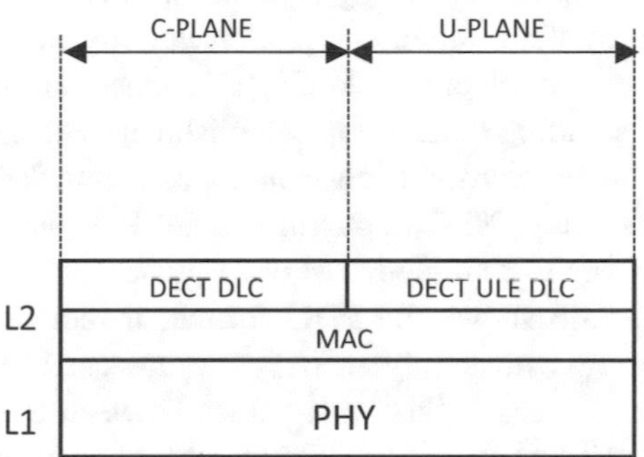

Fig. 3.65 DECT ULE protocol stack

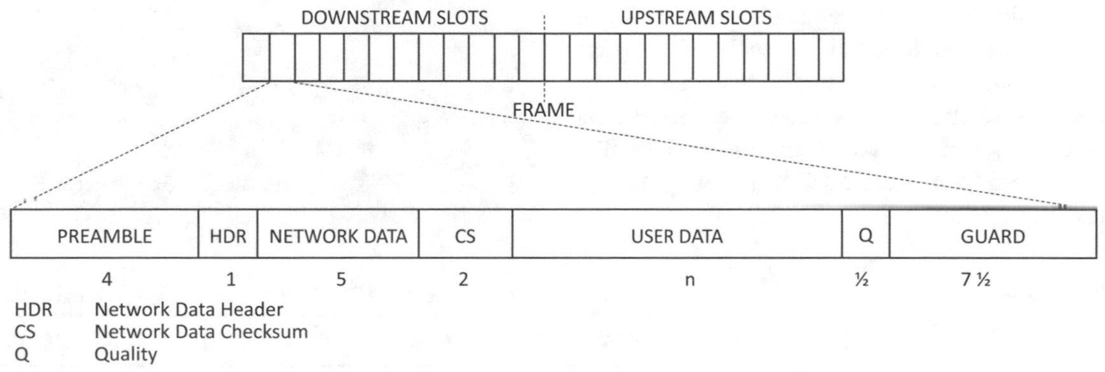

Fig. 3.64 DECT ULE frame

ISM band. In a passive configuration, NFC consists of an interrogator and a tag. The interrogator generates an RF field that induces an electric field and powers the tag. In an active configuration, which is associated with NFC peer-to-peer mode, devices are always powered up and those waiting for traffic disable their transmitters. Three transmission rates are possible, 106 Kbps, 212 Kbps, and 424 Kbps that depend on the modulation scheme and the configuration. For 106 Kbps in passive configuration, 212 and 424 Kbps bitstreams the scheme is a combination of Manchester line coding with 10% ASK modulation. This implies that both one and zero are encoded using a Manchester line code, but zero are attenuated by 10% of the peak signal amplitude. For a 106 Kbps bitstream, in active configuration, bit encoding is a combination of MMC with 100% ASK modulation. This means that ones are encoding using an MMC line code, but zeros are completely attenuated and not transmitted.

3.3.7.2 Link Layer

The NFC link layer is carried out by the *Logical Link Control Protocol* (LLCP). LLCP provides three mechanisms: (1) link management, (2) connection-oriented transmission, and (3) connectionless transmission. Link management enables *packet data unit* (PDU) aggregation and disaggregation as well as link status monitoring. Connection-oriented transmission keeps track of maintaining all connection-related exchanges including connection establishment and tear down. Connectionless transmission handles unacknowledged data exchanges. Most NFC deployments are P2P with a single device talking to a single M2M or IoT gateway.

Figure 3.66 shows the PDU format; it includes a 6-bit DSAP along with a 6-bit SSAP that are used to identify the service access points with values between 0 and 15 to indicate well-known services, values between 16 and 31 to indicate registered local services, and values between 32 and 47 to indicate services assigned by the upper layers. Note that the format of DSAPs and SSAPs is not the same as that of DSAPs and SSAPs associated with IEEE 802.2. The PDU also includes a 4-bit *PDU Type* (PTYPE) that specifies the syntax and semantic of the remaining fields. Additionally, an optional 8-bit sequence field identifies receiving and transmitting sequence numbers. The frame ends with a variable length payload. Note that address values between 32 and 47 can be used to generate IPv6 interface identifiers. IPv6 datagrams are transmitted under NFC by adopting some of the 6LoWPAN techniques.

Summary

There are several physical and link layer technologies that are an integral part of many IoT solutions. Tables 3.17 and 3.18 compare the technologies presented in this chapter including their most important characteristics. Note that CB and CF in the MAC column stand for *contention-based* (CB) and *contention-free* (CF) MAC. While many of these mechanisms are stand-alone protocols, a large proportion of them belong to proprietary stacks like ZigBee and BLE that do not natively support IPv6. This chapter started by briefly reviewing basic physical layer modulation techniques like line codes, passband modulation, spread spectrum, and discrete multicarrier modulation that serve as building blocks of most

Table 3.17 Wireline physical/link layers

Protocol	Medium	Modulation	MAC	Max nominal rate
Ethernet	Twisted pair, fiber optics	PAM	CF	400 Gbps
ITU-T G.9903	Existent electrical wiring	DPSK, QAM	CF	34 Kbps
IEEE 1901.2	Existent electrical wiring	DPSK, QAM	Both	500 Kbps
MS/TP	RS-485 based	Differential signaling	CB	115.2 Kbps

Table 3.18 Wireless physical/link layers

Protocol	Bands	Max coverage	Modulation	MAC	Max nominal rate
IEEE 802.11	ISM/non-ISM	1 km	DSSS, OFDMA	Both	10 Gbps
IEEE 802.15.3	ISM 2.4 GHz	10 m	QAM	CB	55 Mbps
IEEE 802.15.4	ISM 1/2.4 GHz	1 km	DSSS, OQPSK	Both	250 Kbps
BLE	ISM 2.4 GHz	400 m	FHSS, GFSK	CB	3 Mbps
ITU G.9959	ISM 1 GHz	30 m	GFSK	CB	100 Kbps
DECT ULE	ISM 1.8 GHz	500 m	GFSK	CB	1 Mbps
NFC	ISM 13.56 MHz	10 cm	ASK	CF	424 Kbps

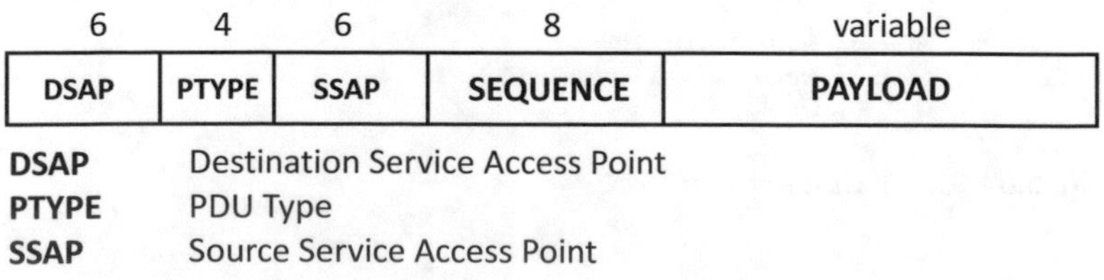

6	4	6	8	variable
DSAP	PTYPE	SSAP	SEQUENCE	PAYLOAD

DSAP Destination Service Access Point
PTYPE PDU Type
SSAP Source Service Access Point

Fig. 3.66 PDU format

of these the technologies. The chapter first explored wireline protocols like Ethernet, ITU-T G.9903, IEEE 1901.2, and MS/TP and then proceeded to analyze wireless mechanisms like IEEE 802.11, IEEE 802.15.3, IEEE 802.15.4, BLE, ITU-T G.9959, DECT ULE, and NFC. In all cases, each of these technologies was presented from the perspective of their modulation and media access algorithms that are respectively associated with physical and link layers.

Homework Problems and Questions

3.1 In Fig. 3.2, how does signal coverage relate to LLNs?

3.2 The readouts generated by an IoT sensor become the following codewords when encoded by a block code: *0000*, *0011*, *0101*, *0110*, *1001*, *1010*, *1100* and *1111*. What is the code rate?

3.3 If a wireline connected sensor transmits the following differentially encoded signal. What does the transmitted bit-stream look like?

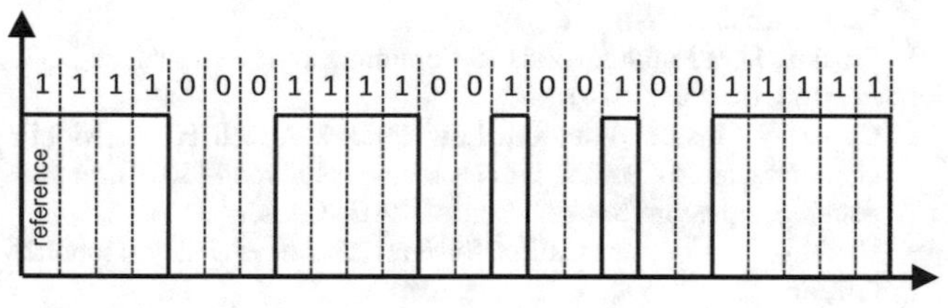

3.4 What do you think is the problem of a scheme where the sensor of Problem 3.2 generates a continuous sequence of *0000* codewords encoded with AMI in a wireline environment?

3.5 If a wireless sensor uses a 64-QAM modulation scheme and the symbol duration is $T = 100\,\mu\text{s}$, what are the symbol R_s and transmission R_b rates?

3.6 An OFDM scheme consists of 16 16-QAM subchannels and 16 64-QAM subchannels with symbol duration $T = 6.4\,\mu\text{s}$, what is the transmission R_b rate?

3.7 Consider two IoT sensors A and B that transmit readouts to sensor C that, in turn, forwards the aggregated traffic to the network core. The signal coverage of A and B is such that they cannot directly communicate with each other. How is this scenario affected by the hidden and exposed station phenomena?

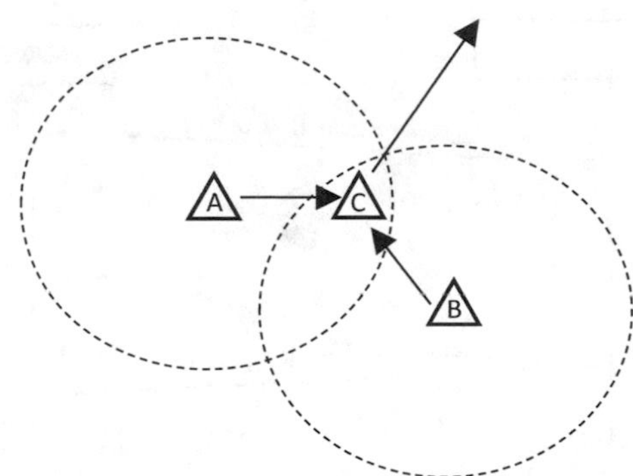

3.8 A sensor generates a single readout once every second. Each readout can take one out of 16384 possible values that are subjected to a block code with rate $r = \frac{7}{8}$. Assuming the overhead due to network, transport, and session layer headers is 48 bytes, what is the protocol efficiency and the nominal transmission rate of the following mechanism?

(a) Ethernet
(b) MS/TP
(c) IEEE 802.15.4 ($S = 0, C = 1, SAM = 10, DAM = 10$)

The efficiency is given by the ratio between the size of the actual sensor data and the total size of the frame. What can be concluded from the efficiency and transmission rate numbers? Ignore preamble and start frame delimeters in the frame length computation.

3.9 One of the rates supported by IEEE 802.11ah is 1.8 Mbps, how is this nominal rate obtained based on the modulation scheme and code rate?

3.10 How long does it take to transmit a full sizc IEEE 802.15.4 frame? How does it compare to a full size Ethernet frame? What are the implications?

3.11 If an IEEE 802.15.4 layer receives a 130-byte datagram from its upper layer, what does the IEEE 802.15.4 layer do?

3.12 Under IEEE 802.15.4, if the address field is 8 bytes long. What are the values of the *SAM*, *DAM*, and *C* fields?

3.13 An IEEE 802.15.4 frame has the following fields *S=0*, *C=0*, *SAM=11*, *DAM=10* and a 40-byte payload. How long does it take to be transmitted?

3.14 Consider the BLE communication flow below. What are the missing transactions?

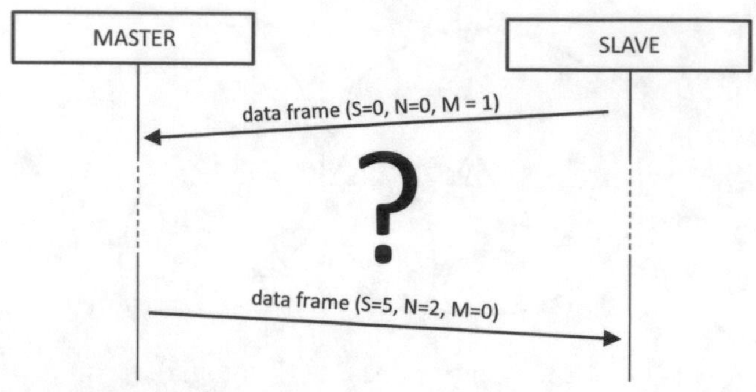

3.15 What are some advantages and disadvantages of the topologies shown in Fig. 3.46?

3.16 Given the BLE state machine in Fig. 3.57, is it possible for a Master device to generate advertising messages?

3.17 What technologies presented in this chapter do not rely on ISM bands?

3.18 To minimize interference in the overcrowded ISM bands, what other technologies can be used together with ITU-T G.9959?

References

1. IEEE standard for information technology– local and metropolitan area networks– specific requirements– part 15.1a: Wireless medium access control (MAC) and physical layer (PHY) specifications for wireless personal area networks (WPAN). IEEE Std 802.15.1-2005 (Revision of IEEE Std 802.15.1-2002) pp. 1–700 (2005)
2. ISO/IEC 18000-3: Parameters for air interface communications at 13,56 MHz. Standard, International Organization for Standardization, Switzerland (2010)
3. ANSI/ISA-100.11a-2011 Wireless systems for industrial automation: Process control and related applications. Standard, International Society of Automation, Research Triangle Park (2011)
4. IEEE standard for local and metropolitan area networks–part 15.4: Low-rate wireless personal area networks (LR-WPANs) amendment 1: Mac sublayer. IEEE Std 802.15.4e-2012 (Amendment to IEEE Std 802.15.4-2011) pp. 1–225 (2012)
5. IEEE standard for low-frequency (less than 500 kHz) narrowband power line communications for smart grid applications - amendment 1. IEEE Std 1901.2a-2015 (Amendment to IEEE Std 1901.2-2013) pp. 1–28 (2015)
6. ZigBee Specification. Standard, The ZigBee Alliance, California (2015)
7. IEC 62591:2016 Industrial networks - Wireless communication network and communication profiles - WirelessHART. Standard, International Electrotechnical Commission, Geneva (2016)
8. IEEE standard for information technology—telecommunications and information exchange between systems local and metropolitan area networks specific requirements—part 11: Wireless lan medium access control (MAC) and physical layer (PHY) specifications. IEEE Std 802.11-2016 (Revision of IEEE Std 802.11-2012) pp. 1–3534 (2016)
9. ETSI TS 102 939. Standard, European Telecommunications Standards Institute, Sophia Antipolis (2017)
10. IEEE standard for ethernet. IEEE Std 802.3-2018 (Revision of IEEE Std 802.3-2015) pp. 1–5600 (2018)
11. IEEE Standards Association: IEEE standard for low-rate wireless networks. IEEE Std 802.15.4-2020 (Revision of IEEE Std 802.15.4-2015), pp. 1–800 (2020)
12. Benedetto, S., Biglieri, E.: Principles of Digital Transmission: With Wireless Applications. Kluwer Academic Publishers, Dordrecht (1999)
13. Bluetooth, S.: Bluetooth 5.2 Core Specification, p. 3256. Bluetooth SIG, Kirkland (2019)
14. Cominelli, M., Patras, P., Gringoli, F.: Dead on arrival: An empirical study of the bluetooth 5.1 positioning system. In: Proceedings of the 13th International Workshop on Wireless Network Testbeds, Experimental Evaluation & Characterization, pp. 13–20 (2019)
15. Gilb, J.P.K.: Wireless Multimedia: A Guide to the IEEE 802.15.3 Standard, 250 p. IEEE Press (2003). ISBN 0-7381-3668-9
16. Glover, I., Grant, P.: Digital Communications. Prentice-Hall, Upper Saddle River (2010)
17. Proakis, J.G.: Fundamentals of Communication Systems, 2nd edn. Pearson, Boston (2014)
18. Gupta, N.: Inside Bluetooth Low Energy. Artech House Mobile Communications Series. Artech House, Norwood (2016). https://books.google.com/books?id=hRoQkAEACAAJ
19. Haykin, S.: Communication Systems, 5th edn. Wiley, Hoboken (2009)
20. International Telecommunication Union: Short range narrow-band digital radio communication transceivers - PHY and MAC layer specifications, ITU-T Recommendation G.9959, January 2015
21. International Telecommunication Union: Narrowband orthogonal frequency division multiplexing power line communication transceivers for G3-PLC networks, ITU-T Recommendation G.9903, May 2013
22. Kurose, J.F., Ross, K.W.: Computer Networking: A Top-Down Approach, 6th edn. Pearson, London (2012)
23. Newman, H.M.: BACnet: The Global Standard for Building Automation and Control Networks. Momentum Press, New York (2013)
24. Perahia, E., Stacey, R.: Next Generation Wireless LANs: 802.11n and 802.11ac, 2nd edn. Cambridge University Press, Cambridge (2013)
25. Rackley, S.: Wireless Networking Technology: From Principles to Successful Implementation. Newnes, USA (2007)
26. Sklar, B.: Digital Communications: Fundamentals and Applications. Pearson Prentice Hall, Upper Saddle River (2017)

Network and Transport Layers

4

4.1 Why IP?

Many of the state-of-the-art technologies used in M2M, CPS, and hybrid IoT scenarios like ZigBee [32], BLE [2], ANT+ [14], Z-Wave [12], and NFC [11], among others, have no native IP support, and in many cases they are not even IP compliant. They provide full stacks with their own physical, link, network, transport, and application layers that do not usually interoperate outside their specific technology. This lack of compatibility is caused by the fact that they are standalone stacks that have evolved into independent standards. For broad interaction in the context of IoT, IP is typically provided by means of border gateways that provide the translation between traffic generated by these technologies and the datagrams that enable some basic IP connectivity.

There are many problems associated with this approach: (1) increased deployment costs due to more complex network topologies that result from the many different standards that must be translated, (2) increased transmission delays due to extra translations that negatively affect real-time performance during sensing and actuation, (3) lack of real end-to-end IP connectivity that enables applications to monitor and access devices in an efficient way, (4) increased device deployment and management costs due to the need of interacting with different overlapping technologies, and (5) packet loss and other network impairments due to routing issues caused by the lack of a common network layer.

The state-of-the-art standards rely on some of the physical and link layer technologies introduced in Chap. 3. These technologies, in turn, exhibit very good properties that, if adapted to provide end-to-end IP support, can enable IoT solutions. Moreover, with a common network layer protocol like IP, many of the issues indicated above can be greatly eliminated or at least mitigated. In recent years, and to this end, there has been a massive effort by IETF to provide adaptation layers that add IP support to these proprietary physical and link layers.

Since this adaptation is performed in the same device, there is no need for extra equipment or expensive network topology changes. Moreover, introducing an adaptation layer is typically performed by means of software updates that require no hardware changes. This comparatively lowers deployment and management costs, reduces both latency as well as packet loss, and therefore improves connectivity and overall QoS.

Enabling devices to natively support IP-based protocols has several advantages; (1) devices can directly connect to IP networks without the need of additional hardware and software modules; (2) changes to IP network and routing infrastructure are minimized; (3) IP-based protocols, that have been around for decades, are proved to work and scale well; (4) standard socket APIs, that allow users to manage and control IP-based protocols, are an industry standard that is available in almost every single software platform; (5) IP-based protocols, standardized and documented by means of IETF RFCs, are open, free, and available to everyone; and (6) IP networks can be managed and monitored through a number of widely available tools and mechanisms.

As already mentioned in Sect. 1.1, traditional IPv4 imposes address size restrictions that makes its use impractical for the volume of traffic and number of devices that IoT is intended to support. It is true that there are mechanisms to overcome these restrictions, in particular NAT, but they are quite cumbersome as they add unnecessary complexity and violate basic principles of the layered architectures. Essentially, NATting uses transport layer ports to extend the range of the network layer addresses. In this context, the best alternative to IPv4 is IPv6. IPv6 relies on 128-bit addresses in order to supply an address space that is large enough to fit billions of devices supporting end-to-end IP connectivity.

M2M, CPS, and hybrid IoT physical and link layers like those of ZigBee, BLE, Z-Wave, and NFC provide comparatively low transmission rates that, in order to minimize latency, limit their maximum frame sizes to values that are not compatible to mainstream IPv6 fragmentation. To address

R. Herrero, *Fundamentals of IoT Communication Technologies*, Textbooks in Telecommunication Engineering, https://doi.org/10.1007/978-3-030-70080-5_4

this issue, IPv6 adaptation is an intermediate mechanism that sits between the link layer and the network layer to address this fragmentation issue as well as other related problems. The most popular IPv6 adaptation technology is 6LoWPAN [26, 19, 15] that interfaces with IEEE 802.15.4 and has been widely extended to other link layer technologies like BLE as 6LoBTLE [22].

Regardless of limitations like frame size, using IPv6 and upper layer protocols directly without adaptation can be demanding for embedded IoT devices due to several reasons: (1) traditional security mechanisms that provide authentication and encryption like *IP Security* (IPSec) [6] and *Transport Layer Security* (TLS) [24, 23] are too complex, (2) the most popular session management application layer protocol transmitted over TCP, HTTP, is also too complex, (3) many IPv6 session and application layer protocols require continuous connectivity (this is a limitation that is not compatible with power duty cycles of many embedded IoT devices), (4) IPv6 multicast transmissions are not always supported by link layers of constrained technologies like IEEE 802.15.4, and (5) IPv6 routing technologies are not prepared to deal with capillary networks that rely on mesh topologies to improve energy consumption.

In order to understand IPv6 adaptation technologies like 6LoWPAN, it is first necessary to understand IPv6 and its benefits.

4.2 IPv6

The most important and relevant change introduced by IPv6 [5, 8] is the support of 128-bit addresses that exponentially increase the potential number of accessible devices. IPv6 address notation is based on eight 16-bit values separated by colons. Each 16-bit value is represented, in turn, by a hexadecimal number with all its leading zeros removed. An IPv6 address can be represented in a short form by replacing any sequence of all-zero 16-bit values with a double colon. For example, the long form 2001:0:0:0:0:10:21:10 can be represented as the short form 2001::10:21:10.

Table 4.1 shows this and other relevant IPv6 address types including their IPv4 example counterparts. Each IPv6 address has a prefix that is used for network identification. Specifically, the prefix indicates how many leading bits determine the network identity. For example, if the address is 2001::10:21:10/64, it means the first 64 bits, that is 2001::, identify the network.

Table 4.1 IPv6 addresses

Long form	Short form	Type	IPv4 equivalent example
2001:0:0:0:0:10:21:10	2001::10:21	Unicast	192.168.21.5
FF01:0:0:0:0:0:0:2105	FF01::2105	Multicast	224.0.0.24
FE80:0:0:0:983D:3F44:2819:CFCC	same	Link-local	169.254.1.10
0:0:0:0:0:0:0:1	::1	Loopback	127.0.0.1
0:0:0:0:0:0:0:0	::	Unspecified	0.0.0.0

The first few bits of an IPv6 address always indicate its format. Figure 4.1 shows the format of unicast addresses. They start with the *001* binary sequence leading to prefixes 2000::/3 and 3000::/3. It follows 45-bit and 16-bit fields known as global routing prefix and subnet identification, respectively. The global routing prefix is assigned to a site, and the subnet identification is assigned to a subnet within a site. The 64-bit *Interface Identification* (IID) that follows identifies the specific device in the subnet. Because devices supporting unicast IPv6 addresses can participate in the global IPv6 infrastructure, unicast addresses are known as *global unicast addresses* (GUAs). Some unicast addresses are known as *Unique Local Unicast Addresses* (ULAs) when they are routable inside a limited area like a site, but they are not expected to be routable in the global Internet.

Figure 4.2 shows the format of multicast addresses. They identify a group of devices such that a single device can belong to multiple groups. The first 8 bits of a multicast address are all ones that are followed by a flag and a scope field. The address ends with a 112-bit group identification field. Figure 4.3 shows the 4-bit flags field with three relevant bits: (1) *T* that indicates that the address is temporary, (2) *P* that indicates that the address is a unicast-prefix based IPv6 multicast address, and (3) *R* that indicates that the rendezvous point address is embedded in the multicast address. Table 4.2 shows the 4-bit scope field that specifies the scope of the multicast group. As in the IPv4 case, some multicast addresses are used to assign specific groups; FF02::1 and FF02::2 identify all devices and all routers, respectively. As the scope is link-local, datagrams are not forwarded by a *border router*.

Figure 4.4 shows the format of link-local addresses. They are used for bootstrapping and connectivity within a single link. They start with the 10-bit binary sequence *1111111010* followed by 54 zeros and the 64-bit IID. The prefix of a link-local address is, therefore, always FE80::/10.

Besides the address size change, IPv6 introduces a different header format that is shown in Fig. 4.5. It includes several fields: (1) a 4-bit version field used to identify the IP version number that always carries a value of 6; (2) an 8-bit traffic class field that provides QoS information; (3) a 20-bit flow label that identifies the datagram as part of a flow of datagrams; (4) a 16-bit payload length that indicates the number of payload bytes that follow this header; (5) an 8-bit next header field that specifies the protocol under which the payload is encoded; (6) an 8-bit hop limit field, also known as *IP Time to Live* (IP TTL), that provides a counter that is decremented as the datagram is forwarded throughout

3
45
16
64
bits
001
global routing prefix
subnet ID
interface ID

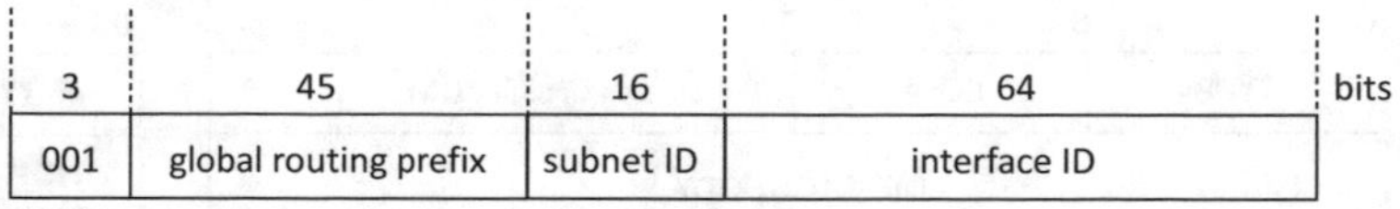

Fig. 4.1 IPv6 unicast addresses

8
4
4
112
bits
11111111
flags
scope
group ID

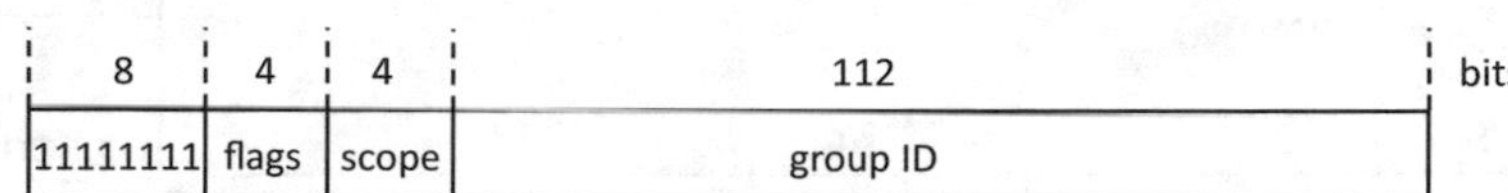

Fig. 4.2 IPv6 multicast addresses

0 1 2 3
0 R P T
R rendezvous point
P unicast-prefix based IPv6 multicast address
T temporary address

Fig. 4.3 IPv6 multicast flags field

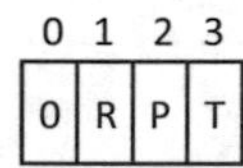

Table 4.2 IPv6 multicast scope field

Hex value	Scope
1	Interface
2	Link
4	Admin
5	Site
8	Organization
E	Global

10
54
64
bits
1111111010
0
interface ID

Fig. 4.4 IPv6 link-local addresses

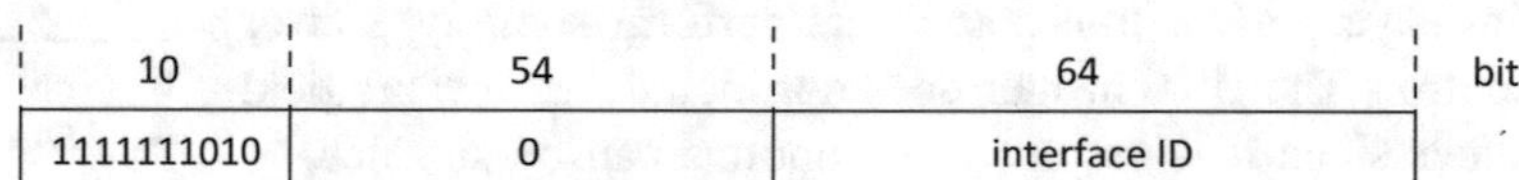

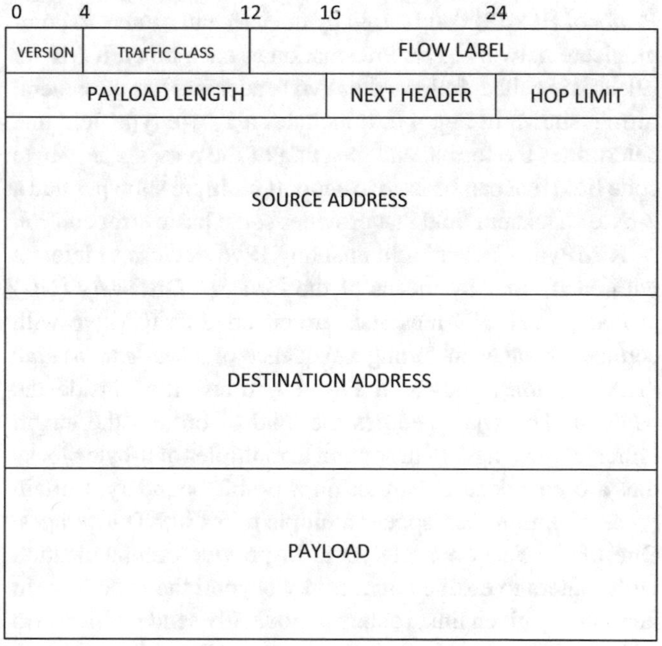

0
4
12
16
24
VERSION
TRAFFIC CLASS
FLOW LABEL
PAYLOAD LENGTH
NEXT HEADER
HOP LIMIT
SOURCE ADDRESS
DESTINATION ADDRESS
PAYLOAD

Fig. 4.5 IPv6 header

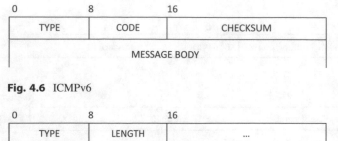

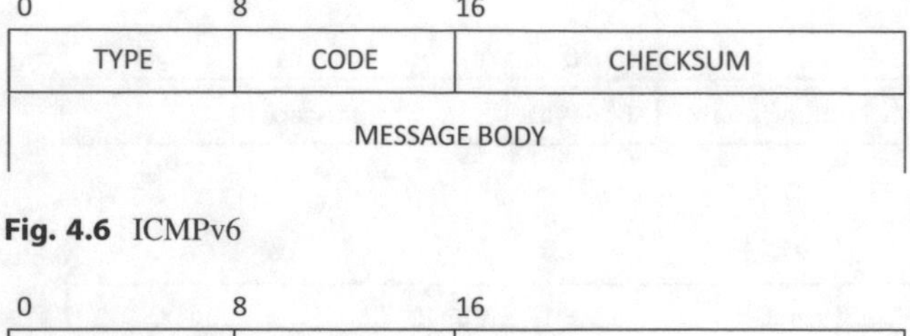

Fig. 4.6 ICMPv6

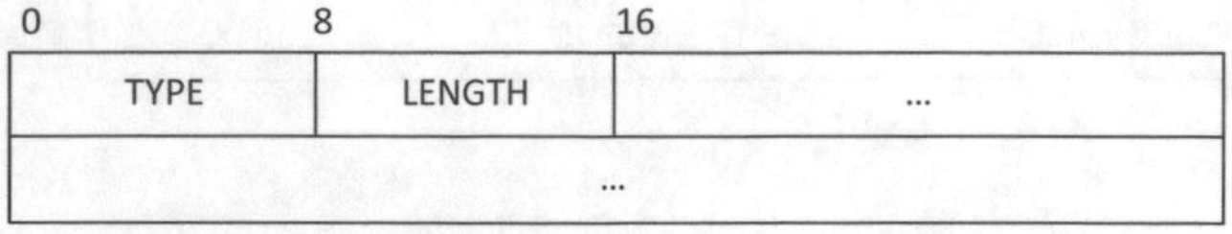

Fig. 4.7 ND message options

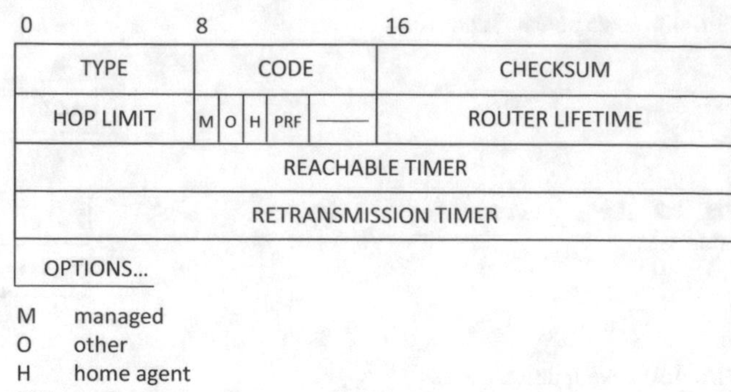

Fig. 4.8 Router advertisement ND message

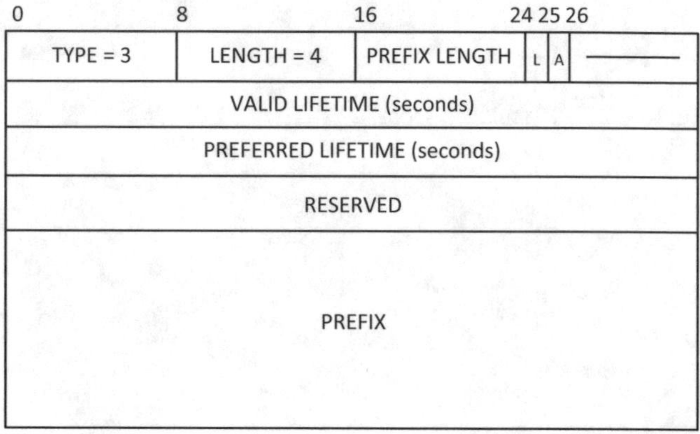

Fig. 4.9 Prefix information

the network; and (7) the aforementioned 128-bit source and destination addresses. Note that when the IP TTL reaches zero, the datagram is dropped from the network. The payload that follows the IPv6 header is processed by the destination device.

An IPv6 header does not include many of the fields found in IPv4 headers. Fragmentation/reassembly fields are not present since IPv6 does not support fragmentation and reassembly at intermediate routers and can only be performed at source and/or destination by means of a special header. A header checksum field is also absent since it is assumed that link-layer and/or transport layers perform some type error control. The IPv6 header does not include an *options* field either since it is assumed that options can be signaled by means of additional headers.

IPv6 borrows from IPv4 the *Internet Control Message Protocol* (ICMP) that is used by devices and routers to communicate network layer information to each other. ICMPv6 [9], as it is called, follows the IPv6 header and has the general format shown in Fig. 4.6. It includes a 1-byte type field that determines the format and meaning of the message, a 1-byte code field that can be used to support multiple subtypes and a 2-byte checksum field that provides some basic error control.

ICMPv6 is essential in enabling IPv6 devices to interact with each other by means of the *Network Discovery* (ND) protocol [27]. ND messages are encoded as ICMPv6 with certain messages including a sequence of *type-length-value* (TLV) options, shown in Fig. 4.7, transmitted inside the payload. The type specifies the kind of option; the length indicates the length of the option in multiples of 8-byte blocks that are encoded, as values, on a 64-bit boundary. Certain types of options can appear multiple times in ND messages. One of the functions of ND is to provide communication with routers to enable connectivity beyond the local link. In fact, on any given link, routers periodically send multicast (to address FF02::1) *Router Advertisement* (RA) ND messages that are used to provision devices with network parameters like prefix and hop limit information. Devices rely on this information to build a list of candidate routers that can be used to reach other networks.

Figure 4.8 shows an ICMPv6 RA ND message. It includes several fields; an 8-bit hop limit field for transmitters in the link, a managed bit that, when set, specifies that addresses are obtained via DHCPv6 or through *Stateless Address Autoconfiguration* (SAA, described later); otherwise, an *other* bit that indicates, when set, that additional configuration information is obtained by means of DHCPv6, a home agent bit used to support mobility, a 2-bit preference field that advertises the priority, between −1 and 1, of the router when compared to others, an 8-bit router lifetime that tells the devices how long the router is available, a 16-bit reachable timer field that indicates how long they should consider a neighbor to be reachable after they have received reachability confirmation, and a 16-bit retransmission timer field that says how long a device should wait before retransmitting neighbor solicitation messages. Options, included as TLVs, encode information that can be used to specify the router link layer address and/or the network MTU size and/or prefix information.

Figure 4.9 shows the prefix TLV; it includes an 8-bit prefix length, an on-link determination bit that indicates whether all addresses covered by the prefix are considered link local, an

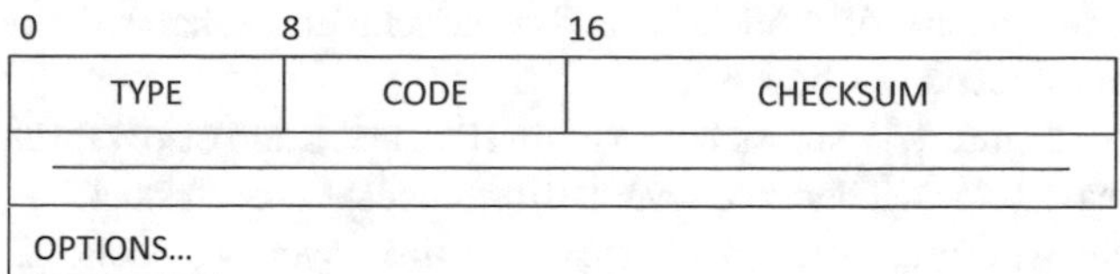

Fig. 4.10 Router solicitation ND message

SAA support bit that specifies that the prefix is to be used for SAA, a 32-bit valid lifetime field that indicates how long the prefix is valid for existing connections, a 32-bit preferred lifetime field that indicates how long the prefix is valid for new connections, and a 128-bit prefix field. The prefix is padded with zeros to comply with the specified length.

If a newly deployed device has not received any RA ND message, it can send out a *Router Solicitation* (RS) ND message to the multicast address (FF02::2) to induce routers to transmit RAs. As shown in Fig. 4.10, RS ND messages are just empty ICMPv6 messages.

The ND protocol also addresses connectivity between devices by means of the *Neighbor Solicitation* (NS) ND message that is primary used as a substitute of the IPv4 *Address Resolution Protocol* (ARP). Specifically, ICMPv6 NS ND messages allow devices to find out the link layer address of other on-link devices. This, in turn, enables devices to find out if other on link devices are reachable by keeping track of the *Neighbor Unreachability Detection* (NUD) cache. In response to NS ND messages, a device sends a *Neighbor Advertisement* (NA) ND message that includes its own link layer address. Note that NA ND messages can also be sent unsolicited to signal link layer address changes. Another type of ND message is the ICMPv6 Redirect ND message that is used by routers to inform other IPv6 devices on the link about a better next hop device.

SAA [21], mentioned above, is a distributed mechanism that allows devices to obtain a unique unicast IPv6 address that serves for proper routing of datagrams. A device relying on SSA dynamically generates its IPv6 address by combining its 64-bit *Interface Identifier* (IID) with the router supplied prefix. The IID is derived from the device link layer address and depends, therefore, on the link layer technology in use. For Ethernet interfaces, this means converting the 48-bit MAC address into a modified IEEE 64-bit EUI-64-based IID. This is done by inserting the 16-bit binary sequence *1111111111111110* in between the third and fourth byte and flipping a bit known as the universal/local bit in the resulting EUI-64 address.

Each SAA generated address can be in one of three states. The *tentative* state is the initial state of an address, when it cannot be used since it is being validated by means of the *Duplicate Address Detection* (DAD) mechanism (described in the next paragraph). The *preferred* state is the state of an address once it has been validated. Addresses in preferred state can be used for connectivity if they are used within the preferred lifetime. The *deprecated* state is the state of an address after its lifetime expires. Addresses in deprecated state are no longer valid and can only be used for preexistent flows and connections. The deprecated state is active during the predefined valid lifetime and enables devices to switch to new addresses that must also be in the tentative state before becoming preferred.

The reason for relying on a state machine and not using IPv6 addresses right away is because link layer addresses are not always unique. In fact, although it is often true that most link layer technologies (like Ethernet) provide unique addresses, this is not always guaranteed. To this end, during the tentative state, a device performs DAD and validates its own generated IPv6 address by transmitting multicast a NS ND message requesting the link layer address of the address. If any other device on the link is already using that IPv6 address, it responds by transmitting an NA ND message that indirectly tells the querying device that the address is not available for use. In this case, the device needs to obtain a new address by some other means. If the request timeouts, the address transitions to the preferred state. The latency introduced while a device is in the tentative state can be avoided if instead of plain DAD, optimistic DAD is alternatively used. Under optimistic DAD, devices start using their generated IPv6 addresses right away even while validation is still in progress. Traffic flow, in this case, is minimized by only allowing the transmission of those datagrams that do not affect the address translation caches of other on-link devices. Note that SAA relies on dynamic lifetime values provided by routers in RA ND messages. If a router changes its lifetime estimations, devices relying on SAA must update their address configuration accordingly.

Example 4.1 Consider an IEEE 802.15.4 device with a 11:22:33:44:55: 66:77:88 MAC (EUI-64 format) address. If the network prefix is 2001:2105::/64, what IPv6 address is derived from this MAC address by means of SAA?

Solution The IID is given by the MAC address; therefore the IPv6 address generated by means of SAA is 2001:2105::1122:3344:5566:7788.

4.3 6LoWPAN

For the many reasons mentioned in previous Sections, IPv6 cannot be directly used with most of the physical and link layer technologies that are part of IoT. Sitting in between network and link layers, 6LoWPAN and 6LoWPAN-like mechanisms can be used to adapt and translate IPv6 datagram into a format that is suitable for transmission over LLNs. Essentially 6LoWPAN gives constrained and low-power devices native IPv6 support at a cost of minimal overhead. This IPv6 support, of course, enables connectivity to other Internet-based hosts and routers. Although 6LoWPAN was originally designed to provide full IPv6 support under IEEE 802.15.4, many of its features have been adopted by other physical and link layer technologies like BLE and ITU-T G.9959. In fact, the term 6LoWPAN network typically refers to an LLN that relies on 6LoWPAN principles.

The 6LoWPAN architecture, shown in Fig. 4.11, is integrated by several WPANs that are characterized by being access stub networks with traffic that is never transmitted to other networks. All devices in a WPAN have IPv6 addresses that share the same prefix usually retrieved through RA ND messages. There are three types of WPANs: (1) simple WPANs that have connectivity to the IP core by means of edge routers, (2) ad hoc WPANs that locally connect devices and have no access to the IP core, and (3) extended WPANs that have connectivity to the IP code by means of several edge routers along a backbone link.

An edge router resides in between the access and core networks and plays the role of a traditional IoT gateway. It handles traffic in and out of the WPAN by performing 6LoWPAN adaptation, ND for interaction with devices on the same link, and other types of operations like IPv4-to-IPv6 translations for communication with other entities in the IP core. Devices in a WPAN can behave like either hosts or routers depending on source and destination addresses of the datagrams. A sensor can act as a host when it generates readouts, and it can act as a router when it forwards traffic from other devices in a mesh scenario. Most devices only have a single interface (i.e. wireless) that is used to receive and transmit datagrams. This is even true when the devices play the role of routers and forward datagrams on the same interface in which they were received as a mechanism to extend coverage. Note that this is quite different to what traditional Internet routers do, where traffic received on one interface is forwarded to a different one.

Access to core communication in the context of WPANs follows the standard rules of any other IP communication. Each WPAN device is identified by its unique IPv6 address, and it can send and receive datagrams. As processing capabilities are limited in constrained devices, the type of transport and application traffic is also limited. While protocols like UDP are suitable in this situation, TCP and HTTP [1] fail due to computational complexity and excessive resource consumption; TCP relies on sophisticated state machines, while HTTP imposes high memory requirements due to the size of its messages. This means that although TCP and HTTP can be used in WPANs, this is not typically recommended. More details of application and session layer mechanisms are introduced in Chap. 5.

Figure 4.12 shows both a plain IPv6 stack and 6LoWPAN-based stack. There are several differences; (1) each stack has, as expected, different physical and link layer protocols, (2) IPv6 cannot run directly over IEEE 802.15.4 as 6LoWPAN adaptation is needed, (3) TCP cannot be used over 6LoWPAN due to performance issues, and (4) CoAP over UDP, as opposed to HTTP over TCP, is used on the 6LoWPAN-based stack. Note that the 6LoWPAN-based stack has the 6LoWPAN adaptation layer placed in between the IPv6 and the IEEE 802.15.4 link layers. Also note that in many implementations, UDP and IPv6 layers are integrated directly into 6LoWPAN with APIs that interface with network and transport parameters. Because of this, 6LoWPAN adaptation is typically shown as part of the IPv6 layer.

Figure 4.13 shows a basic translation between an Ethernet wireline core and an IoT IEEE 802.15.4 wireless access that usually occurs at an WPAN edge router. Incoming IPv6 traffic from the core, on the Ethernet interface, is encoded as 6LoWPAN datagrams before being transmitted over the IEEE 802.15.4 interface. Similarly, incoming 6LoWPAN traffic on the IEEE 802.15.4 interface is decoded as IPv6 datagrams before transmitted over the Ethernet interface. Note that this bidirectional translation is quite efficient, and depending on the 6LoWPAN compression scheme, it can be either stateless or stateful. More details of the 6LoWPAN compression mechanisms are discussed later in this chapter.

Although 6LoWPAN is intended to work over IEEE 802.15.4, it can work over other link layer technologies with some small modifications. Specifically, 6LoWPAN requires that link layer protocols support framing, presented in Sect. 3, unicast transmission, and unique addresses that can be used, in turn, to derive unique IPv6 addresses by means of SAA. In addition, because IPv6 fragments cannot be smaller than 1280 bytes, 6LoWPAN performs its own fragmentation to adapt datagram transmission to link layer mechanisms with small MTUs.

It is therefore desirable for link layer frames to be as large as possible with payloads of at least 60 bytes to minimize the number of fragments that 6LoWPAN needs to track. Moreover, for fixed network packet loss, it is always good to minimize the number of fragments per datagram since the more fragments the higher the probability that a datagram will get dropped by the network. Note that 6LoWPAN compression can be used to efficiently compress IPv6 and UDP headers in order to maximize link layer payload size and therefore minimize the number of fragments per datagram. In the context of other link layers, it is also important, although not required, that the adaptation layer provides reliability by means of error detection and correction. Additionally, the adaptation layer must support security through encryption and authentication.

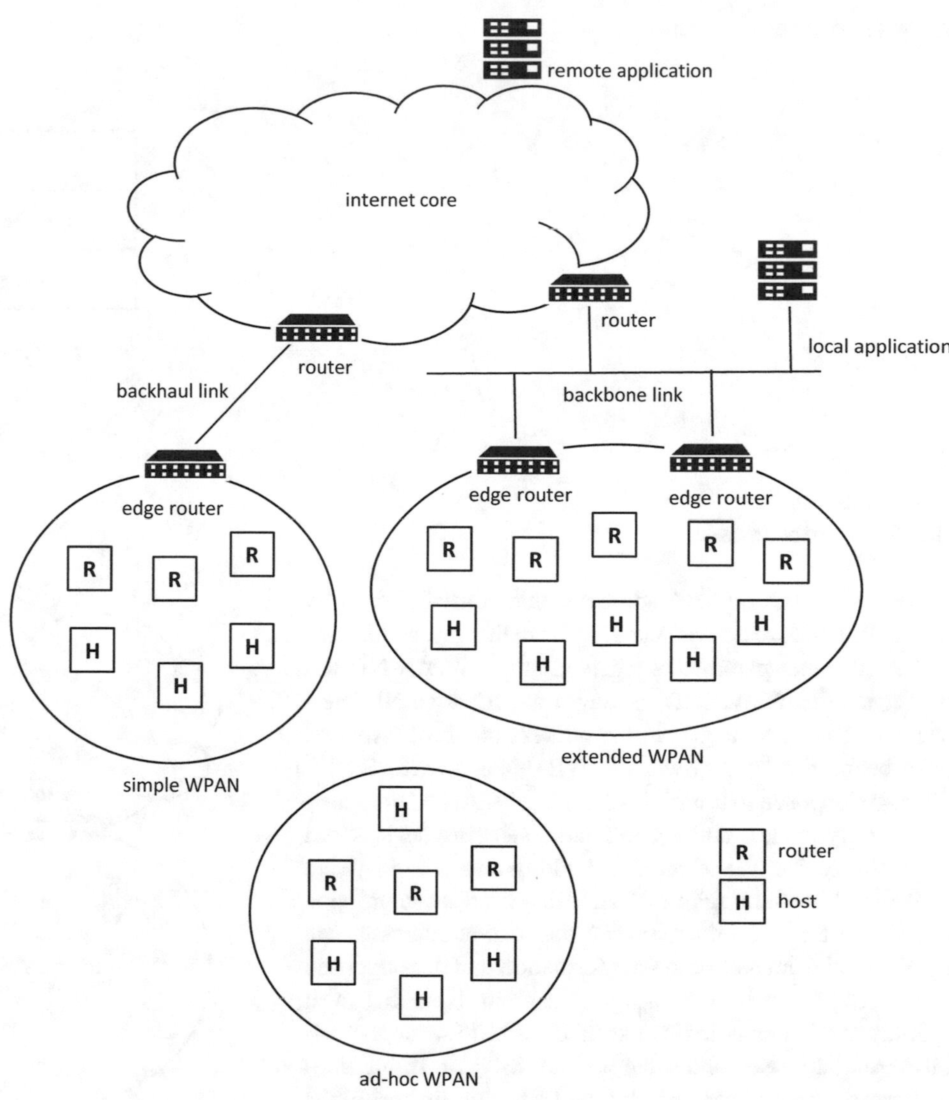

Fig. 4.11 6LoWPAN architecture

Fig. 4.12 6LoWPAN stack

IPv6 Stack				6LoWPAN Stack	
HTTP			application	CoAP	
TCP	UDP	ICMP	transport	UDP	ICMP
IPv6			network	IPv6 6LoWPAN	
ETHERNET			link	IEEE 802.15.4	
ETHERNET			physical	IEEE 802.15.4	

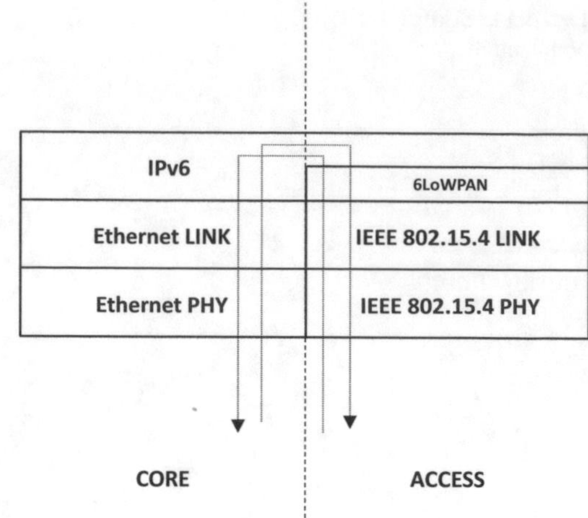

Fig. 4.13 6LoWPAN translation

4.3.1 Addresses

6LoWPAN relies on IPv6 addresses that consist of a prefix and an IID that are derived via SAA from link layer addresses and edge router parameters received in 6LoWPAN ND RA messages. 6LoWPAN ND messages are ICMPv6 ND messages that are exchanged in the context of 6LoWPAN and have been optimized to work in WPANs as per IETF RFC 6775. As opposed to traditional ND, 6LoWPAN ND supports power duty cycles, and it can handle sleeping devices and to minimize the use of multicast addressing. The mapping is essential in enabling 6LoWPAN IPv6 header compression as most of the 128-bit source and destination addresses can be obtained from well-known information that does not need to be transmitted in every single datagram. Note that most link layer IoT technologies like IEEE 802.15.4 support 64-bit MAC addresses and configurable, 8-bit or 16-bit short addresses typically assigned by the PAN coordinator. In the context of 6LoWPAN, either one of these types of addresses can be used to form the 64-bit IIDs.

As mentioned in Sect. 2.3, IoT devices are intended to be deployed, provisioned, and configured with minimal (and if possible, without) human intervention. This has to do with the volume of sensors and actuators as well as the remote nature of many of the locations where they are deployed. Initial bootstrapping is performed by the link layer that besides assigning a link local address, attempts to obtain a unicast globally routable address by means of SAA. In this scenario, the 6LoWPAN edge router exchanges 6LoWPAN ND messages that not only support SAA but also enable it to keep track of all devices in the link. Specifically, the edge router creates a registry of devices and assigns simpler IPv6 addresses that have the overall effect of simplifying the network operation.

Figure 4.14 shows an end-to-end IP network that combines a traditional IPv6 core with an 6LoWPAN access. The IPv6 core network is associated with the prefix 2001:10::/64, while the 6LoWPAN access network is associated with the prefix 2001:11::/64. The access network is composed of five devices that interface with an edge router. These devices can behave like routers or hosts depending on their location and their traffic processing capabilities. This is because end devices do not have direct connectivity to the edge router and rely on mesh routing to reach it. The edge router, in turn, connects to the core network and can access a server running on 2001:10::2105.

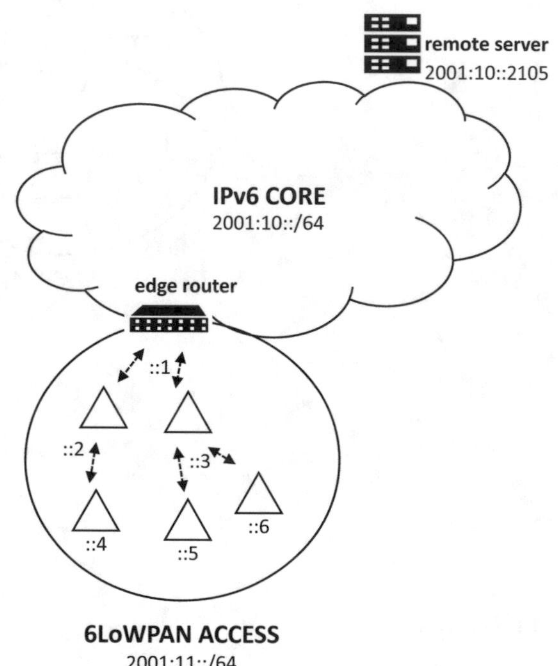

Fig. 4.14 6LoWPAN example

Bootstrapping starts when the edge router begins to advertise the 2001:11::/64 prefix to the devices by means of 6LoWPAN ND RA messages. The devices use this prefix and their own IID derived from the 64-bit IEEE 802.15.4 MAC address to generate IPv6 addresses. This is further simplified when devices, also relying on 6LoWPAN ND, register with the edge router and receive a 16-bit IID that is combined with the prefix to generate a less complex IPv6 address. The edge router assigns its own access interface to address 2001:11::1 and randomly assigns addresses 2001:11::2 through 2001:11::6 to the devices. This opens the door for 6LoWPAN compression since it is a lot more efficient to encode a 16-bit address than a 64-bit IID. Moreover, for traffic within the WPAN, datagrams only need to specify the 16-bit source and destination addresses since the network prefix is well-known to all devices (i.e. traffic from ::4 to ::6). For datagrams going from one device to its closest neighbor, there is no need to include any IPv6 address since their IEEE 802.15.4 MAC addresses provide all the routing information needed (i.e. traffic from ::5 to ::3). In other words, if an 6LoWPAN stack finds a header that has neither source nor destination address, it assumes that the traffic is within neighbors. When datagrams flow from one device to the external server, the datagram includes a source 16-bit address (i.e. ::4) and the destination 128-bit address (i.e. 2001:10::2105). The edge router looks at its registry and converts the 16-bit address into the corresponding global 128-bit unicast address of the device before forwarding the datagram to destination. In general, the edge router performs any type of header compression and decompression in the process of translating network and transport layers back and forth.

4.3.2 Header Format

The 6LoWPAN layer maps fields of IPv6 and upper layer headers into their compressed versions that become the 6LoWPAN header. This compression can be either stateless or stateful depending on whether the compressed information embedded in the datagram is enough to recover the original IPv6 and transport layer headers. To be specific, Fig. 4.15 shows the difference between stateless and stateful compressions. Under stateless compression, 6LoWPAN compresses each datagram header based on intra-datagram redundancy alone. On the other hand, under stateful compression, 6LoWPAN compresses each datagram header based on both intra- and inter-datagram redundancy. Stateless compression is simple, and, as opposed to stateful compression, it does not require a mechanism to keep track of the state of inter-datagram redundancy. Unfortunately, simplicity comes to a price of lower compression rates. Stateful compression relies on associating redundant information in uncompressed IPv6 headers with a context identifier that is transmitted in the datagrams instead. Consequently, stateful compression requires coordination to make sure that contexts are always correctly synchronized.

In most cases IPv6 addresses are not included in 6LoWPAN headers because they can be derived one way or another. As already stated, if a 128-bit destination address is from a device in the same WPAN, it can be compressed as a 16-bit address. This would be a case of stateless compression since each address is mapped based on the 128-bit address contained in that datagram. If stateful compression were in place, then the 128-bit address that is common to all datagrams would be associated with a context identifier, and it would not need to be transmitted. This, of course, requires coordination between transmitter and receiver. Usually, 6LoWPAN ND provides some mechanisms for context information dissemination.

Both IPv4 and IPv6 work well with fast link layer protocols like Ethernet and IEEE 802.3 that natively support many of the regular IP requirements including the support of large MTU sizes and multicast addresses. In those scenarios, there is no need for any adaptation mechanism. Other more constrained link layer standards like IEEE 802.15.4 support very small MTU sizes and do not allow multicast addressing due to restrictions linked to low transmission rates and low-power consumption. As many times mentioned, 6LoWPAN provides an adaptation mechanism that overcomes these and other problems.

One of these problems is because link layer power limitations also prevent devices from directly communicating with other devices on the same link. In many cases, a device relies on intermediate devices to transmit frames to the destination. This mesh topology requires additional information like original and final MAC addresses that can be inserted by means of 6LoWPAN adaptation. Another problem is that because link layer protocols like IEEE 802.15.4 were designed to work within proprietary stacks, they do not include a type field that can be used to identify the upper layer datagram. This is, again, a big difference when compared to fast link layer protocols like Ethernet. Specifically, Ethernet includes the ethertype field to indicate if the upper layer is, for example, IPv4, IPv6, or ARP. 6LoWPAN introduces a special scheme to solve this issue. This scheme together with fragmentation and header compression provides the most important features of 6LoWPAN.

To address the lack of a link layer type field, 6LoWPAN starts every datagram with an 8-bit field named dispatch value that serves as the datagram type indicator [18]. Although the dispatch value is 8 bits long, most of the time only a few of the most significant bits are used to indicate the type, while the remaining bits are used to signal additional header information. Table 4.3 shows the dispatch values that are used to assign the different types of 6LoWPAN headers; for example, a dispatch value starting with *10* indicates a mesh header, while one starting with *01000001* specifies an uncompressed IPv6 header. Note that a dispatch value *01000000* is used to indicate that the dispatch value extends beyond one byte. All other dispatch values not shown in the table are reserved values.

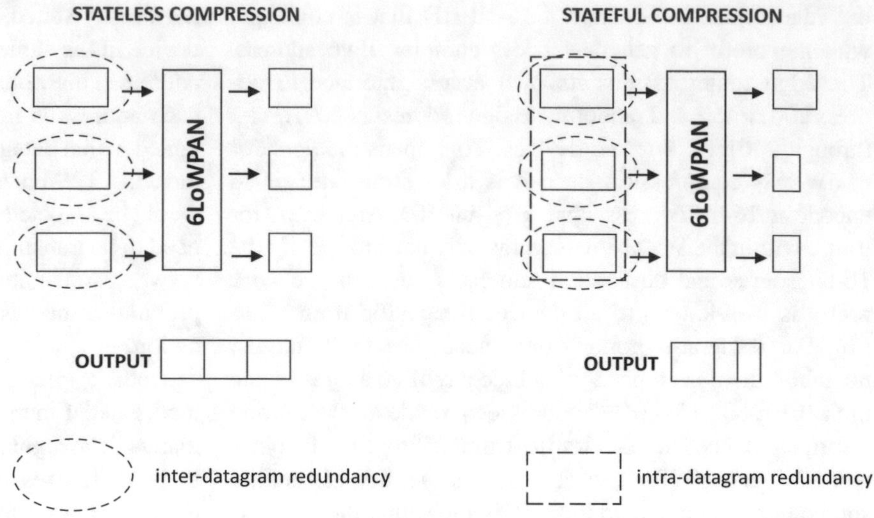

Fig. 4.15 Stateless vs stateful compression

Table 4.3 Dispatch values

Pattern		Header type	Order
00		Not a LoWPAN frame (NALP)	
01	000001	Uncompressed IPv6	4th
	000010	HC1 stateless compression	
	010000	BC0 broadcast	2nd
	1	IPHC stateful compression	4th
	111111	Extension Escape (ESC)	
10		MESH header	1st
11	000	FRAG1 initial fragment	3rd
	100	FRAGN subsequent fragments	
	1010	RFRAG recovery fragments	

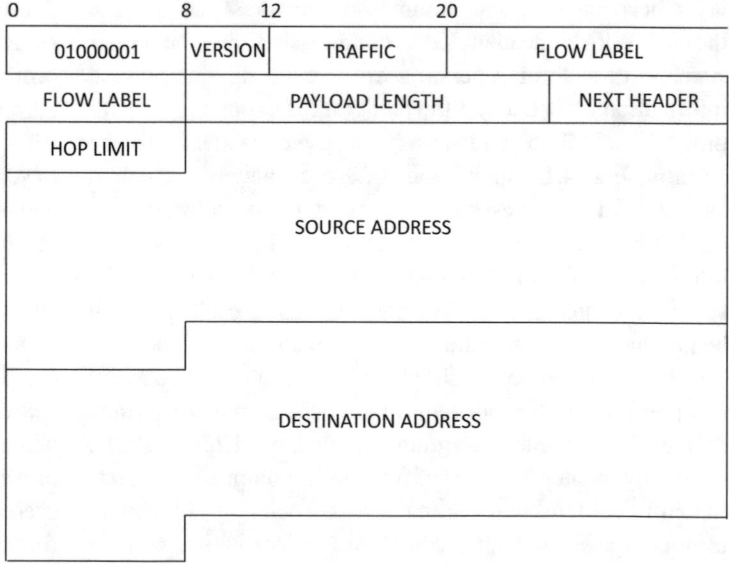

Fig. 4.16 Uncompressed IPv6 header

Figure 4.16 shows an IPv6 header that is transmitted uncompressed by means of 6LoWPAN by pre-appending a *01000001* dispatch value. Note that most IPv6 header fields are placed within a 32-bit boundary that makes them accessible by means of a single clock transition in 32-bit and 64-bit processors. The extra 8-bit dispatch value disrupts the 32-bit boundary of some fields like the flow label and forces an extra clock transition that lowers processing speed. Fortunately, many embedded IoT devices rely on 8-bit and 16-bit processors that are less affected by this displacement. Note that the 6LoWPAN uncompressed header can be also used to transmit IPv4 datagrams and other types of packets. In this case, 6LoWPAN provides fragmentation that enables the end-to-end transmission of full IPv4 stacks over IEEE 802.15.4. Essentially, the gateway only transforms physical and link layers and preserves all the upper layers. Of course, the drawback is the fact that because IPv4 datagrams are not compressed, they are inefficiently transmitted.

Note that multiple dispatch values and headers can be encoded in a single datagram based on the order given in Table 4.3. For example, if a datagram is compressed and then fragmented, the fragmentation dispatch value and header must go before the compression dispatch value and header since a given fragment cannot be decompressed until it is properly reassembled. Essentially, the order of the different 6LoWPAN headers enables synchronization between the encoder and decoder when many of them are included in a single datagram. The first header must be, if present, the mesh header that carries the original and final MAC addresses along with a hop count. It follows, if present, the broadcasting header that contains hop-by-hop link layer information. The third header is, if present, the fragmentation header in order to support transmission of datagrams larger than the link layer MTU size. The packet ends with the uncompressed or compressed network/transport layer header if present. Note that this order is consistent with the order of plain headers in regular IPv6 datagrams.

6LoWPAN relies on the dispatch value to specify the different headers, while IPv6 relies on the next header field to do the same. Moreover, after the dispatch value, the 6LoWPAN protocol adds the compressed fields associated with the header and then continues with the uncompressed fields that are directly copied from the IPv6 header. These uncompressed fields are known as in-line because they are sent in-line after the compressed fields. Examples of in-line fields include IPv6 addresses associated with external hosts that cannot be easily compressed based on context and implied information.

To understand how compression works under 6LoWPAN, it is interesting to look at a simple example. Figure 4.17 shows the headers that are used when application traffic is encapsulated by means of UDP over IPv6. The UDP and IPv6 headers are 8 and 40 bytes long, respectively, accounting for a total of 48 bytes that can be found on every single datagram regardless of the values of the fields. Figure 4.18, on the other hand, shows the maximum compression that can be achieved when adapting the same headers with 6LoWPAN. The 6-byte 6LoWPAN header starts with a 2-byte stateful *IP Header Compression* (IPHC) dispatch value and associated compressed fields. It follows a 1-byte *Next Header Compression* (NHC) UDP compressed header that includes a 1-byte field to identify both source and destination UDP ports. Note that uncompressed UDP ports are always 2 bytes long. The 6LoWPAN header ends with the 2-byte UDP checksum field that is transmitted in-line and therefore uncompressed. The overall compression rate is 6–48 or 1–6.

4.3.3 Routing and Forwarding

One of the roles of a network layer is to move datagrams from a sending to a receiving device. Two main functions are identified: (1) forwarding that involves moving a datagram from an incoming to an outgoing link in a device or router and (2) routing* that involves determining a route or path that datagrams must follow from source to destination. Routing is a network wide decision that results from executing a routing algorithm that defines a *forwarding information base* (FIB). Forwarding, in turn, is a device decision derived from the FIB.

Figure 4.19 shows how encapsulation works under traditional IPv6 routing. A router device with a FIB has two interfaces on two different links associated with two different subnets, 2001:10::/64 and 2001:11::/64. A frame received on interface 0 is processed and, based on the destination IPv6 address found in the FIB, is forwarded to the outgoing interface 1. Figure 4.20 shows a datagram being sent from device 1 at address 2001:10::10 to device 2 at address 2001:11::10. First, device 1 looks at its FIB table to determine that in order to reach 2001:11::10, it must first send the datagram to the router on interface 0 at address 2001:10::1. Then device 1 looks in its ND table to find out the MAC address corresponding to IPv6 address 2001:10::1. The resulting 11:22:33:44:55:66 MAC address is used as a destination link layer address for the datagram sent by device 1. The router receives the datagram, and it looks at its FIB table to determine that the datagram must be sent directly over interface 1 to reach destination IPv6 address 2001:11::10. The router looks at its ND table to obtain the MAC address corresponding to IPv6 address 2001:11::10. The resulting 11:12:13:14:15:16 MAC address is used as a destination link layer address for the datagram sent by the router.

When 6LoWPAN is in place, communication between devices is by means of capillary networking where devices acting as routers have only one interface. Datagrams enter the router and leave the router on the same interface. In this

Fig. 4.17 UDP over IPv6

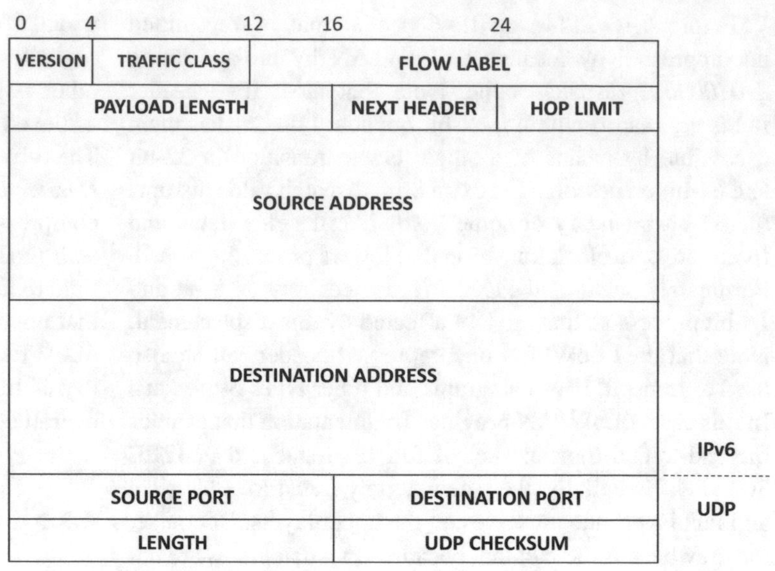

Fig. 4.18 UDP over IPv6 over 6LoWPAN

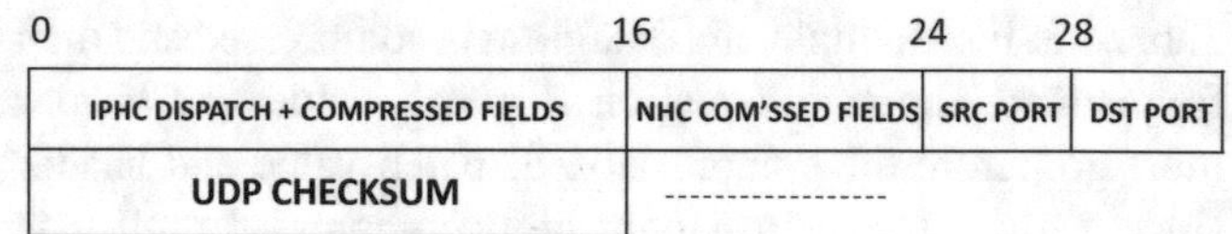

Fig. 4.19 IPv6 encapsulation

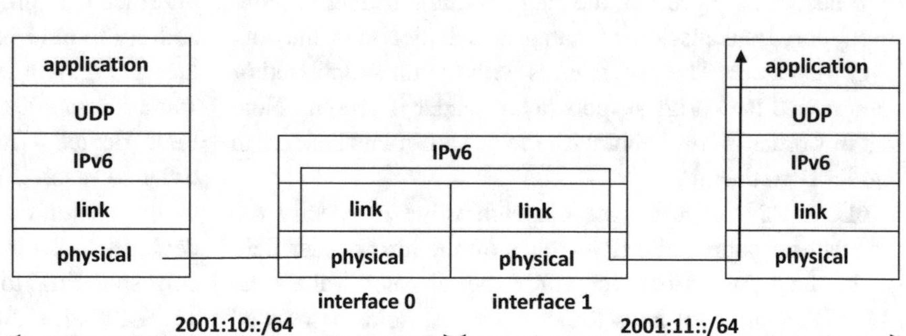

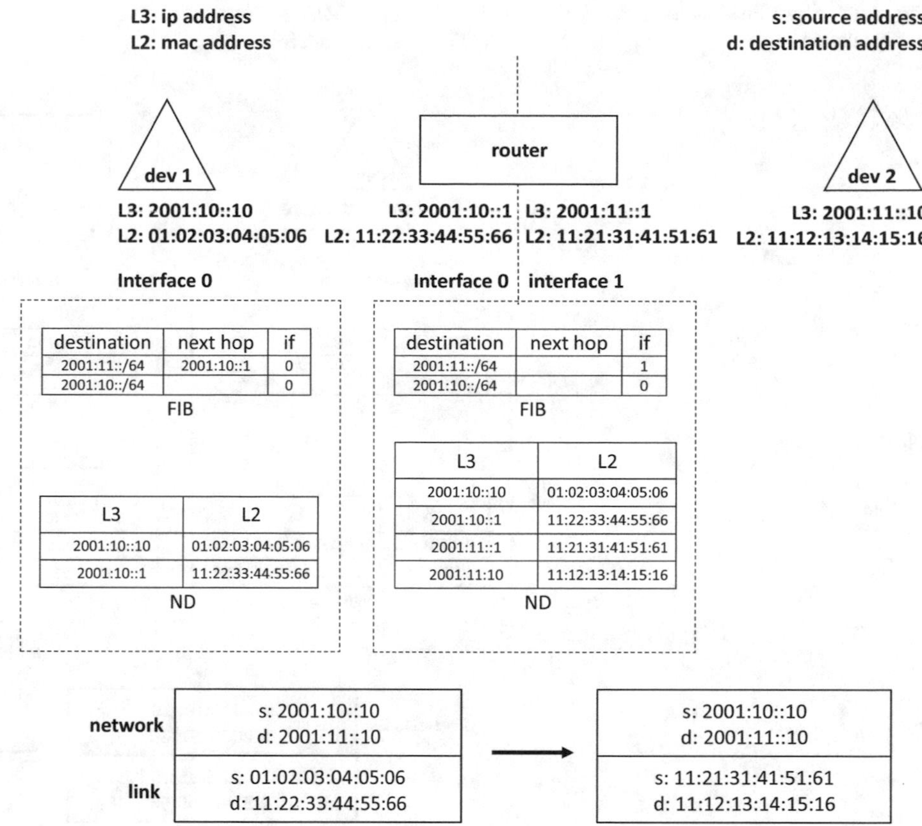

Fig. 4.20 IPv6 routing

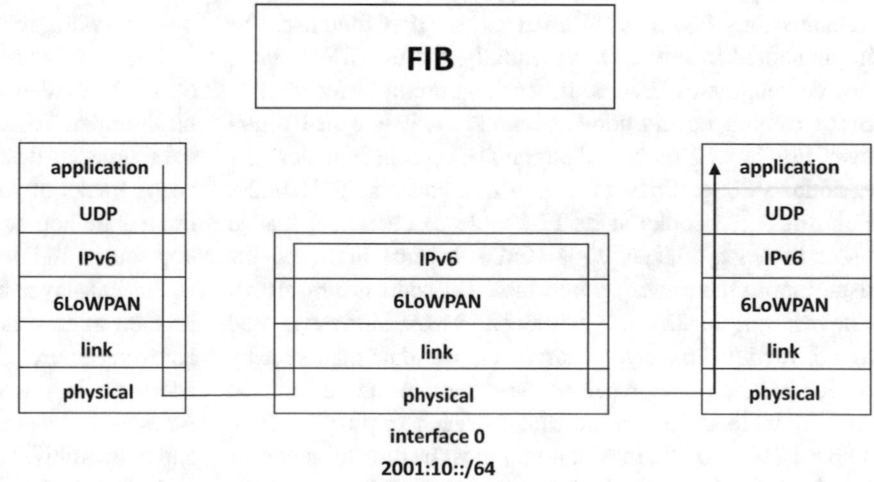

Fig. 4.21 6LoWPAN route-over

Fig. 4.22 6LoWPAN route-over encapsulation

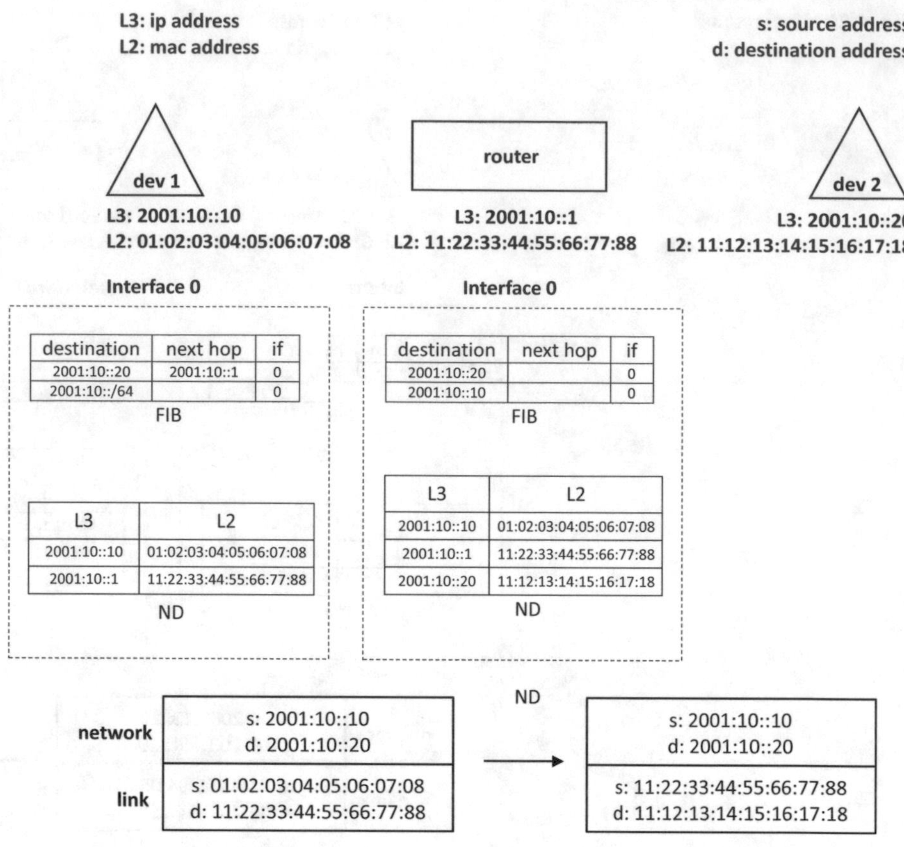

scenario, not all devices on a single link can talk to each due to the power limitations that prevent long-distance radio transmissions. Figure 4.21 illustrates how IPv6 routing can be implemented in this case. Essentially, the 6LoWPAN layer just decompresses IPv6 addresses that are used for routing. Since routing occurs above 6LoWPAN, it is called route-over. Figure 4.22 shows a datagram being sent from device 1 at address 2001:10::10 to device 2 at address 2001:10::20. First, device 1 looks at its FIB table to determine that in order to reach address 2001:10::20, it must first send the datagram to the router on interface 0 at address 2001:10::1. The resulting 11:22:33:44:55:66:77:88 MAC address is used as destination link layer address for the datagram sent by device 1. The router receives the datagram, and it looks at its FIB table to determine that to reach destination IPv6 address 2001:10::20, the datagram must be directly sent over interface 0. The router looks at its NB table to obtain the MAC address corresponding to IPv6 address 2001:10::20. The resulting 11:12:13:14:15:16:17:18 address is used as a destination link layer address for the datagram sent by the router.

The alternative to route-over routing is to rely on the link layer to perform multi-hop frame mesh forwarding and enable local link connectivity. This is shown in Fig. 4.23 where the link layer not only keeps track of source and destination MAC addresses for immediate hop communication but also original source and destination MAC addresses for end-to-end support. In the context of IEEE 802.15.4, mesh forwarding is introduced as part of IEEE 802.15.5. Link layer mesh forwarding is invisible to both 6LoWPAN and IPv6.

Figure 4.24 shows how to bring mesh forwarding to the 6LoWPAN adaptation layer. This approach is also called mesh-under. In this case, 6LoWPAN must keep track of original source and destination MAC addresses since the link layer keeps tracks of source and destination MAC addresses for immediate hop communication. Essentially at every single-hop source and destination, MAC addresses are overwritten at the link layer by means of the original source and destination addresses carried in the 6LoWPAN mesh header. Knowing source, destination, and original and final MAC addresses is not only needed for mesh forwarding but also for other services provided by 6LoWPAN like fragmentation and reassembly.

Figure 4.25 shows an example of 6LoWPAN mesh forwarding; frames go from device 1 to device 4 through devices 2 and 3. Each hop shows the frame source, destination, and original and final MAC addresses.

Figure 4.26 shows a 6LoWPAN mesh header. The header defines two 1-bit fields, *O* and *F* that, respectively, indicate whether the original and final MAC address is 16-bit PAN coordinator assigned or 64-bit based. The header also includes a *hops left* field that indicates how many times the packet can be forwarded before it is dropped by the network. As the datagram traverses a network, and it is forwarded by different

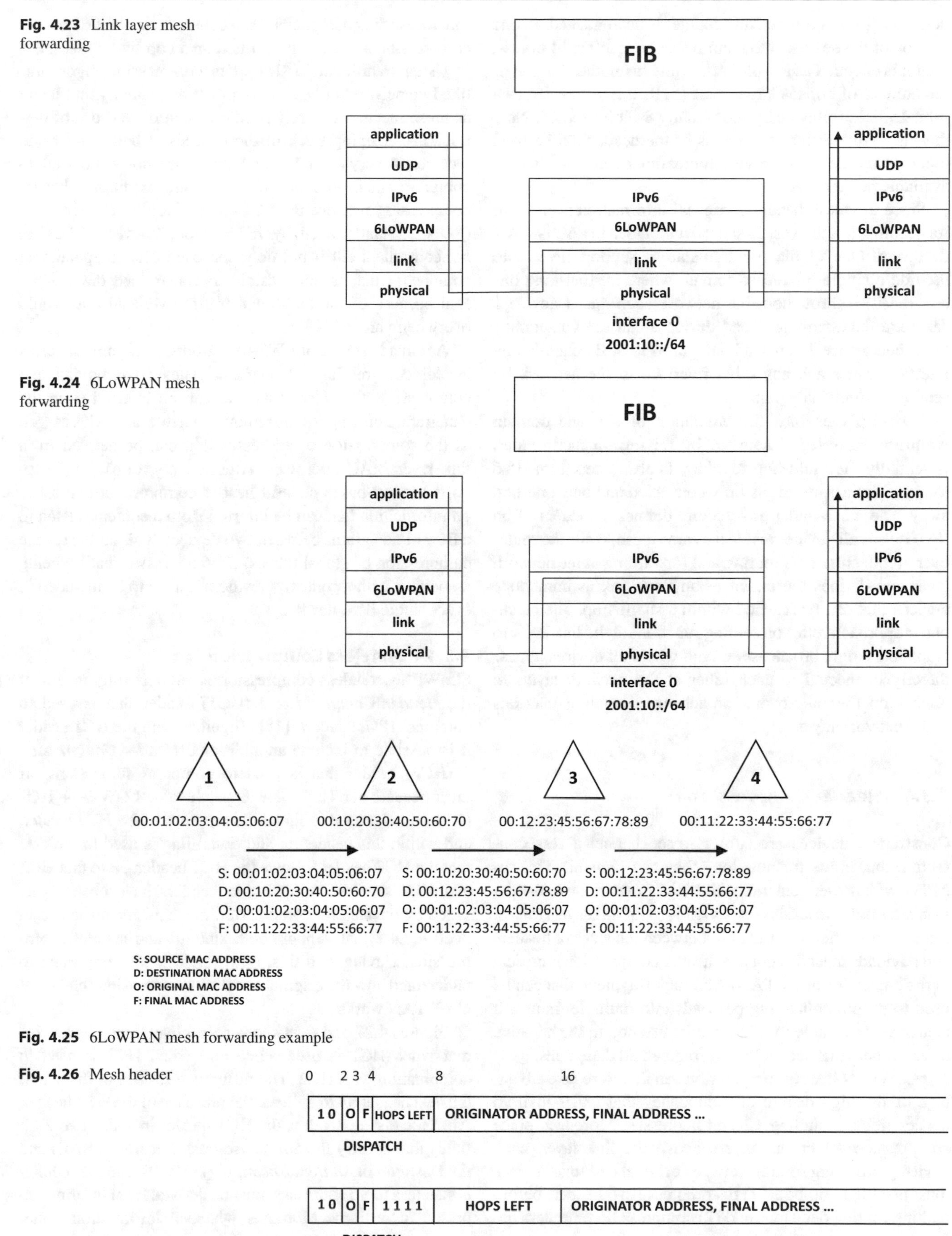

Fig. 4.23 Link layer mesh forwarding

Fig. 4.24 6LoWPAN mesh forwarding

Fig. 4.25 6LoWPAN mesh forwarding example

Fig. 4.26 Mesh header

devices acting as routers, this counter is decremented. In one version of the header, if the number of hops left is 14 or less, then it is encoded as a 4-bit field, while in another version, if the number of hops is larger than 15, then it is encoded as a 4-bit 15 field followed by the actually 8-bit hop count. Note that these two different versions of the header can be used together in order to improve compression by maximizing the available header space.

Since the mesh header carries all information needed for forwarding, it is the very first header included in a 6LoWPAN datagram. One of the requirements to support IPv6 is to provide a single broadcast domain where all transmissions are transitive throughout the network such that if device 1 can send datagrams to 2 and device 2 can send datagrams to 3, then device 1 can send datagrams to 3. Essentially any interface can reach any other interface in the network by sending a single datagram.

6LoWPAN enables the emulation of a single domain within an 6LoWPAN network by relying on mesh-under. Essentially the multi-hop topology is abstracted from IPv6 to make devices appear as fully connected and only one-hop away. The mesh-under architecture defines the extent of an IPv6 link as all devices inside the same mesh, while the route-over architecture puts routing and forwarding at the network layer and defines the extent of an IPv6 link as immediate devices that can be reached within a single hop. The mesh-under approach relies on routing functions at the link layer to emulate a single broadcast domain where all devices appear directly connected to each other at the network layer. In opposition, the route-over approach puts all routing functions at the network layer.

4.3.4 Header Compression

Constrained devices are subjected to channel access contention that limits for how long they can transmit. This together with power constraints that lower maximum transmission rates put restrictions on payload size of link layer frames. Because for efficiency the ratio between the sizes of headers and payloads must be kept low, header compression is critical in the context of 6LoWPAN. Although fragmentation can be used to accommodate big payloads into multiple frames, it is always preferable to minimize its use due to the negative effect of network packet loss in fragmented datagrams.

6LoWPAN header compression can therefore take advantage of the high degree of redundancy that exists in IPv6 header fields including 128-bit addresses. Moreover, since compression affects all layers above the link layer, each device performing routing between original and final devices must be able to do both (1) decompress IPv6 headers before making routing decisions and (2) compress IPv6 headers before forwarding datagrams. Consequently, 6LoWPAN header compression is usually performed on a hop-by-hop basis.

Using traditional lossless data compression algorithms like Lempel-Ziv for header compression is highly inefficient as those mechanisms rely on the formation and use of dictionaries. Keeping track of dictionaries at both sender and receiver is very complex and that does not scale well to compress small amounts of data. Using traditional header compression frameworks like *Robust Header Compression* (ROHC), standardized by IETF through several RFCs, is not convenient either not only because of its computational complexity that is not feasible in constrained devices but also because of the additional traffic overhead that would overwhelm any LLN.

As already stated, 6LoWPAN introduces a simple stateless header compression scheme that relies on removing and compressing all information that can be inferred from the datagram. This implies the removal of irrelevant fields as well as the compression of addresses that can be derived from link layer MAC addresses. Alternatively, 6LoWPAN supports context-based stateful header compression that relies on information that can be inferred from a context related to a flow of datagrams between two devices. This addresses the compression of global unicast IPv6 addresses that by being associated with a context, they do not need to be included in every single IPv6 header.

4.3.4.1 Stateless Compression

6LoWPAN stateless compression is supported by means of the *Header Compression 1* (HC1) header that is used to compress IPv6 headers [18]. Together with the HC1 header it is possible to include an additional *Header Compression 2* (HC2) header that is used to compress some transport layer headers like UDP. The dispatch value LOWPAN_HC1 (*01000010*) is used to signal the presence of the HC1 header, and within this header an additional flag is used to indicate that the HC2 header follows the HC1 header. Note that HC1 and HC2 stateless compression headers have been superseded by the stateful compression mechanisms presented in Sect. 4.3.4.2 that support both stateful and stateless compression. Having said this, however, it is still important to understand how the original stateless compression scheme of 6LoWPAN works.

Figures 4.27 and 4.28 show, respectively, the header format when HC1 is used alone and when HC1 is used in combination with HC2. The difference resides in the use of the next header bit to indicate the presence of the HC2 header. After the dispatch value, the HC1 header includes two 2-bit fields that specify the *Source Address Encoding* (SAE) and the *Destination Address Encoding* (DAE). Because encoding is stateless, IPv6 addresses must be derived from information present in the frame. Moreover, whatever information cannot

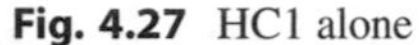

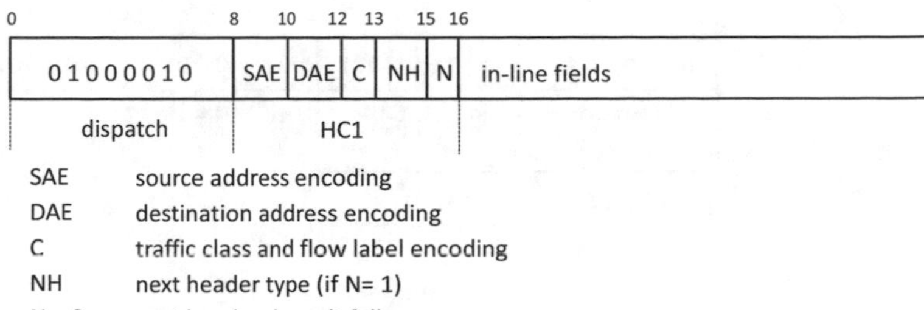

Fig. 4.27 HC1 alone

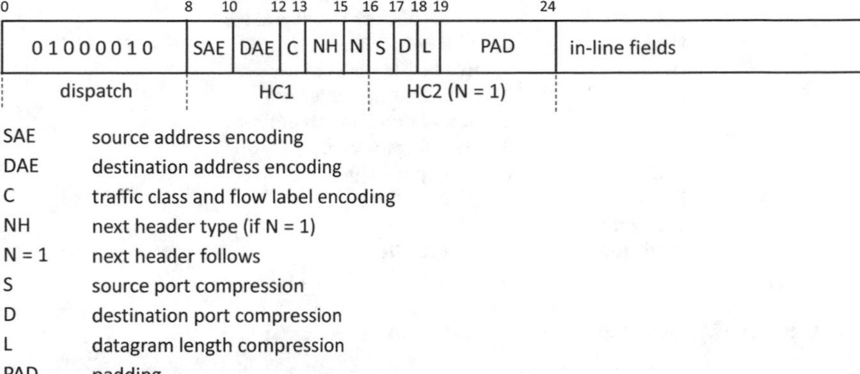

Fig. 4.28 HC1 with HC2

Table 4.4 HC1 address encoding

SAE/DAE value	64-bit prefix	64-bit IID
00	In-line	In-line
01	In-line	Derived
10	Link-local	In-line
11	Link-local	Derived

Table 4.5 HC1 next header

Value	Meaning
00	In-line
01	UDP
10	ICMP
11	TCP

be derived from the frame, it must be sent in-line after the compressed headers.

Table 4.4 illustrates the different possible values for DAE and SAE. The network prefix is either transmitted in-line or assumed to be link-local as FE80::/64. Similarly, the IID is either transmitted in-line or derived from the link layer addresses. The 6LoWPAN header then includes a bit that is used to flag whether the traffic class and flow label fields are transmitted or not. Specifically, the IPv6 8-byte traffic control and 20-byte flow label are rarely used, so they can be removed from the header when this bit is set. If this bit is not set, then the traffic control and flow label are transmitted in-line.

The IPv6 next header field is compressed right after as a 2-bit value in accordance with Table 4.5. As previously stated, the last bit of the HC1 header specifies whether an HC2 header follows. Note that IPv6 fields like version and payload length are not transmitted since they can be inferred from the packet itself. The IPv6 8-bit hop limit is always transmitted in-line. Note that non-compressed in-line fields are transmitted in the order in which they appear in the original IPv6 header.

HC2 that compresses the UDP header starts with two 1-bit fields that indicate, when set, that source and destination ports are compressed. If not set, the corresponding ports are sent in-line. When compressed, ports are encoded by the four least significant bits of the port range between 61616 and 61631. The following field specifies if the UDP payload length is removed from the header and inferred from the frame or transmitted in-line. The UDP checksum is transmitted in-line. Figure 4.29 shows an example of a combination of HC1 and HC2 headers when encoding UDP over IPv6 relying on link-local addresses. Note that the resulting headers are about 56 bits long.

► **6LoWPAN Implementations** There are many commercial and open source 6LoWPAN implementations with support ranging from common constrained RTOS to general purpose Linux distributions. The list below shows the most popular open platforms that support IEEE 802.15.4.

Contiki	Supported by means of the well-known open source LwIP protocol stack.
TinyOS	Supported by means of the Berkeley *Low Power IP* (BLIP) protocol stack.
OpenWSN	Natively supported by the OS.
Zephyr	Natively supported by the OS.
Mbed	Natively supported by the OS.
Linux	Kernel modules available for specific hardware.

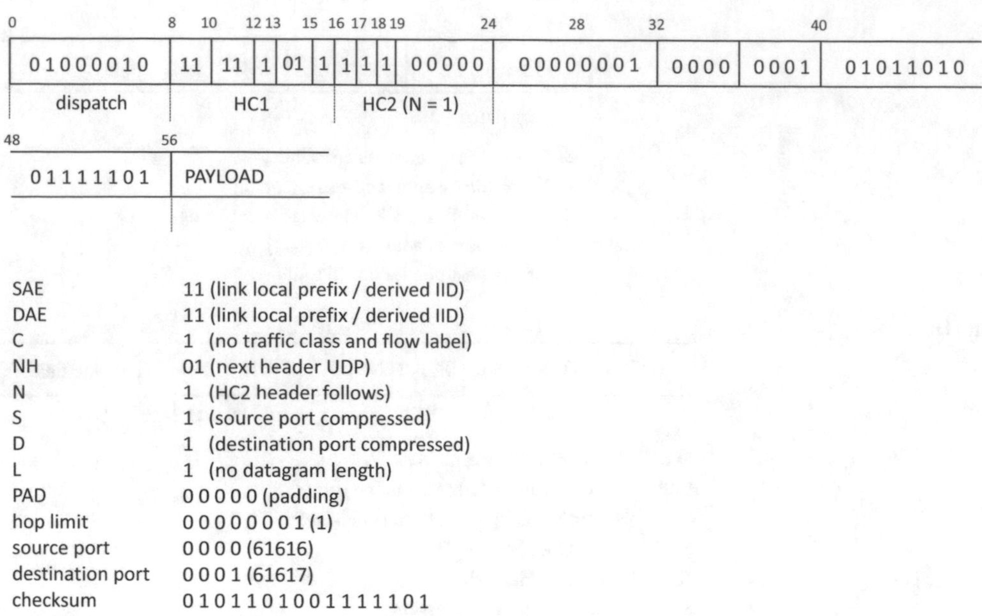

Fig. 4.29 6LoWPAN Datagram HC1/HC2 (link-local addresses)

4.3.4.2 Stateful Compression

Stateless compression that is standardized as IETF RFC 4944 "Transmission of IPv6 Packets over IEEE 802.15.4 Networks" has been superseded by stateful compression standardized as IETF RFC 6282 "Compression Format for IPv6 Datagrams over IEEE 802.15.4-Based Networks" [28]. The main difference between stateful and stateless compression is that the former addresses some of the limitations of the latter by enabling the compression of parameters that are common to datagrams associated with streams of data generated by devices. Because these common parameters can be thought of as part of a context, stateful compression is also known as context-based compression. For example, for a session between a sensor and an external application, all datagrams sent by the device carry the same unicast global IPv6 addresses. If the sensor and the application agree to assign the destination address to a given context, then compression can be accomplished. Specifically, datagrams can carry a context identifier that is encoded with a lot fewer bits than those needed to encode unicast addresses.

The shared context negotiation is not specified by IETF RFC 6282, and it is intended to be agreed by means of other mechanisms like ND. When the 6LoWPAN stack of a device wants to compress traffic based on a context, it needs to make sure that the destination 6LoWPAN stack context information is fully synchronized with that of the sender. Usually, and to prevent context synchronization problems due to connectivity and power cycle limitations, context information is divided into multiple slots that can be modified independently.

In general, context-based synchronization problems can be detected and prevented by higher-layer protocols like UDP or TCP that calculate checksums based on pseudo headers. These headers include several fields including IPv6 source and destination addresses. When the checksum is incorrect, these datagrams are dropped and retransmitted when transport is TCP based.

As in the case of stateless compression, stateful compression includes the aforementioned IPHC as well as the optional next header NHC header compression schemes. Figures 4.30 and 4.31, respectively, show the header format when IPHC is used alone and when IPHC is used in combination with NHC. Under stateful compression, only three bits of the 8-bit dispatch value are used to signal the presence of the IPHC header. The remaining five bits of the dispatch value are used to encode actual header fields.

The IPHC header starts with a 2-bit TF field that encodes traffic class and flow label as in-line fields as indicated in Fig. 4.32. If the field is encoded as *11* traffic class, flow label are assumed to be all zero. All other values of TF are used to encode combinations of the 2-bit *Explicit Congestion Notification* (ECN), the 6-bit *Differentiated Services Code Point* (DSCP), and the 20-bit flow label. Note that variable padding is used to make sure that encoding is within an 8-bit boundary. The 1-bit flag N follows to indicate whether an NHC header is also included. If this bit is not set, it implies that the 8-bit IPv6 next header field follows in-line the IPHC header.

The 2-bit hop limit field, shown in Table 4.6, is used to specify, by means of variable length coding, the hop limit of the original IPv6 header. Border routers and gateways

Fig. 4.30 IPHC alone

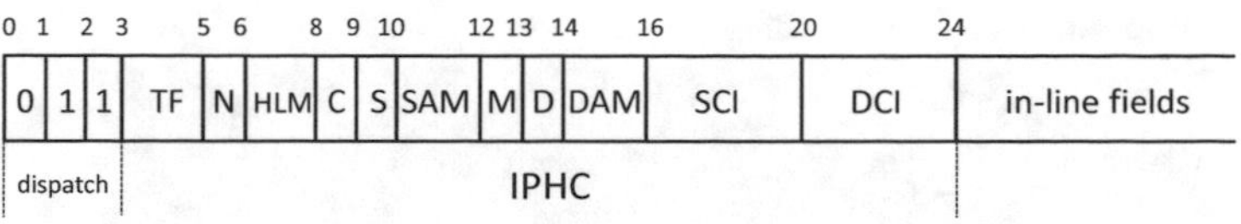

TF traffic class and flow label encoding
N = 0 NHC doesn't follow
HLM hop limit encoding
C include context information fields
S context based source address encoding
SAM source address mode
M multicast destination address
D context based destination address encoding
DAM destination address mode
SCI source context identifier (if C = 1)
DCI destination context identifier (if C = 1)

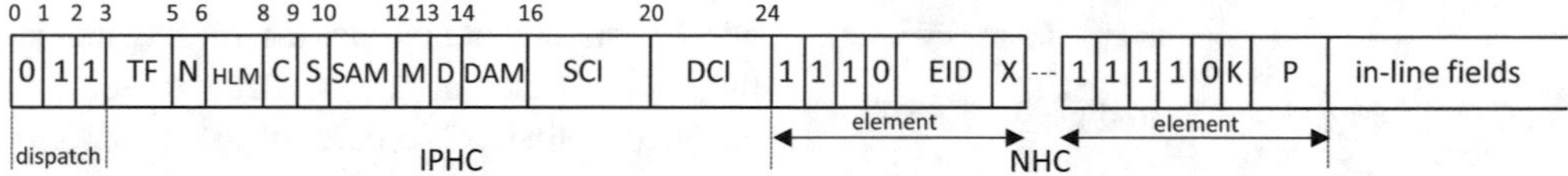

TF traffic class and flow label encoding
N= 1 NHC follows
HLM hop limit encoding
C include context information fields
S context based source address encoding
SAM source address mode
M multicast destination address
D context based destination address encoding
DAM destination address mode
SCI source context identifier (if C = 1)
DCI destination context identifier (if C = 1)
EID IPv6 extension header identifier
X more NHC elements
K UDP checksum removal
P UDP source and destination port encoding

Fig. 4.31 IPHC with NHC

Fig. 4.32 Traffic class encoding

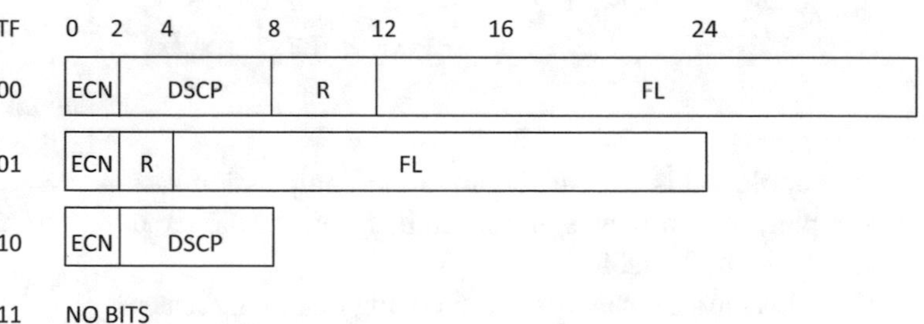

TF traffic class and flow label encoding
ECN explicit congestion notification
DSCP differentiated services code point
FL flow label
R reserved field

Table 4.6 Hop limit encoding

Value	Meaning
00	1 hop
01	64 hops
10	255 hops
11	In-line

Table 4.7 Unicast S/SAM and D/DAM Values

Context-based	Address mode	In-line bits	Meaning
0	00	128	Full unicast address
0	01	64	FE80:0:0:0:inline (link-local + 64-bit IID)
0	10	16	FE80:0:0:0:0:0:0:inline (link-local + 16-bit IID)
0	11	0	FE80:0:0:0:link-layer (link-local + link-layer address)
1	01	64	Context[0...63]:inline (context + 64-bit IID)
1	10	16	Context[0...111]:inline (context + 16-bit IID)
1	11	0	Context[0...127] (context)

Table 4.8 Multicast D/DAM values

Context-based	Address mode	In-line bits	Meaning
0	00	128	Full unicast address
0	01	48	FFxx::xx..xx
0	10	32	FExx::xx..xx
0	11	8	FE02::00xx
1	00	48	FFxx:context[0...31]:xx..xx

can manipulate this field to improve the compression rate by mapping certain ranges of hop limits to one of the fixed values shown in the table.

The 1-bit source context-based encoding flag (S) indicates whether the following 2-bit *Source Address Mode* (SAM) is context based or not. The following 1-bit flag (M) specifies if the destination address is multicast. Next, a 1-bit destination context based encoding flag (D) specifies whether the following 2-bit *Destination Address Mode* (DAM) is context based or not. Table 4.7 shows how SAM and DAM fields are encoded for unicast address support. If the destination address is multicast, Table 4.8 shows how the combination of the destination address bits encode the address.

When context encoding is set, the 4-bit source and destination context identifiers indicate prefix used by the source

Table 4.9 EID

000	IPv6 hop-by-hop options
001	IPv6 routing
010	IPv6 fragment
011	IPv6 destination options
100	IPv6 mobility header
111	Nested IPv6 header, IPHC encoded

and destination addresses, respectively. If a NHC header is present, several 8-bit elements follow. One type of element starts with a *1110* sequence and includes a 3-bit *Extension Identifier* (EID) that is mapped as illustrated in Table 4.9. The EID is followed by another 1-bit flag that indicates whether another element follows.

Another type of element starts with a *11110* and identifies a UDP header that includes a 1-bit flag that specifies whether the UDP checksum is to be transmitted or not. If it is not encoded in-line, it is recalculated on reception by the 6LoWPAN layer when the IPv6 header is generated. The following 2-bit field (P) is used to define how source and destination ports are encoded. This field is encoded in accordance to Fig. 4.33. Essentially, the source and destination ports can be sent in-line as 16-bit uncompressed fields or encoded as an 8-bit field that indicates the offset between 61440 and 61695 or, just as in the HC2 case, encoded as a 4-bit field that indicates the offset between 61616 and 61631.

Figure 4.34 shows an example of a combination of IPHC and NHC headers when encoding UDP over IPv6 relying on link-local addresses. Note that the resulting headers are about 48 bits long, this is one byte less than when HC1 and HC2 are used instead as shown in Fig. 4.29. In addition, it is possible to further reduce two bytes of the NHC header by removing the checksum and relying on lower layer protection. On the other hand, Fig. 4.35 shows an example of a combination of IPHC and NHC headers with global unicast addresses based on a default context and 16-bit short addresses that are encoded in-line instead.

4.3.5 Fragmentation

Link layer MTU sizes are small enough that fragmentation is usually required to support the transmission of very large datagrams. One issue with fragmentation is that network packet loss gets amplified whenever a fragment gets lost. Specifically, for a datagram to be reassembled correctly, all fragments must arrive at their destination. The datagram loss probability, P, is given by

$$P = 1 - (1 - p)^n$$

Fig. 4.33 NHC port encoding

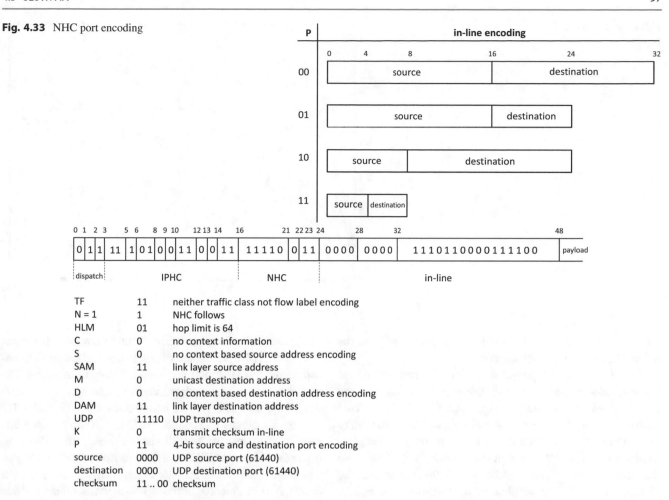

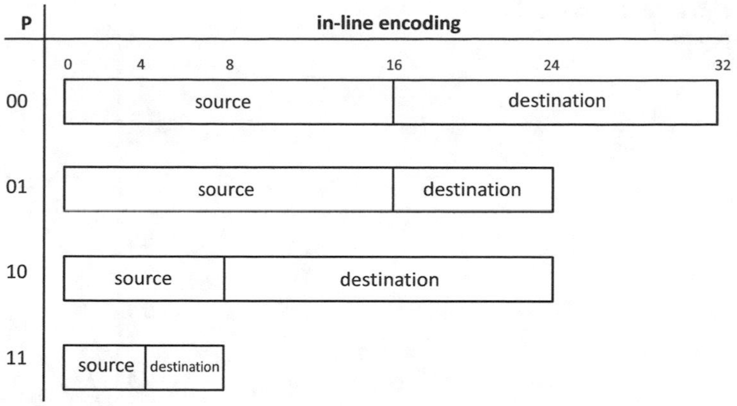

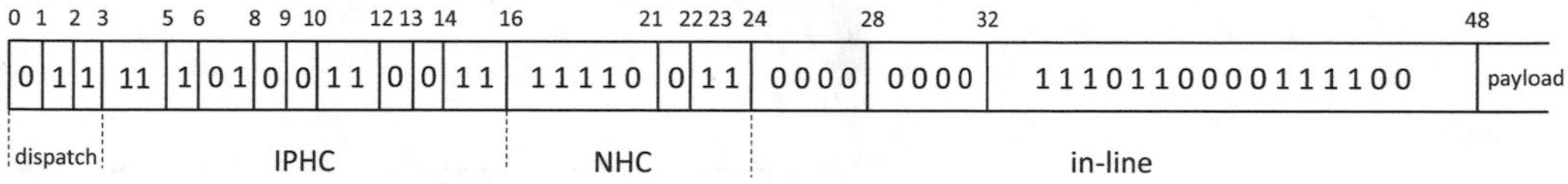

P	in-line encoding
00	source (16 bits, 0–16) | destination (16 bits, 16–32)
01	source (16 bits) | destination (8 bits)
10	source (8 bits) | destination (16 bits)
11	source (4 bits) | destination (4 bits)

0 1 2 3 5 6 8 9 10 12 13 14 16 21 22 23 24 28 32 48

0	1	1	11	1	01	0	0	11	0	0	11	11110	0	11	0000	0000	1110110000111100	payload

dispatch | IPHC | NHC | in-line

TF	11	neither traffic class not flow label encoding
N = 1	1	NHC follows
HLM	01	hop limit is 64
C	0	no context information
S	0	no context based source address encoding
SAM	11	link layer source address
M	0	unicast destination address
D	0	no context based destination address encoding
DAM	11	link layer destination address
UDP	11110	UDP transport
K	0	transmit checksum in-line
P	11	4-bit source and destination port encoding
source	0000	UDP source port (61440)
destination	0000	UDP destination port (61440)
checksum	11 .. 00	checksum

Fig. 4.34 6LoWPAN datagram IPHC/NHC (link-local addresses)

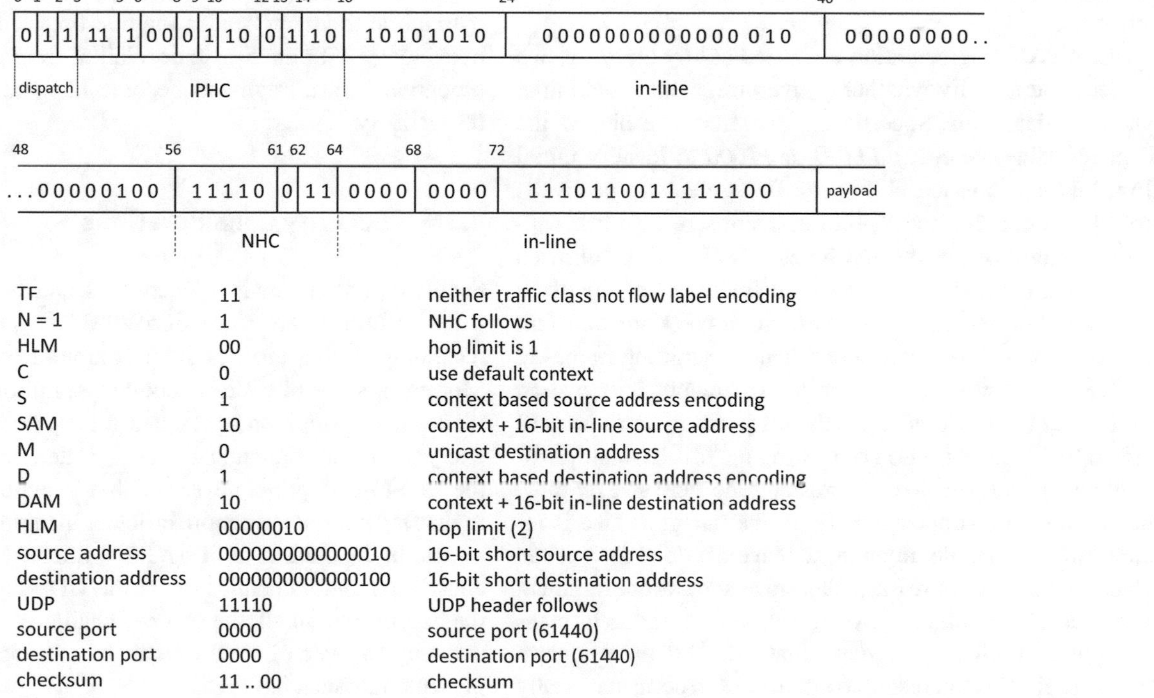

Fig. 4.35 6LoWPAN datagram (global addresses)

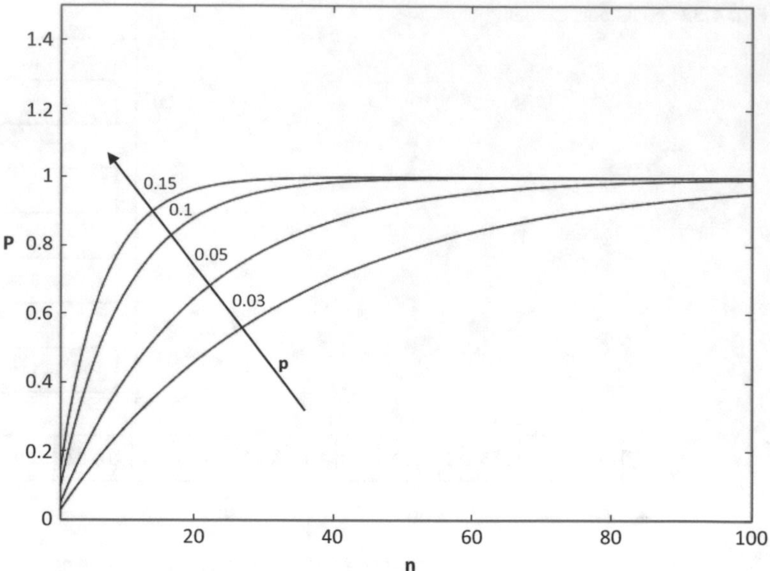

Fig. 4.36 Fragmentation problem

where n and p are the total number of fragments and the network packet loss, respectively.

The datagram loss probability (P) as a function of the number of fragments (n) for different levels of network packet loss (p) is plotted in Fig. 4.36. Clearly, for a fixed network packet loss (p), the overall datagram loss probability (P) increases as the number of fragments per datagram increase (n). This implies that the best decision is to avoid fragmentation whenever possible and, if this is unavoidable, to minimize the total number of fragments per datagram. This is particularly important in the context of IoT where network packet loss is usually a lot higher than in mainstream networks.

6LoWPAN fragmentation is carried out by means of two headers that identify whether a given fragment is the initial one in a datagram. Specifically, the first five bits of the dispatch value are either *11100* or *11000* to identify initial (Fig. 4.37) and non-initial (Fig. 4.38) fragments, respectively. Both headers include a 11-bit datagram size field that supports a length of up to 2048 bytes. The length is followed by a 16-bit tag that is unique for all fragments of any given datagram. For the worst-case scenario, it takes around 4 min for the tag counter to roll over when transmitting frames at IEEE 802.15.4 rates. For non-initial fragments, it follows an 8-bit offset field that indicates the fragment offset in units of eight bytes. The restriction of making the offset a multiple of eight has the advantage of decreasing the field size by three bits. Note that as opposed to IPv4, the datagram size is included in every single fragment so that embedded devices can allocate a buffer to store the datagram as soon as one fragment is received. The alternative of storing all fragments in maps and queues until the fragment that includes the datagram size is received is too resource intensive and computationally complex for constrained devices.

Figure 4.39 shows an example of 6LoWPAN fragmentation. A 310-byte chunk of application data is first encapsulated via UDP and becomes a 318-byte message. The message is then encapsulated via IPv6 and becomes a 358-byte datagram. 6LoWPAN compresses both IP and UDP headers by means of stateful IPHC and NHC headers that end up reducing the overall datagram size to 337 bytes. Since the maximum IEEE 802.15.4 payload size is 127 bytes and fragmentation headers are either 4 or 5 bytes long, the datagram is fragmented into two 120-byte fragments and one 77-byte fragment. Note that 120-byte fragments are that size to preserve the 8-byte offset boundary limitations. With the 4-byte initial fragment header, the first fragment becomes 124 bytes long. With the 5-byte non-initial fragment header, the second and third fragments become 125 and 82 bytes long, respectively.

4.3.6 Security Considerations

Security is paramount in IoT networks, and this is particularly true for 6LoWPAN since 6LoWPAN is one of the main technologies that enables IPv6 communications in LLNs. Moreover, since 6LoWPAN compresses information in network and transmission layers, it is in a key position to provide encryption and authentication. As stated in Sect. 3.3.3.2, the end-to-end principle states that providing security at higher layers is always more efficient than doing so at lower layers. In the context of 6LoWPAN, this means that placing encryption and authentication at the UDP layer is probably the best option. In all, the security challenges, requirements, and threats over 6LoWPAN can be summarized as those presented in Sect. 4.3.6.

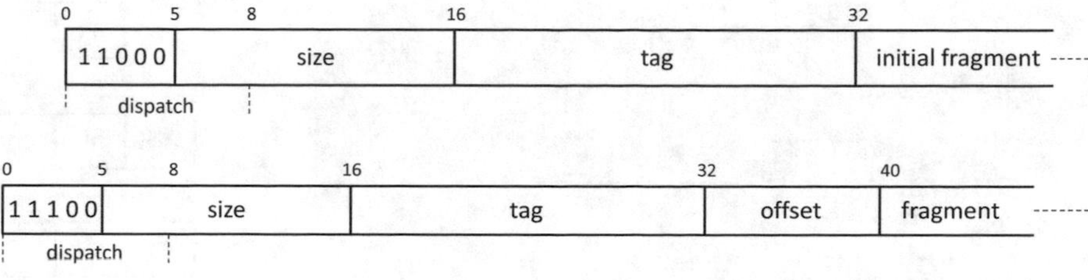

Fig. 4.37 6LoWPAN fragmentation (initial fragment)

Fig. 4.38 6LoWPAN fragmentation (non-initial fragment)

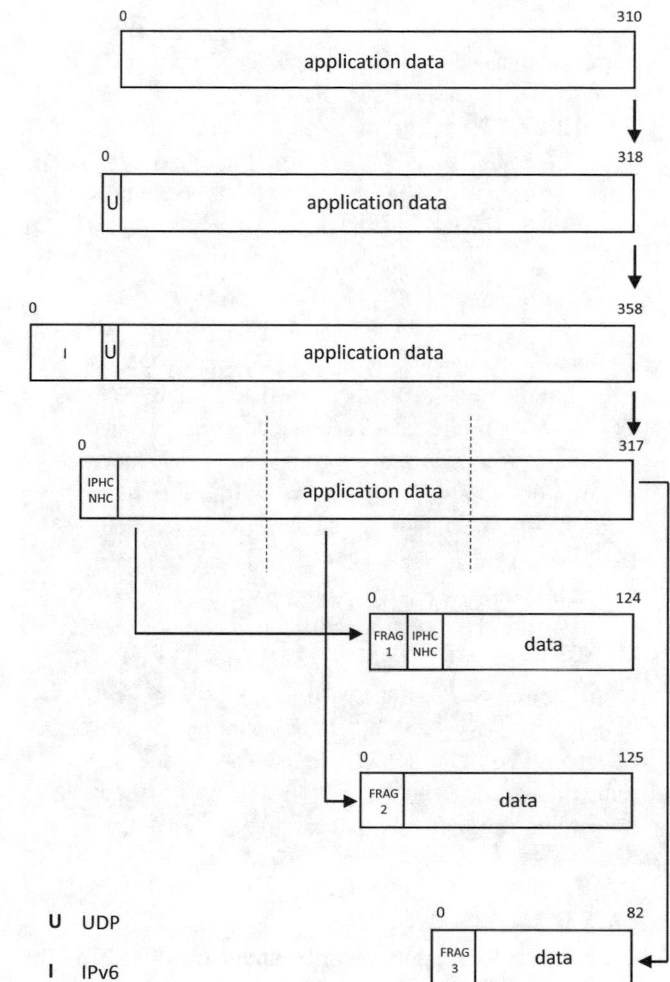

Fig. 4.39 6LoWPAN fragmentation example

Example 4.2 Given the following 53-byte IPv6 datagram:

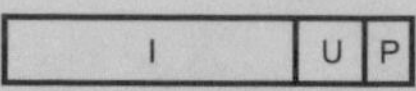

I	IPv6 header (40 bytes)
U	UDP header (8 bytes)
P	Payload (5 bytes)

Assuming best-case scenario stateful 6LoWPAN compression (i.e. smallest possible IPHC and NHC header sizes), how long does it take to transmit the IEEE 802.15.4 frame? Also assume that IEEE 802.15.4 frame headers are a small as possible. How does it compare to a scenario where uncompressed 6LoWPAN is used instead?

Solution The best-case scenario of stateful 6LoWPAN compression consists of an IPHC header that is 3-byte long and a NHC header that is 1-byte long. The corresponding best-case scenario header size of IEEE 802.15.4 is 10 bytes long because it included a minimum size 4-byte address field and no security. Because the best-case scenario transmission rate of IEEE 802.15.4 is $R = 250$ Kbps, the frame transmission takes $T = \frac{L}{R} = 608$ μs where $L = 8 \times (3 + 1 + 5 + 10) = 152$ bits. Under uncompressed 6LoWPAN, the 1-byte dispatch is added to the datagram, therefore $L = 8 \times (53 + 1 + 10) = 512$ bits and $T = \frac{L}{R} = 2.048$ ms. Essentially, by relying on stateful compression, the transmission time is more than three times shorter. This also has an effect on power consumption and battery life preservation.

4.3.6.1 IPSec/IKE

One candidate to support security under 6LoWPAN is the IPSec/*Internet Key Exchange* (IKE) combo [6, 13]. They are well-known network security protocols that are used to establish a secure channel between two mutually authenticated nodes. IPSec provides encryption and authentication, while IKE is used to set *security associations* (SAs) between entities. Note that IPSec can be used with or without IKE because the latter is typically used to exchange cryptographic keys and other security parameters that can be set up relying on other mechanisms otherwise. IKE is transmitted over UDP as request/response transactions consisting of two stages: (1) *IKE_SA_INIT* that is used to exchange security policies like algorithms and perform a *Diffie-Hellman* to exchange session keys for both endpoints to communicate with each other and (2) *IKE_AUTH* that is used for mutual node authentication. Both endpoints are authenticated, and a simple IPsec child SA is established.

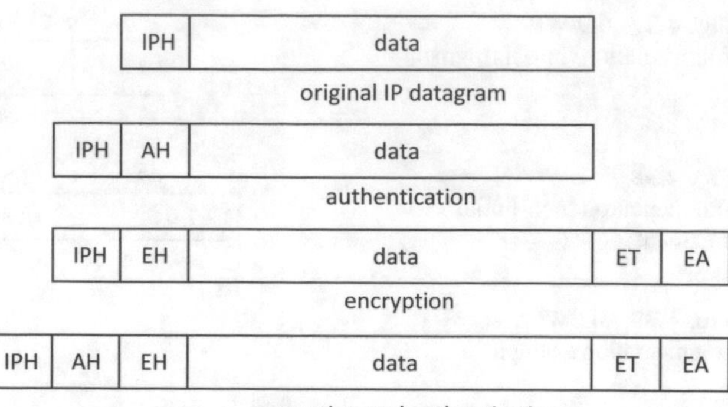

IPH	IP header
AH	Authentication Header
EH	Encapsulating Security Payload Header
ET	Encapsulating Security Payload Trailer
EA	Encapsulating Security Payload Authentication

Fig. 4.40 IPSec/IKE

IPSec includes two headers: (1) *Authentication Header* (AH) that provides data integrity, data origin authentication, and protection against replays and (2) *Encapsulating Security Payload* (ESP) that provides confidentiality and ESP payload authentication. IPSec supports two modes of operation, *transport mode* used in node-to-node operations where routing is not changed and *tunnel mode* used in encapsulation scenarios of *Virtual Private Networks* (VPNs).

Figure 4.40 shows an example of transport mode including an IP datagram, its authentication by means of AH, its encryption by means of ESP and authentication, as well as encryption when combining both AH and ESP headers. Note that ESP encrypts data and provides authentication over the ESP header, trailer, and data. On the other hand, AH authenticates the whole datagram including the IP header. Several 6LoWPAN extensions have been proposed to support the compression of IPSec AH and ESP headers in the context of stateful compression as 6LoWPAN_NHC_AH and 6LoWPAN_NHC_ESP. There have also been some proposals to adapt IKE to the 6LoWPAN context. In all cases, especially when considering IKE, these protocols have a comparatively higher overhead than other alternatives. Because of this, IPSec/IKE is not the recommended option for 6LoWPAN security.

4.3.6.2 HIP

Another candidate to provide security in 6LoWPAN is the *Host Identity Protocol** (HIP) that separates the identity of a host from the location role of its IP address [20]. Based on public key infrastructure (PKI) HIP assigns universal host identities that exist beyond different addressing domains associated with NAT. Moreover, HIP support for end-to-end identities simplifies and makes more efficient mobility as well as it ensures that node locations are kept anonymous as

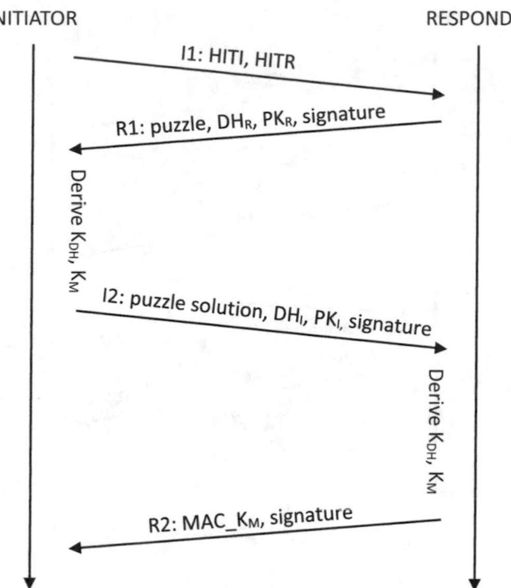

Fig. 4.41 HIP handshake

identify is disassociated from IP addresses. HIP also provides end-to-end encryption.

HIP starts with a secure handshake called *HIP Base Exchange* (HIP BEX) that is used to establish the security context between an initiator and a responder (Fig. 4.41). As IPSec, HIP information is transmitted directly over the IP layer. As part of HIP-BEX, the initiator starts by transmitting a message I1 that contains the initiator identity HITI and the responder identity HITR. The responder replies by transmitting a message R1 that includes a puzzle, its Diffie-Hellman public value (DH_R), its host identity (public key PK_R), and a signature. With the R1 information, the initiator can derive the Diffie-Hellman public key (K_{DH}) and the master key (K_M). The initiator then transmits message I2 that has the puzzle solution, its Diffie-Hellman public value (DH_I), its host identity (public key PK_I), and a signature. With the I2 information, the responder can derive the Diffie-Hellman public key (K_{DH}) and the master key (K_M). The responder then replies by transmitting a message R2 that has a message authentication code computed using the Diffie-Hellman public key and a signature.

There have been several proposals to reduce the complexity of HIP in order to make it more suitable for 6LoWPAN scenarios. Some approaches like *Lightweight HIP* (LHIP) reduce complexity by compromising security. Another approach known as *HIP Diet Exchange* (HIP DEX) removes the need of signatures. In all cases, however, HIP overhead is too high to consider the protocol as a viable security solution in the context of 6LoWPAN.

4.3.6.3 DTLS

Datagram Transport Layer Security (DTLS) is a security mechanism that is based on the well-known TLS protocol. DTLS was designed to be as robust as TLS and provide similar security levels [25, 23]. As opposed to TLS, however, DTLS does not require reliable transport to work. Specifically, DTLS relies on UDP for transport, and it is, therefore, more suitable for use in 6LoWPAN and IoT scenarios. Reliability which is critical in the context of security is addressed by DTLS by adding (1) loss handling, (2) datagram reordering, and (3) message size management. Note that under TLS, these features are handled by stream-based TCP transport.

DTLS is made of two layers; (1) the low layer called *DTLS Record Protocol* that ensures that connections are private by means of symmetric encryption and message authentication codes and (2) the high layer that supports the *DTLS Handshake Protocol*, the *DTLS Change Cipher Spec Protocol*, the *DTLS Alert Protocol*, and the *DTLS Application Data Protocol*. The DTLS Handshake Protocol enables the negotiation of security parameters. The DTLS Change Cipher Spec Protocol is used to signal the transitions in ciphering strategies and to indicate what records are protected by the negotiated parameters. The DTLS Alert Protocol is used to terminate a session by indicating an error condition. Finally, the DTLS Application Data Protocol ensures that messages are fragmented and transmitted by the DTLS Record Protocol.

Figure 4.42 shows a DTLS handshake. Each transmission from the client to the server or from the server to the client is called a *flight*. There are two entities: a client that initiates the session and a server that accepts the session establishment request. Note that these roles are inherited from TLS and TCP where a client and servers are essential to establishing secure connections. The client starts by sending a Client Hello message that is replied by the server with a Client Hello Verify that includes a random cookie. These two optional messages are used to protect the server against DoS attacks. The client then resends the *Client Hello* message but includes the cookie in order to indicate that it can process messages from the server. The Client Hello message signals the protocol version and cipher suites supported by the client. The server then replies by sending a Server Hello message that indicates the selected cipher suite from the client's list and its own ITU-T X.509 certificate. If the server is configured to support mutual authentication the Server Hello message also includes a certificate request. The server ends the transaction by transmitting a Server Hello Done message. The client then sends the certificate if requested by the client and the Client Key Exchange message encrypting a master secret using the

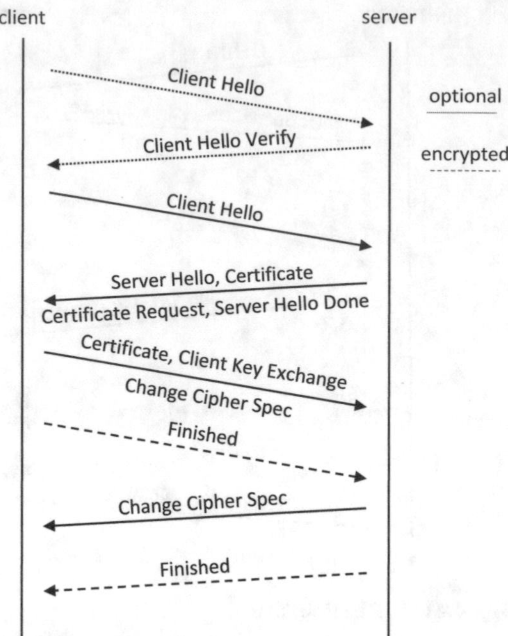

Fig. 4.42 DTLS handshake

server's public key obtained from its certificate. The master secret is used by both endpoints to derive encryption and authentication session keys. The client verifies the server's certificate against a known Certificate Authority (CA), and, under mutual authentication, the server verifies the client's certificate against a known CA. The client then transmits a Change Cipher Spec message to indicate all subsequent messages are encrypted. The client transmits a Finished message to indicate the client is done with the DTLS handshake. Similarly, the server transmits its own Change Cipher Spec and Finished messages.

DTLS breaks the data stream into DTLS records, appends a MIC to each record for integrity checking, and then encrypts both the record and the MIC. Figure 4.43 shows a DTLS record that consists of type, version, epoch, sequence number, and data length fields as well as a data section and the MIC. Note that the first five fields are not encrypted. The type field indicates whether the record is a handshake message or a message that contains application data. The epoch field is a counter value that is incremented on every cipher state change. The sequence number field is a counter value that is incremented with each record.

In the context of 6LoWPAN networking, DTLS is not fully optimal as neither DTLS nor TLS were designed to work with constrained networks and devices. One of these issues is the fact that many of the messages exchanged during handshaking are very large in nature as they carry certificates and other security parameters. This requires endpoints to have large buffers to store messages in case they need to be retransmitted. Moreover, certificates are typically represented by a chain of trust that includes intermediate certificates that need to be transmitted along the way. In 6LoWPAN networks, with a 127-byte MTU, this is a limiting factor that leads to fragmentation, and, therefore, it increases the chances of datagram loss. Essentially a 127-byte MTU yields about 70 bytes of payload after authentication and headers are put in place. Fortunately, DTLS addresses this issue by introducing a new mechanism by which instead of transmitting the whole ITU-T X.509 certificate chain, public keys are sent. Once received, the public keys are then validated by endpoints by means of an out-of-band mechanism. DTLS also introduces handshake level fragmentation that can be used to prevent IP fragmentation of handshake messages. Handshake level fragmentation, as opposed to the IP one, has the advantage of being controlled by the DTLS Application Data Protocol and supports retransmissions at the fragment level. Of course, there is a network overhead associated with this due to the extra datagrams being transmitted. Regardless, and as stated before, memory requirements in all scenarios are large, as fragments associated with certificates need to be stored for later reassembly at the receiver.

Although DTLS provides handshake level fragmentation, it does not support application data fragmentation. This means that for applications to prevent fragmentation, they must attempt to fit each message and all headers in a single datagram. There have been several proposals to minimize the DTLS header length in order to increase the space available for message transmission including removing the version field in DTLS record carrying application data since it has already been negotiated during handshake and removing the length field of the last record in a given datagram since it can be derived from its length.

One way to prevent any type of fragmentation is by reducing the number of bytes to be transferred so that fewer

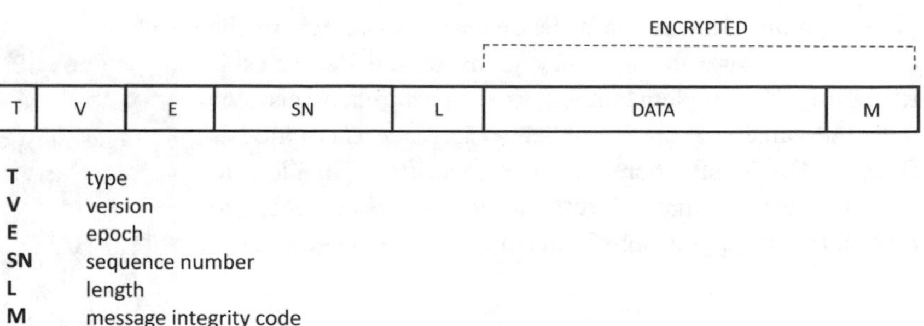

Fig. 4.43 DTLS record

fragments can be potentially lost. One approach to accomplish this is by exchanging, out of band, large chunks of data. To this end, the TLS Cached Information Extension is a mechanism that enables an endpoint to omit the transmission of information, such as a certificate, that is already known to its peer. Another way to reduce the number of transmitted bytes is by introducing header compression. There have been multiples proposal for DTLS header compression under 6LoWPAN including the use of IETF RFC 7400 that introduces *6LoWPAN-GHC* a generic header compression mechanism under 6LoWPAN [3]. Other proposals include transmitting only the least significant bits of the sequence number, using self-delimiting fields instead of fixed sized fields as with Huffman coding and using single bit fields instead of multiple type fields to encode handshake messages. IETF RFC 7925 that addresses the security profiles under TLS and DTLS for IoT states that compression at the TLS/DTLS layer is not needed since application-layer protocols are highly optimized, and the compression algorithms at the DTLS layer increases code size and complexity [29].

Another critical element are timers; DTLS recommends that protocol implementations include an initial timer of 1 s and double the value each time a message is retransmitted up to a maximum value of 60 s. In the context of DTLS, an initial timer of 1 s may lead to bogus retransmissions as it may not give enough time for embedded IoT devices to perform message transmissions. In 6LoWPAN networks, therefore, the initial retransmission timer value may need to be adjusted depending on hardware requirements. IETF RFC 7925 recommends an initial timer value of 9 s.

The number of handshake messages and their complexity have memory implications that can be challenging in 6LoWPAN networks. When compared to TLS, DTLS is even more complex because it incorporates the initial cookie exchange intended to prevent DoS attacks. There have been several proposals addressing this issue including introducing client puzzles to prevent DoS attacks in a similar way to HIP DEX. Another possibility is for the client to create DTLS connections before they are really needed in order to minimize the time it takes to send secure datagrams once they become available. This approach, however, requires servers to be powerful enough to support and maintain multiple simultaneous DTLS sessions.

Although a DTLS handshake stage consumes a lot of memory due to the buffers required for retransmissions, once a session is established, memory requirements are a lot less stringent. In general, a single session must have enough memory to store session keys and sequence numbers. This is, in general, used to dimension the number of supported DTLS sessions in 6LoWPAN scenarios. With most embedded devices and *real-time operating systems* (RTOSs), DTLS connections are allocated statically as part of pools of predefined size. Because of the memory implications of DTLS, determining when a DTLS connection terminates is important. DTLS connection closure consists of a *closure alert* that is not retransmitted even if packet loss occurs. This, again, can lead to memory exhaustion due to stale connections that result from lost alerts in the context of lossy 6LoWPAN networks. One way to guarantee that DTLS connections are terminated is by keeping track of them such that when a device starts running out of resources, the *least frequently used* (LFU) connections are removed first. IETF RFC 7925 recommends using a mechanism introduced in IETF RFC 6520 that introduces a heartbeat mechanism to verify whether a given DTLS connection is still active [31].

Since a single flight is associated with one or more messages like Server Hello and Certificate Request in Fig. 4.42, several fragments may end up being transmitted. This is aggravated by the fact that many messages are independent and can be transmitted in parallel. In this case, multiple fragments may simultaneously arrive at the receiver causing input buffer overflows that could lead to datagram loss. One way to prevent this is by making sure that the DTLS protocol implementation transmits independent messages in a flight sequentially. Another problem with fragmentation is the fact that message integrity check requires the whole record to be processed even if it has been corrupted. Additionally, if a message is to be retransmitted, the cleartext or unencrypted version of the message must be stored in order to be re-encrypted with every single retransmission. The idea behind preserving a cleartext message is to make sure that every retransmission is different since sequence numbers are different, thus providing replay protection. This also increases memory requirements and prevents the implementation from implementing *in-place* encryption. This mechanism consists of encrypting data in the same memory space where the cleartext message is. One way to overcome this issue is to modify the implementation to support retransmissions with the same sequence number so in-place encryption is feasible. Obviously, there are some security implications to this. Another way, that may result in slower performance, is to rely on cryptographic algorithms that use a lot less memory.

4.3.7 TCP and 6LoWPAN

6LoWPAN HC2 and NHC headers provide ways to compress an UDP header in a stateless and stateful way, respectively. Although there have been several attempts to introduce a compression scheme for TCP under 6LoWPAN, no formal specification exists to this day. TCP is a transport protocol that, as opposed to UDP, provides a connection-oriented service that characterizes for enabling reliable data transfer. In order to accomplish this, TCP relies, when configured accordingly, on a *Automatic Repeat reQuest* (ARQ) mechanism with selective acknowledgments. A TCP stack not only requires computational complexity that is not always available in embedded devices, but it also introduces retransmissions that result in additional end-to-end delay. Because LLNs are prone to packet loss, retransmissions are quite common in the context of IoT. Moreover, retransmissions cause, in many cases, more problems as they increase the volume of traffic that, with limited network capacity, results in collisions. These collisions, in turn, cause more packet loss. By the time datagrams arrive at their destination, the application has no more use for them and ends up dropping them and causing additional application layer packet loss. Therefore, network packet loss is amplified by retransmissions that cause an increased application layer packet loss.

Another issue of TCP transport is that it assumes that packet loss is caused by congestion due to intermediate devices, like routers, dropping packets when their queues are full. IoT packet loss, however, is mostly due to low SNR situations where signals are heavily corrupted by noise. For scenarios of low-volume traffic, the TCP approach attempts to further decrease the rate as it slows down the transmission of packets to cope with the congestion. In many of these scenarios, however, it is convenient to rely on very fast retransmissions that provide a FEC-like mechanism to overcome the loss introduced by the channel.

Example 4.3 Consider a sensor that transmits 3-byte readouts every 20 ms directly over a TCP stream (no session/application layer protocol). What is the expected additional delay due to the Nagle's algorithm (when it is enabled)? Assume that the algorithm will buffer 60 bytes before transmitting.

(continued)

(continued)

Solution Clearly, around $N = \frac{60 \text{ bytes}}{3 \text{ bytes}} = 20$ messages must be buffered before they can be transmitted. The delay of each transmission is given by $d_i = (N - 1 - i) \times 20$ ms where $i = 0, \ldots N - 1$. The expected additional delay is, therefore, given by $E = \sum_{i=0}^{N-1} \frac{d_i}{N} = 1 \text{ ms} (19 + 18 + 17 \ldots + 1 + 0) = 190$ ms. Because a sensor readout is transmitted every 20 ms, it can be assumed that transmissions are uniformly distributed.

Another important issue with TCP is that in real-time sensing applications, traffic needs to be sent as fast as it is generated. Through the Nagle's algorithm, TCP natively buffers packet payloads until it has enough data to make it efficient to send a datagram. This extra delay affects latency, and, as stated in the previous paragraph, it can also result in application packet loss. Fortunately, some applications can disable Nagle's algorithm in their TCP stacks to improve real-time performance. In either case 6LoWPAN is not intended to rely on TCP for session management, and therefore its use is only supported as uncompressed traffic.

4.4 6Lo

The support of IPv6 over IEEE 802.15.4, by means of 6LoWPAN adaptation, has opened the door for other physical and link layer LLNs technologies like BLE (described in Sect. 3.3.4), DECT ULE (described in Sect. 3.3.6), NFC (described in Sect. 3.3.7), IEEE 802.11ah (described in Sect. 3.3.1), IEEE 1901.2 (described in Sect. 3.2.3), MS/TP (described in Sect. 3.2.4), and ITU-T G.9959 (described in Sect. 3.3.5) to enable IPv6 connectivity through their own adaptation schemes. These mechanisms are comprehensively called 6Lo technologies as they are the focus of standardization of the IETF *IPv6 over Networks of Resource Constrained Nodes* (6Lo) working group [7].

These stacks rely on IPv6 adaptation that is either partially or fully derived from 6LoWPAN functionality. Specifically, these technologies modify some of the 6LoWPAN features to better fit their physical and link characteristics. Figure 4.44 illustrates the different LLN families with the proposed and standardized adaptation mechanisms. The figures indicates the different topologies associated with each of the technologies and the corresponding standardized or proposed adaptation protocol: plain 6LoWPAN for IEEE 802.15.4, modified 6LoWPAN for ITU-T G.9959 [4], LoBAC for MS/TP [16], 6LoPLC for IEEE 1901.2 [10], 6LoWPAN for DECT ULE [17], 6LoBTLE 6LoWPAN for BLE [22], 6LoWPAN for NFC, and 6Lo IEEE 802.11ah [7]. These

Fig. 4.44 6Lo technologies

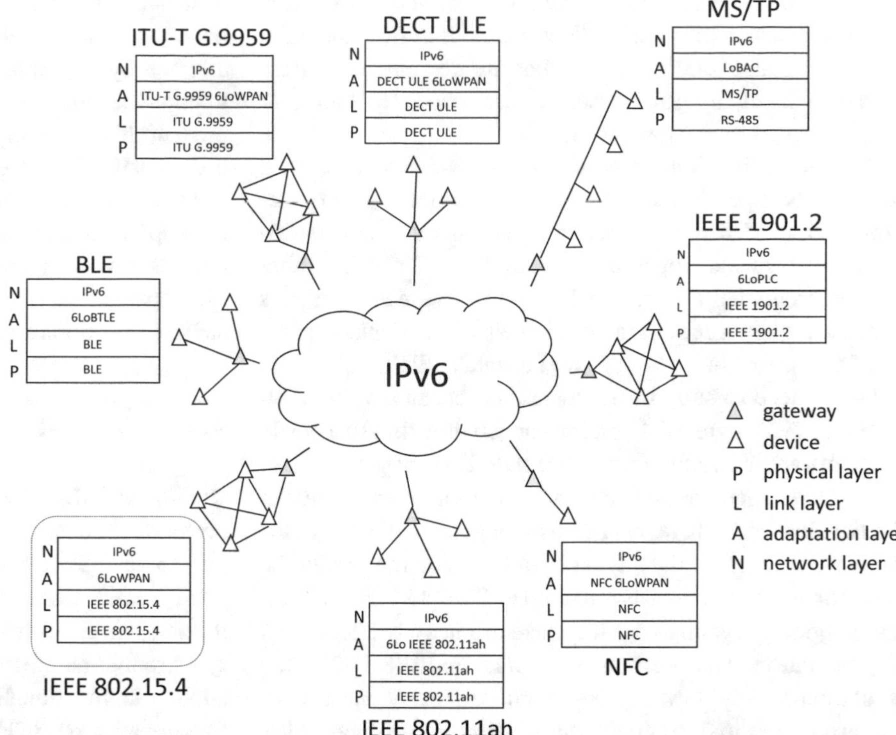

Table 4.10 6Lo Technologies

Technology	Adaptation layer	Topology
ITU-T G.9959	ITU-T G.9959 6LoWPAN	M
MS/TP	LoBAC	M
DECT ULE	DECT ULE 6LoWPAN	S
IEEE 1901.2	6LoPLC	M
NFC	NFC 6LoWPAN	S
IEEE 802.11ah	6Lo IEEE 802.11ah	S
IEEE 802.15.4	6LoWPAN	M

technologies are compared in Table 4.10 where the topology column indicates whether they support *multi-hop* (M) or *single-hop* (S) communication.

Many of these technologies are not just physical and link layer technologies. Like ZigBee with IEEE 802.15.4, BLE, NFC, and DECT ULE prescribe full stacks that include their own network, transport, and application layers. In these scenarios, 6Lo replaces these upper layers with IETF IP-based protocols. In all cases, however, the different physical and link layers face different challenges associated with issues ranging from low transmission rates to small MTU sizes including lack of link layer fragmentation and asymmetries between uplink and downlink mechanisms.

As previously indicated in Sect. 4.3.3, 6LoWPAN provides two main modes of operation to support a multi-hop topology: mesh-under where the multi-hop path appears as a single link to the network layer and route-over where physical hops are IP hops. The only 6Lo technologies that support a mesh topology are IEEE 1901.2 and ITU-T G.9959 that reuse 6LoWPAN functionality. ITU-T G.9959 supports both modes of operation, while IEEE 1901.2, under 6LoWPLC, only supports mesh-under forwarding.

Address compression is also key to both 6LoWPAN and 6Lo. Specifically, the IID, obtained from link layer addressing, is not only used to derive IPv6 addresses by means of SAA but also to provide their compression. All adaptation mechanisms standardized and proposed for the 6Lo technologies described in this section rely on address compression. One drawback of this compression, however, is the fact that the universal and static nature of IIDs can lead to security attacks. In this context, different proposals that target this issue under 6Lo have been proposed.

Header compression, as provided by 6LoWPAN, can be stateless or stateful to take advantage of the inter- and intra-flow information correlation. The difference between one approach and the other is extra computational complexity in favor of stateful compression. All 6Lo technologies, other than IEEE 1901.2 under 6LoWPLC which is based on IEEE 802.15.4, modify 6LoWPAN header compression to adapt to specific needs of the link layer. ITU-T G.9959, MS/TP LoBAC, IEEE 802.11ah, and NFC introduce minor changes associated with the address length of the link layer addresses. Single-hop star topologies like BLE and DECT ULE are slightly less complex from a header compression point of view since the gateway, when receiving a datagram, can safely assume that the source address is that of the originator. Moreover, when a device receives a datagram from the gateway, it can also safely assume that the destination address is that of itself. In either case, this leads to IPv6 source or destination address suppression.

Fragmentation is another important issue associated with 6LoWPAN. Specifically, 6LoWPAN introduces fragmentation because IEEE 802.15.4 does not include any mechanism to truncate upper layer datagrams. Since the other 6Lo technologies do support link layer fragmentation, this is not typically a requirement for their IPv6 adaptation layers. Moreover, under IPv6, MS/TP and IEEE 802.11ah do not even need to support fragmentation because with 2032-byte and 7951-byte MTU sizes, respectively, they can easily transport any minimum size 1280-byte IPv6 fragment. Also related to fragmentation is the encapsulation overhead due to the physical and link layers headers. In general, the larger the headers, the smaller the payload available and the greater the need for fragmentation. For example, BLE and DECT ULE that support very small payload size are heavily affected by fragmentation. The overhead of MS/TP and IEEE 802.11ah is comparatively low as they both can carry most IPv6 datagrams without fragmentation. The other technologies, ITU-T G.9959, IEEE 1901.2, NFC, and IEEE 802.15.4 suffer some fragmentation for large datagrams. In all scenarios, the fragmentation overhead at either layer, link, or adaptation is very low.

All 6Lo adaptation methods, other than that of IEEE 1901.2, rely on either 6LoWPAN ND or traditional IPv6 ND for neighbor discovery. Specifically, IEEE 1901.2 relies on DHCPv6 for address autoconfiguration. IPv6 multicasting is particularly important in IoT as it enables applications to reach multiple devices through the transmission of a single datagram. Of all 6Lo technologies, only IEEE 802.11ah natively supports multicast and broadcast transmissions. Moreover, BLE and DECT ULE only support unicast transmissions. Regardless, for IPv6 multicast to work properly, it is critical that the adaptation layer is able to present to the upper layer multicast support. This implies that 6Lo adaptation follows 6LoWPAN in the fact that each multicast datagram is transmitted by the adaptation layer to each device in the multicast group as individual unicast datagrams. This has to be done very efficiently as it can lead to depleted batteries.

Security is also a key factor with 6Lo but fortunately most adaptation mechanisms like 6LoWPAN rely heavily on link layer security. Since all these technologies, including IEEE 802.15.4, provide some sort of security at the link layer this is not usually a problem. Also, most applications support end-to-end security that, as explained in Sect. 4.3.6, is always more efficient than security placed at lower layers. Due to the limited resources, the biggest challenge is to make sure that upper and lower layer security mechanisms do not inefficiently overlap. In this context, it is important to understand what threats can affect 6Lo technologies and how they can be prevented. Neighbor discovery is usually affected by DoS attacks where an intruder pretends to be a legitimate device by sending fake ND messages. One way to prevent this is by relying on router priorities while limiting ND transmission rates. Fragmentation is another issue that can lead to security attacks. Specifically, an attacker can send incomplete datagrams with pending fragments that cause devices to waste memory resources for long periods of time. The intruder can also send fake and duplicate fragments that cause corruption of incoming datagrams.

4.5 6TiSCH

IPv6 over TSCH [30] is an umbrella term for a group of protocols intended to provide IPv6 support over the IEEE 802.15.4e TSCH media access mechanism presented in Sect. 3.3.3.3. Because traditional 6LoWPAN and associated routing mechanisms are not designed to handle the synchronous nature of the TSCH link layer, 6TiSCH provides additional functionality that compensates for this missing functionality. 6TiSCH introduces a thin layer above the MAC functionality of TSCH called *6TiSCH Operational Sublayer* (6top) that binds asynchronous 6LoWPAN to synchronous TSCH.

Specifically, 6LoWPAN provides IPv6 adaptation for traditional CSMA/CA-based IEEE 802.15.4, but it is not prepared to deal with TSCH-based IEEE 802.15.4e. 6TiSCH sits in between 6LoWPAN and IEEE 802.15.4e. Figure 4.45 illustrates the layer differences between traditional CSMA/CA and TSCH stacks.

6TiSCH supports dynamic scheduling by means of two separate entities; (1) the *6top Protocol* (6P) that can be used by neighbors to negotiate the addition and removal of cells from the schedule and (2) a *Scheduling Function* (SF) that triggers 6P negotiations. The 6P, which is standardized under IETF RFC 8480, provides several commands that neighbors can execute to support schedule cell management including adding, removing, and relocating cells from a schedule as well as listing cells and signaling SFs.

6P negotiation is carried out by means of 2-step or 3-step transactions. 2-step transactions consist of one device transmitting a request to its neighbor that replies by transmitting a response. This is usually used when a device wants to add, remove, or relocate a given cell. Three-step transactions add an extra confirmation that is transmitted by device that initiated the transaction. This is used when one device requests adding one or more cells to the schedule and expects the receiving neighbor to offer a list of available cells for the originator to select from. This is illustrated in Fig. 4.46 where device 1 sends a request to add two cells and the receiving neighbor device 2 offers four cells (2,3), (3,5), (1,2), (5,1)

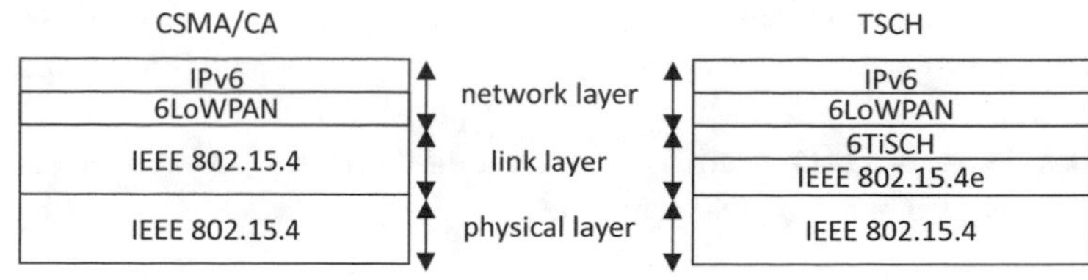

Fig. 4.45 6TiSCH stack

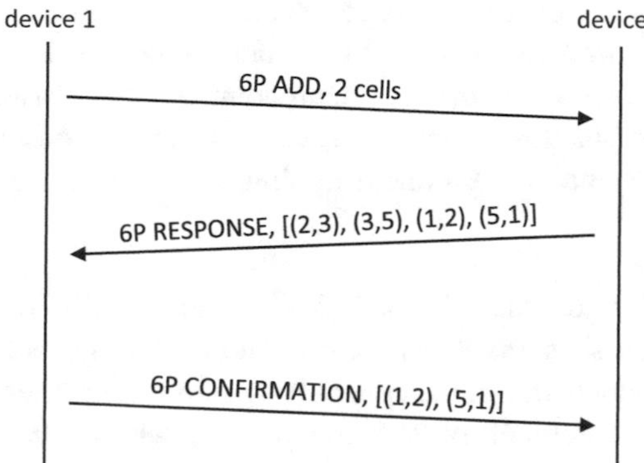

Fig. 4.46 6P 3-step transaction

specified by the TSCH timeslot and channel offsets. Then the originator device 1, confirms by selecting cells (1,2) and (5,1) for communication. All these commands are carried by IEs in the IEEE 802.15.4 frame. Each IE includes not only the command information but also an 8-bit sequence number that is used to keep track of transactions while detecting for inconsistencies like missing acknowledgments. When an inconsistency is detected, though, an SF may send a command to clear all cells between the neighbors, or it may send a command to list all cells and attempt to delete and add individually. Because devices store the expected sequence numbers of their neighbors, they can detect problems when the sequence number received in a command does not match the expected one. Again, this situation is fixed by transmitting a clear command that resets the sequence number.

The default SF supported by 6TiSCH is called *minimal SF* (MSF). MSF defines two types of cells: autonomous cells that provide proactive cell scheduling without any type of negotiation and negotiated cells that rely on 6P to install and remove cells from the schedule in a reactive way. Autonomous cells can be, in turn, autonomous RX cells that are permanently installed in the schedule or autonomous TX cells that are installed on demand. If a frame is to be sent out and there is no cell installed, an autonomous TX cell is briefly installed, the frame is sent out, and then the cell is removed right away. The cell itself is derived from hashing the destination address of the frame. Since hashing can lead to collisions where the same cell is allocated as an autonomous RX and an autonomous TX cell resolution may be needed. In general, autonomous cells are kept in the schedule even after 6P commands to clear the schedule are submitted. Negotiated cells are added or deleted based on how fast neighbors exchange frames. MSF keeps track of cell usage that is defined as the ratio between elapsed cells and cells that are used to transmit frames to the neighbor. If the cell usage ratio is high, MSF adds cells, and similarly, if the ratio is low, MFS removes cells. All of this is done, of course, by means of 6P commands. In this scenario MSF may introduce collisions that are detected and resolved by means of cell relocations. If an MSF negotiation transaction fails, depending on the failure type, different recovery actions are recommended; to do nothing *or* to clear the schedule *or* to clear the schedule and put the device in quarantine dropping all frames intended to it *or* to abort the transaction, wait for a random amount of time and restart the transaction.

Summary

One critical requirement to enable IoT connectivity is to support end-to-end IPv6 networking. This chapter started by introducing the main features of IPv6 including procedures like SAA that are key in allowing devices to self-initialize and configure. IPv6 support in the context of IoT is provided by adaptation mechanisms like 6LoWPAN that introduce network and transport layers over constrained physical and link layer technologies. The chapter explored 6LoWPAN detailing its main characteristics and features including its addressing model, header format, routing and forwarding schemes, stateful and stateless compression support, as well as fragmentation. The chapter also delved into security considerations of IoT network layers by presenting the most relevant security protocols: IPSec/IKE, HIP, and DTLS. Finally, it presented several other IPv6 adaptation mechanisms that fall under the umbrella of 6Lo and support other IoT physical and link layer technologies like BLE and DECT ULE among others.

Homework Problems and Questions

4.1 Consider an Ethernet device with MAC (EUI-48 format) address 11:22:33:44:55: 66 and an IEEE 802.15.4 device with MAC (EUI-64 format) address 11:22:33:44:55:66: 77:88. If the network prefix is 2001::/64, for the following IPv6 address types, what addresses are derived for both Ethernet and IEEE 802.15.4?

(a) SAA based unicast
(b) Link-local

4.2 Why is DAD needed, if SAA relies on using MAC addresses that are typically unique?

4.3 What is the point of using fewer of the eight available bits for dispatch encoding?

4.4 Why does not the initial 6LoWPAN fragment include the offset field?

4.5 Given the following 88-byte IPv6 datagram…

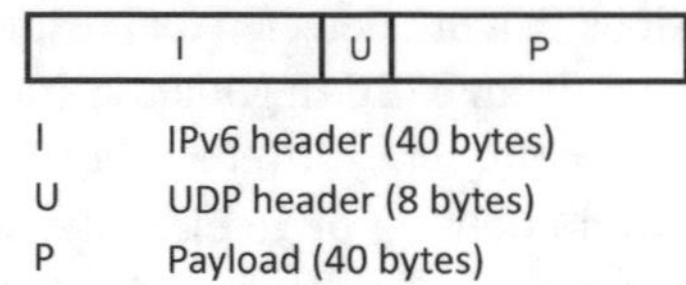

What are the best compression rates for the scenarios listed below? Indicate all assumptions. Note that the compression rate is defined as the ratio between the size of the compressed frame and that of the original datagram. Assume minimum 6LoWPAN compressed network and transport headers.

(a) Stateless 6LoWPAN compression
(b) Stateful 6LoWPAN compression

4.6 Why is the 6LoWPAN fragment offset field a multiple of eight?

4.7 Repeat Problem 4.5 but considering the following 68-byte IPv4 datagram…

I U P

I IPv4 header (20 bytes)
U UDP header (8 bytes)
P Payload (40 bytes)

4.8 Repeat Problem 4.5 but consider an 800-byte payload instead. Assume that the lower layer is IEEE 802.15.4, supporting payload sizes no larger than 118 bytes long.

4.9 In what 6LoWPAN scenarios is stateless compression more convenient than stateful compression?

4.10 If an IoT sensor generates 172-byte readouts every 20 ms, what is its transmission rate? Each readout is transmitted over UDP/IPv6 and compressed by means of 6LoWPAN. Consider best and worst stateful compression scenarios. Indicate all assumptions.

4.11 What is the main difference between the 6LoWPAN S/SAM and SAE fields?

4.12 If an IEEE 802.15.4 IoT sensor generates N-byte readouts at a rate of R readouts per second. Assuming each readout is transmitted over UDP/IPv6 and compressed by means of 6LoWPAN. For the maximum nominal transmission rate and best stateful compression, what is the relationship between R and N? Assume a minimum size IEEE 802.15.4 header.

4.13 Consider the IEEE 802.15.4/6LoWPAN route-over aggregation scenario shown below. The sensors generate 40-byte readouts transmitted every 100 ms. Assuming best-case scenario stateful 6LoWPAN compression, what are the transmission rates for each of the sensors?

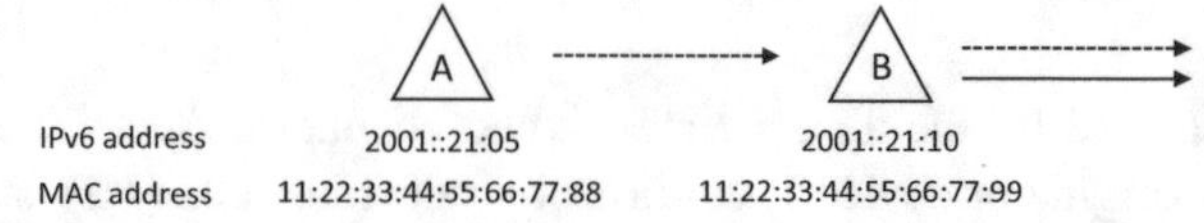

4.14 In an IoT scenario where real time readouts are transmitted over TCP by a sensor, what is a drawback of disabling Nagle's algorithm?

4.15 What does a 6LoWPAN stack that supports DTLS security look like?

4.16 Given the following scenario, if sensor A sends a datagram to the application, what does the IEEE 802.15.4 frame generated by the sensor look like? What does the Ethernet frame generated by the edge router look like? Include all headers and associated fields. Assume a 20-byte payload. Indicate all assumptions including source and destination UDP ports.

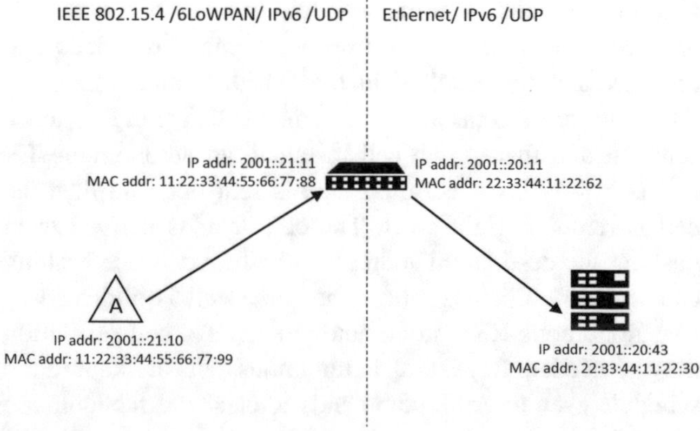

4.17 For the same conditions, what does typically result in lower latency 6LoWPAN over IEEE 802.15.4 or over IEEE 802.15.4e? Why?

4.18 How is a 20-byte payload transported over TCP statefully compressed by 6LoWPAN and transmitted over IEEE 802.15.4? What headers and fields do the IEEE 802.15.4 frame include?

4.19 A sensor generates 500-byte readouts, if the network frame loss is 5%, what is the datagram loss for the following two scenarios?

(a) IEEE 802.15.4
(b) Ethernet

References

1. Belshe, M., Peon, R., Thomson, M.: Hypertext Transfer Protocol Version 2 (HTTP/2). RFC 7540 (2015). https://doi.org/10.17487/RFC7540. https://rfc-editor.org/rfc/rfc7540.txt
2. Bluetooth, S.: Bluetooth 5.2 core specification, p. 3256 (2019)
3. Bormann, C.: 6LoWPAN-GHC: Generic Header Compression for IPv6 over Low-Power Wireless Personal Area Networks (6LoWPANs). RFC 7400 (2014). https://doi.org/10.17487/RFC7400. https://rfc-editor.org/rfc/rfc7400.txt
4. Brandt, A., Buron, J.: Transmission of IPv6 Packets over ITU-T G.9959 Networks. RFC 7428 (2015). https://doi.org/10.17487/RFC7428. https://rfc-editor.org/rfc/rfc7428.txt
5. Deering, D.S.E., Hinden, B.: Internet Protocol, Version 6 (IPv6) Specification. RFC 8200 (2017). https://doi.org/10.17487/RFC8200. https://rfc-editor.org/rfc/rfc8200.txt
6. Frankel, S., Krishnan, S.: IP Security (IPsec) and Internet Key Exchange (IKE) Document Roadmap. RFC 6071 (2011). https://doi.org/10.17487/RFC6071. https://rfc-editor.org/rfc/rfc6071.txt
7. Gomez, C., Paradells, J., Bormann, C., Crowcroft, J.: From 6lowpan to 6lo: Expanding the universe of ipv6-supported technologies for the internet of things. IEEE Commun. Mag. **55** (2017). https://doi.org/10.1109/MCOM.2017.1600534
8. Graziani, R.: IPv6 fundamentals: a straightforward approach to understanding IPv6. Pearson Education (2012)
9. Gupta, M., Conta, A.: Internet Control Message Protocol (ICMPv6) for the Internet Protocol Version 6 (IPv6) Specification. RFC 4443 (2006). https://doi.org/10.17487/RFC4443. https://rfc-editor.org/rfc/rfc4443.txt
10. Ikpehai, A., Adebisi, B.: 6loplc for smart grid applications. In: 2015 IEEE International Symposium on Power Line Communications and Its Applications (ISPLC), pp. 211–215 (2015)
11. ISO/IEC 18000-3: Parameters for air interface communications at 13,56 MHz. Standard, International Organization for Standardization, Switzerland (2010)
12. ITU, T.S.S.: Itu-t g.9959 : Short range narrow-band digital radiocommunication transceivers - phy, mac, sar and llc layer specifications. Tech. rep., International Telecommunication Union (2015)
13. Kaufman, C., Hoffman, P.E., Nir, Y., Eronen, P., Kivinen, T.: Internet Key Exchange Protocol Version 2 (IKEv2). RFC 7296 (2014). https://doi.org/10.17487/RFC7296. https://rfc-editor.org/rfc/rfc7296.txt
14. Khssibi, S., Idoudi, H., Van Den Bossche, A., Val, T., Saidane, L.A.: Presentation and analysis of a new technology for low-power wireless sensor network (2013)
15. Kim, E., Kaspar, D., Gomez, C., Bormann, C.: Problem Statement and Requirements for IPv6 over Low-Power Wireless Personal Area Network (6LoWPAN) Routing. RFC 6606 (2012). https://doi.org/10.17487/RFC6606. https://rfc-editor.org/rfc/rfc6606.txt
16. Lynn, K., Martocci, J., Neilson, C., Donaldson, S.: Transmission of IPv6 over Master-Slave/Token-Passing (MS/TP) Networks. RFC 8163 (2017). https://doi.org/10.17487/RFC8163. https://rfc-editor.org/rfc/rfc8163.txt
17. Mariager, P.B., Petersen, J.T., Shelby, Z., van de Logt, M., Barthel, D.: Transmission of IPv6 Packets over Digital Enhanced Cordless Telecommunications (DECT) Ultra Low Energy (ULE). RFC 8105 (2017). https://doi.org/10.17487/RFC8105. https://rfc-editor.org/rfc/rfc8105.txt
18. Montenegro, G., Hui, J., Culler, D., Kushalnagar, N.: Transmission of IPv6 Packets over IEEE 802.15.4 Networks. RFC 4944 (2007). https://doi.org/10.17487/RFC4944. https://rfc-editor.org/rfc/rfc4944.txt
19. Montenegro, G., Schumacher, C., Kushalnagar, N.: IPv6 over Low-Power Wireless Personal Area Networks (6LoWPANs): Overview, Assumptions, Problem Statement, and Goals. RFC 4919 (2007). https://doi.org/10.17487/RFC4919. https://rfc-editor.org/rfc/rfc4919.txt
20. Moskowitz, R., Heer, T., Jokela, P., Henderson, T.R.: Host Identity Protocol Version 2 (HIPv2). RFC 7401 (2015). https://doi.org/10.17487/RFC7401. https://rfc-editor.org/rfc/rfc7401.txt
21. Narten, D.T., Thomson, D.S.: IPv6 Stateless Address Autoconfiguration. RFC 2462 (1998). https://doi.org/10.17487/RFC2462. https://rfc-editor.org/rfc/rfc2462.txt
22. Nieminen, J., Savolainen, T., Isomaki, M., Patil, B., Shelby, Z., Gomez, C.: IPv6 over BLUETOOTH(R) Low Energy. RFC 7668 (2015). https://doi.org/10.17487/RFC7668. https://rfc-editor.org/rfc/rfc7668.txt
23. Oppliger, R.: SSL and Tls: Theory and Practice, 2nd edn. Artech House, Norwood (2016)
24. Rescorla, E.: The Transport Layer Security (TLS) Protocol Version 1.3. RFC 8446 (2018). https://doi.org/10.17487/RFC8446. https://rfc-editor.org/rfc/rfc8446.txt
25. Rescorla, E., Modadugu, N.: Datagram Transport Layer Security Version 1.2. RFC 6347 (2012). https://doi.org/10.17487/RFC6347. https://rfc-editor.org/rfc/rfc6347.txt
26. Shelby, Z., Bormann, C.: 6LoWPAN: The Wireless Embedded Internet. Wiley, New York (2010)
27. Simpson, W.A., Narten, D.T., Nordmark, E., Soliman, H.: Neighbor Discovery for IP version 6 (IPv6). RFC 4861 (2007). https://doi.org/10.17487/RFC4861. https://rfc-editor.org/rfc/rfc4861.txt
28. Thubert, P., Hui, J.: Compression Format for IPv6 Datagrams over IEEE 802.15.4-Based Networks. RFC 6282 (2011). https://doi.org/10.17487/RFC6282. https://rfc-editor.org/rfc/rfc6282.txt
29. Tschofenig, H., Fossati, T.: Transport Layer Security (TLS) / Datagram Transport Layer Security (DTLS) Profiles for the Internet of Things. RFC 7925 (2016). https://doi.org/10.17487/RFC7925. https://rfc-editor.org/rfc/rfc7925.txt
30. Vilajosana, X., Watteyne, T., Chang, T., Vučinić, M., Duquennoy, S., Thubert, P.: Ietf 6tisch: a tutorial. IEEE Commun. Surv. Tutorials **22**(1), 595–615 (2020)
31. Williams, M., Tüxen, M., Seggelmann, R.: Transport Layer Security (TLS) and Datagram Transport Layer Security (DTLS) Heartbeat Extension. RFC 6520 (2012). https://doi.org/10.17487/RFC6520. https://rfc-editor.org/rfc/rfc6520.txt
32. ZigBee Specification. Standard, The ZigBee Alliance, USA (2015)

5 Application Layer

5.1 Architectures

The IETF layered architecture indicates that the application layer also performs session and presentation functions that enable the management of different sessions as well as encryption and authentication mechanisms. Most so-called IoT application layer protocols provide session management such that application layer-specific functionality is typically carried out by some other mechanism that encodes both sensor and actuation traffic. This is a similar situation to web traffic transmission that relies on HTTP for session management and relies on *Hypertext Markup Language* (HTML) for application data encoding. When considering session management, there are two well-known topologies request/response and publish/subscribe that are shown in Figs. 5.1 and 5.2, respectively [3, 15].

The request/response model, also known as client/server model, bases its interaction between application and device by means of requests and responses. The device, sensor or actuator, acts as a server, while the application acts as a client. The client typically requests sensor data or sends actuation commands, while the server responds by transmitting sensor readouts or actuation acknowledgments. The model is based on synchronous polling where for each request there is a response. On the other hand, the publish/subscribe model relies on a central broker that queues and delivers messages between applications and devices. First, the application subscribes to a specific topic supported by the device. The device publishes device messages by sending them to the broker. The broker stores the messages and forwards them via notifications to the application that is interested in them. This model is based on asynchronous observation of devices that do not rely on polling. The advantage of this mechanism is that it improves latency and network utilization, but it requires the use of a broker that is a single point of failure.

5.2 Request/Response

5.2.1 REST Architecture

IoT connectivity at the session level can take advantage of the existing request/response architecture that is an integral part of the web. Moreover, not only session layer mechanisms like HTTP but other technologies like HTML, JavaScript, Node JS, Ajax, and PHP can be leveraged to provide connectivity between devices and applications. This approach has led to what it is called the Web of Things (WoT) that relies on embedding a simple web server on devices to take advantage of the web infrastructure. There is no need for topology or client changes since all updates are pushed to the devices. In general, request/response mechanisms associated with web technologies like HTTP comply quite well with the REST distributed architecture. REST architectures formalize a series of requirements and interfaces that are necessary for client/server interaction [8]. Not all REST schemes rely on the HTTP protocol, and not all HTTP scenarios support REST architectures. Other protocols like CoAP, presented later in this chapter, also support REST deployments.

For an architecture to be REST based, it must comply with six requirements or principles. Freedom is given on the implementation of each component if it meets these requirements. The six requirements are the following:

(1) The client/server requirement that states that a Uniform Interface separates clients and servers. This implies that storage is associated with servers, while user interaction is a function of the client. If the interface between a client and a server is preserved, they can evolve independently. Moreover, clients and servers can be upgraded and downgraded without affecting each other.

(2) The stateless requirement that states that communication between client and servers is stateless and that each

R. Herrero, *Fundamentals of IoT Communication Technologies*, Textbooks in Telecommunication Engineering, https://doi.org/10.1007/978-3-030-70080-5_5

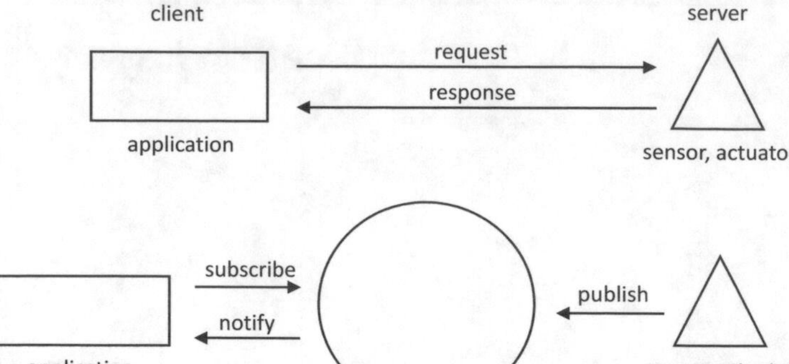

Fig. 5.1 Request/response

Fig. 5.2 Publish/subscribe

request must contain all the information needed by the server to process it and respond. No content stored on the server can be used to process the request. Session state, if any, is kept on the client side.

(3) The cacheable requirement that states that to minimize overall network throughput and guarantee network efficiency, requests can be cached at the clients. Server responses indicate whether a given message is cacheable or not depending on how likely it is to change.

(4) The Uniform Interface requirement that, besides separating the client/server architectures, enables independent protocol evolution.

(5) The layered system requirement that states that the architecture must be composed of several hierarchical layers with visible interfaces but without a formally specified implementation.

(6) The code on demand requirement that allows client code to be extended by dynamically downloading and executing scripts from the server. This allows the simplification of the client infrastructure by minimizing the preinstalled code and expediting software updates.

REST APIs provide the basic functionality to access network resources by means of their representational states. REST APIs rely on the CRUD mechanism; *C* stands for *create*, *R* stands for *read*, *U* stands for *update*, and *D* stands for *delete*. Essentially a client must be able to create, read, update, and delete an object representation. This is ideal for an IoT environment where objects are devices that are (1) created as sensors or actuators, (2) read when they are sensors, (3) updated when they are actuators, or (4) deleted. In the context of the Internet, REST APIs rely on HTTP to provide session management. As shown in Fig. 5.3, different HTTP methods are used; POST is used to create a new device (sensor or actuator), GET is used to retrieve a sensor readout, PUT is used to update the state of an actuator, and DELETE is used to delete a device.

Table 5.1 shows an example of specific HTTP methods and how they are used in the context of individual devices as well as collections of them. More details of the HTTP protocol are presented in the following section. Note that REST requests rely on resource identification by means of Universal Resource Identifiers (URIs) that follow the method. Clients refer to resource identifiers but handle representations of those resources and not the resources themselves. These representations are data structures that can be formatted via HTML, the more generic Extended Markup Language (XML), or JavaScript Object Notation (JSON). If a client has enough permissions, it can use its own resource representation to update the resource on the server accordingly. Moreover, the client can also delete the resources from the server.

One issue with traditional REST architectures is that the CRUD mechanisms rely on the client always initializing the interaction with the server. In the context of IoT, this implies that the application must periodically send requests to obtain readouts from devices. For each readout, therefore, there is a full duplex transaction between client and server. There are two problems associated with this situation: (1) extra latency due to the time it takes for the client to poll the sensor and (2) extra traffic due to the additional datagrams needed to poll the sensor. To address this issue, the reference architecture for IoT, introduced by the *European Telecommunications Standards Institute* (ETSI), extends REST by means of two additional operations: (1) notify and (2) execute. The notify operation, also known as observation, is triggered upon a change in the representation of a resource and results in a notification sent to the client in order to monitor changes of the resource in question. The execute operation enables a client to request the execution of a specific task on the server. In the context of the CRUD set, notify is implemented by an UPDATE operation transmitted from the server toward the requesting client. Similarly, execute is implemented by an EXECUTE operation from the client to the server that includes task information and parameters.

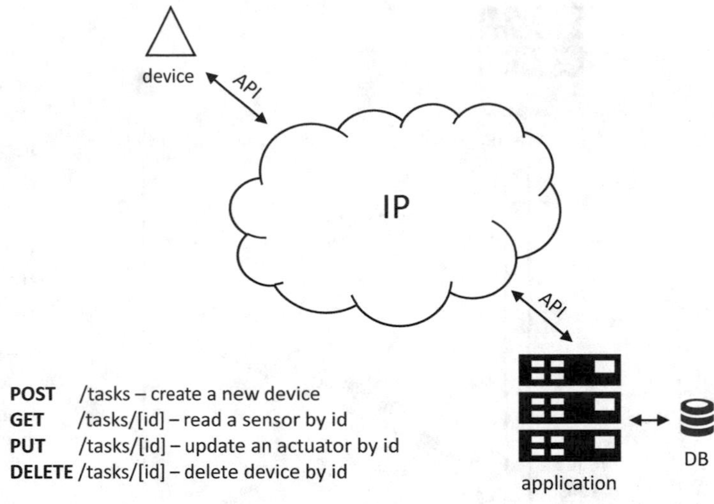

Fig. 5.3 REST APIs with HTTP

Table 5.1 REST API Methods

HTTP method	Entire collection, e.g. /devices/	Specific item, e.g. /devices/id/
GET	200 (OK)—list of devices	200 (OK)—single device 404 (Not Found) if ID is not found or invalid
PUT	404 (Not Found)	200 (OK) or 204 (No Content)—update device information 404 (Not Found) if ID is not found or invalid
POST	201 (Created)—Location header with link to /devices/id/ containing the new ID.	404 (Not Found)—Generally not used
DELETE	404 (Not Found)	200 (OK) 404 (Not Found)—Generally not used

5.2.2 HTTP

HTTP provides session layer management of web applications including client and server support. HTTP traffic relies on TCP transport that cannot be typically compressed by means of 6LoWPAN [9]. For reasons explained in Sect. 4.3.7, since TCP transport is not usually recommended in the context of IoT, there is no standard way to compress TCP headers. Moreover, HTTP messages are *American Standard Code for Information Interchange* (ASCII) encoded, and therefore they are typically too big to fit in a single frame of most link layer IoT technologies like IEEE 802.15.4. In consequence, the lack of TCP header compression added to the large size of the messages results in network fragmentation that leads to elevated network loss. HTTP is used to access web documents that can be HTML content, scripts, or media files that are accessed by means of a subset of *Uniform Resource Identifiers* (URIs) known as *Uniform Resource Locators* (URLs). An HTML document consists of an XML structure used to reference several other web objects via their corresponding URLs. A URL consists of service type, a hostname, and a path. For example, for the http://www.congress.gov/advanced-search/legislation URL, *http://* is the service type, www.congress.gov is the hostname, and */advanced-search/legislation* is the path.

Figure 5.4 shows a client/server interaction where an application sends an HTTP request to obtain a representation of a temperature readout object from a thermometer sensor acting as a server. The sensor receives the requests and sends an HTTP response containing the representation of a temperature readout object. This interaction is done without the server keeping any state information and thus complying with REST architecture principle of statelessness. Since HTTP relies on TCP for transport, before any traffic can be transmitted, a connection must be established. Figure 5.5 shows an example of a connection established between a client/application and a server/sensor/device. Connection es-

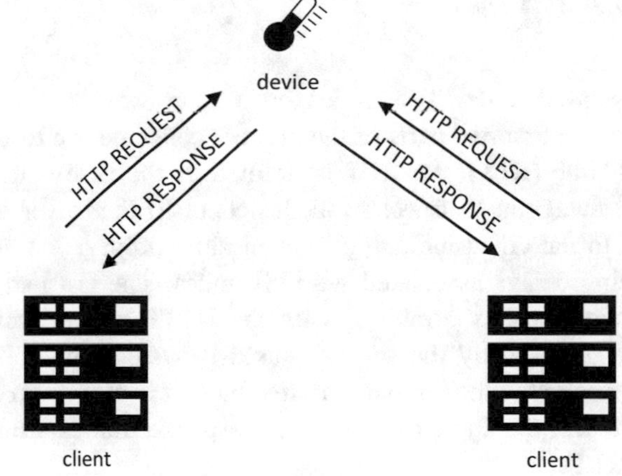

Fig. 5.4 HTTP request/response

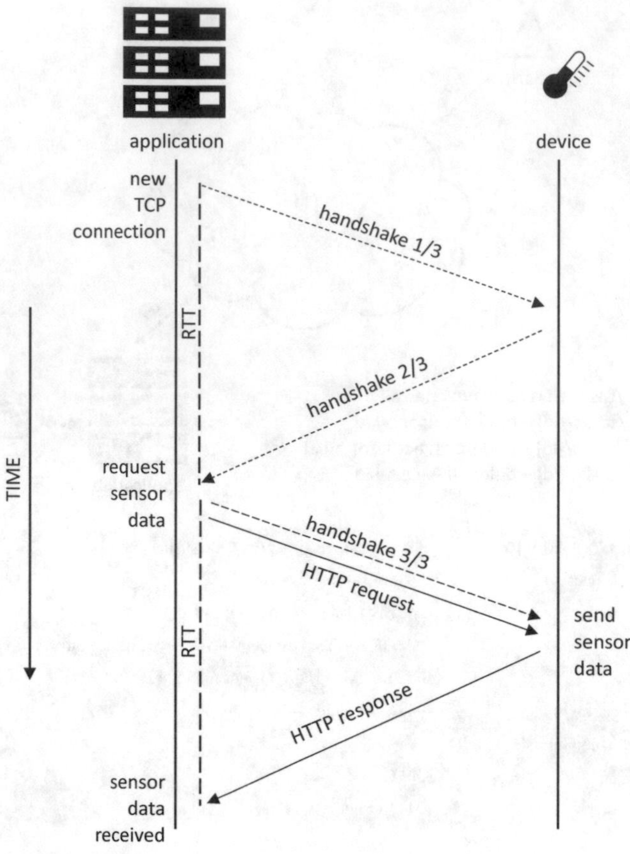

Fig. 5.5 HTTP setup

tablishment under TCP relies on a three-way handshake where the first two parts of the process consume one Round Trip Time (RTT). An RTT is defined as the time it takes for a datagram to travel from the client to the server and back to the client including transmission, propagation, and queuing delays associated with all nodes. The last part of the handshake is combined with the HTTP request which when received by the sensor causes it to send the HTTP response including temperature readouts. It is clear from the figure that the approximate server response time is around two RTTs.

Depending on configuration, HTTP may rely on a single persistent TCP connection to transport all requests and responses, or it may rely on multiple connections such that it associates a single request/response transaction with a single non-persistent TCP connection. Figure 5.6 shows a client requesting two consecutive sensor readouts on both scenarios. With a persistent connection, the client first initiates a TCP connection and transmits the first HTTP request as part of the last part of the three-way handshake. When the sensor receives the request, it reads the temperature value and responds with an HTTP response. Neither the client nor the sensor tears down the connection. A bit later, the client issues a new HTTP request to get an additional readout that, in turn, is replied by the server. With a non-persistent connection, as before, the client initiates a TCP connection and transmits the first HTTP request as part of the last part of the aforementioned three-way handshake. When the sensor responds, and after receiving the readout, the client tears down the connection. Later, when the client tries to obtain a new readout from the server, it creates a new connection before it transmits the HTTP request. The sensor responds by transmitting an HTTP response, which, when received by the client, triggers the latter to tear down this second connection.

HTTP transported by means of persistent TCP connections has become more efficient with newer versions of HTTP. This is shown in Fig. 5.7. The very first version 1.0 of HTTP, formally known as HTTP/1.0, only allowed clients to send requests in a stop-and-wait fashion. In other words, if a client transmits a request, it must wait for the server response before it can transmit another request. The result of this behavior is excessive latency. HTTP 1.1 (formally known as HTTP/1.1) introduces the concept of pipelining, where the client can transmit many simultaneous requests that are processed and answered by the server in the order in which they are received. HTTP 1.1 exhibits lower latency than HTTP 1.0. Although all the requests may arrive at the server at the same time, some of them may take longer to be processed than the others. For example, a request that retrieves the temporal average of temperature readouts takes longer to be processed than a request that retrieves a single readout. Therefore, an earlier request may delay the transmission of the response to a later request that may have been already processed. This is known as head-of-line (HOL) blocking because the processing of a request can prevent other responses from being transmitted. HTTP 2.0 [4] (formally known as HTTP/2.0) addresses this issue by introducing multiplexing that enables the client to simultaneously transmit requests so that they can be answered by the server in the order in which they are processed. As a consequence, HTTP 2.0 exhibits lower latency than HTTP 1.1. HTTP 3.0 (formally known as HTTP/3.0) [5] introduced as an IETF draft in September 2020 attempts to improve latency by relying on *Quick UDP Internet Connection* (QUIC) transport instead of TCP. QUIC is a transport layer protocol that lowers connection setup as well as transmission latency and consequently improves congestion performance [12].

Non-persistent connections are responsible for extra latency introduced when each new connection is established. New connections also introduce additional computational complexity and memory requirements because they rely on allocated buffers and state variables. Persistent connections do not have these restrictions; however, they are always active even when they are not needed. So, if the client does not need to request sensor data, the connection remains causing a waste of computational and memory resources.

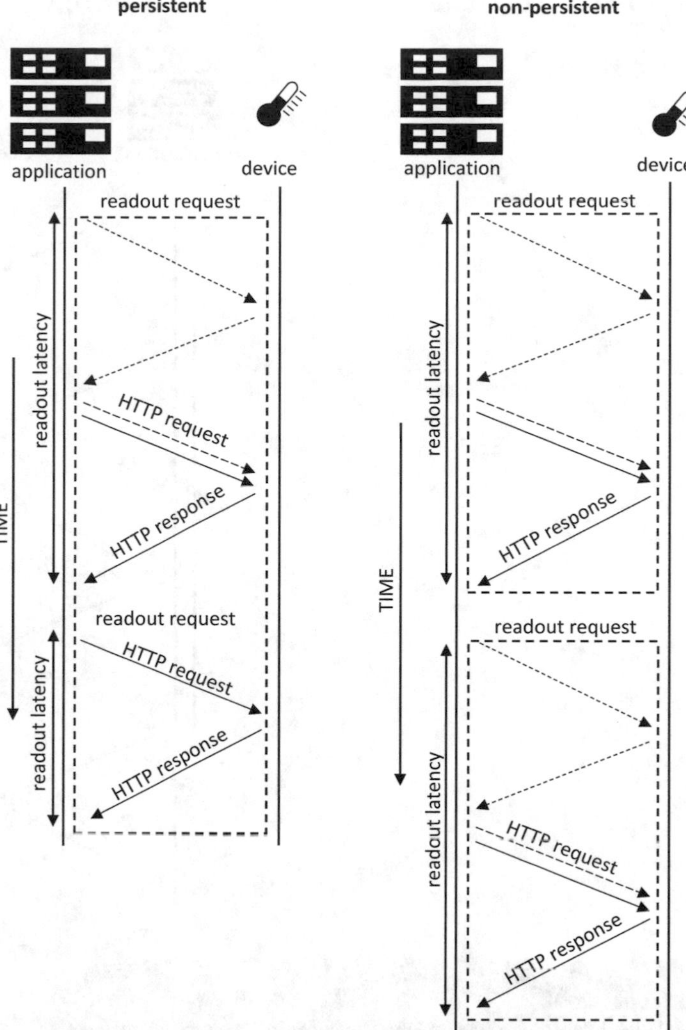

Fig. 5.6 Persistent vs non-persistent connections

5.2.2.1 HTTP Messages

Example 5.1 Consider an end-to-end HTTP scenario where an application requests a readout from a sensor:

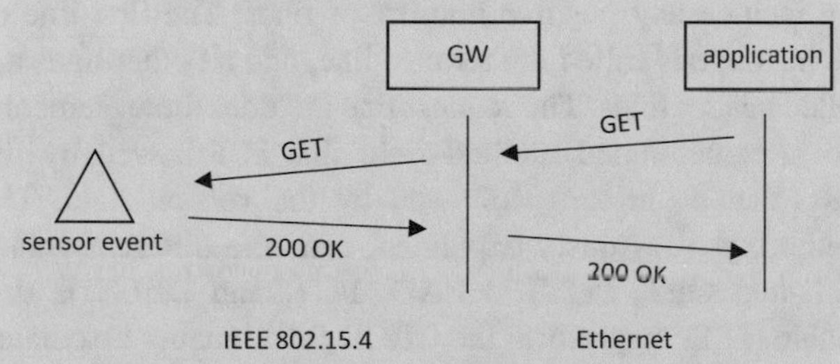

Assume the following: (a) the IEEE 802.15.4 transmission rate is 250 KBps, (b) the Ethernet transmission rate is 1 Gbps, (c) the only delay is the transmission delay, (d) the TCP header is

(continued)

(continued)

$L_{TCP} = 20$ bytes long, (e) the HTTP GET request is $L_{GET} = 200$ bytes long, (f) the HTTP 200 OK response is $L_{200\,OK} = 100$ bytes long, (g) the IPv4 header is $L_{IPv4} = 20$ bytes long, (h) the 6LoWPAN header is $L_{6LoWPAN} = 10$ bytes long, (i) the Ethernet header is $L_{eth} = 16$ bytes long, and (f) the IEEE 802.15.4 header is $L_{IEEE\ 802.15.4} = 10$ bytes long. How long does it take for the application to receive the readout?

Solution The overall delay is given by $T = T_1 + T_2 + T_3 + T_4$ where T_1 is the delay associated with the transmission of the TCP SYN frame to establish the connection (as part of the initial RTT), T_2 is the delay associated with transmission of the

(continued)

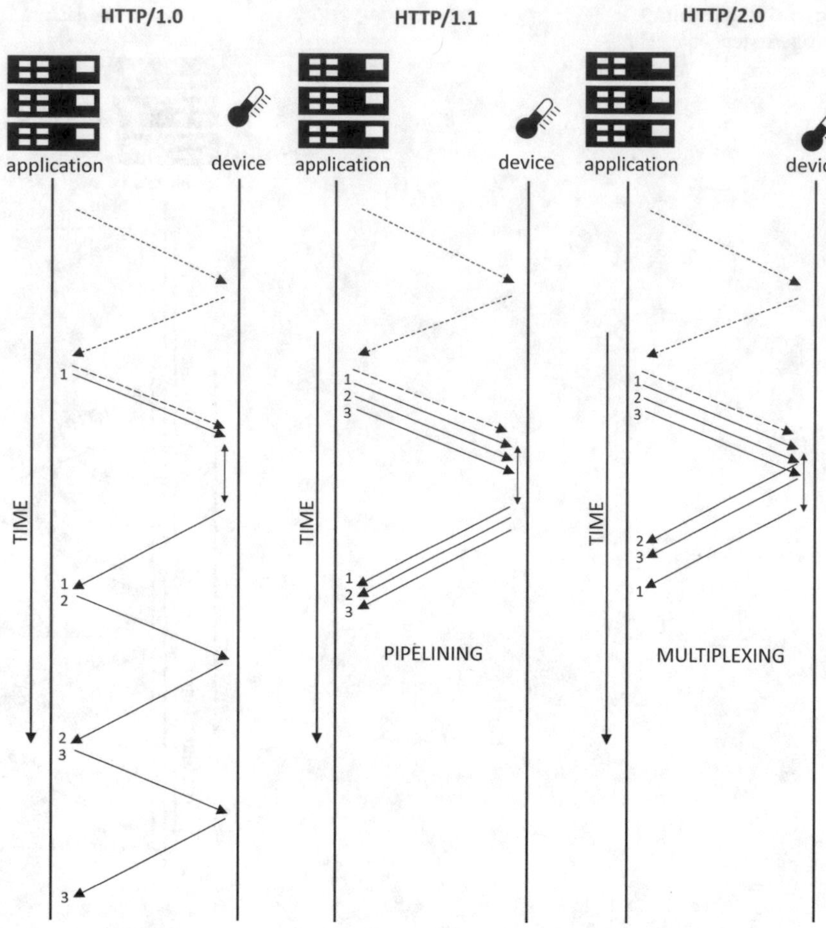

Fig. 5.7 HTTP 1.0 vs HTTP 1.1 vs HTTP 2.0

(continued)
TCP SYN/ACK frame to acknowledge the connection establishment (as part of the initial RTT), T_3 is the delay associated with the transmission of the HTTP GET request, and T_4 is the delay to transmit the HTTP 200 OK. T_1 and T_2 are given by $T_1 = T_2 = \frac{8\times(L_{\mathrm{eth}}+L_{\mathrm{IPv4}}+L_{\mathrm{TCP}})}{10^9} + \frac{8\times(L_{\mathrm{IEEE\ 802.15.4}}+L_{\mathrm{6LoWPAN}}+L_{\mathrm{TCP}})}{250000} = 1.28$ ms. T_3 is given by $T_3 = \frac{8\times(L_{\mathrm{eth}}+L_{\mathrm{IPv4}}+L_{\mathrm{TCP}}+L_{\mathrm{GET}})}{10^9} + \frac{8\times(L_{\mathrm{IEEE\ 802.15.4}}+L_{\mathrm{6LoWPAN}}+L_{\mathrm{TCP}}+L_{\mathrm{GET}})}{250000} = 7.68$ ms. T_4 is given by $T_4 = \frac{8\times(L_{\mathrm{eth}}+L_{\mathrm{IPv4}}+L_{\mathrm{TCP}}+L_{\mathrm{200\ OK}})}{10^9} + \frac{8\times(L_{\mathrm{IEEE\ 802.15.4}}+L_{\mathrm{6LoWPAN}}+L_{\mathrm{TCP}}+L_{\mathrm{200\ OK}})}{250000} = 4.48$ ms. The total delay is, therefore, $T = 14.72$ ms.

HTTP messages are ASCII encoded, and they can therefore be read by humans on the wire by means of network sniffers and analyzers. There are basically two types of HTTP messages that comply with the client/server architecture interaction: (1) request messages and (2) response messages. The message below is an example of an HTTP request transmitted by a client attempting to retrieve a temperature readout from a sensor:

```
GET /sensor102/temperature HTTP/1.1
Host: www.l7tr.com
Connection: close
User-Agent: PXO Sensor
```

The request consists of four lines with each line followed by ASCII carriage return and line feed characters. The very last line of the message is also followed by an additional sequence of ASCII carriage return and line feed characters. Although this message is four lines long, HTTP messages can include any positive number of lines. The first line of the message is called the request line, and all other lines are called header lines. The request line includes three elements: the aforementioned method field that is followed by the URL that is, in turn, followed by the version field. The method, as previously explained, can take different values including GET, POST, HEAD, PUT, and DELETE that enable HTTP to perform the CRUD functionality associated with REST architectures. Most HTTP transactions involve GET requests like the one in the example above. The */sensor102/temperature* URL specifies the full path to the server resource being accessed. The first header in the example is the *Host* header that points to the www.l7tr.com server. Although this header may seem redundant, it is critical when

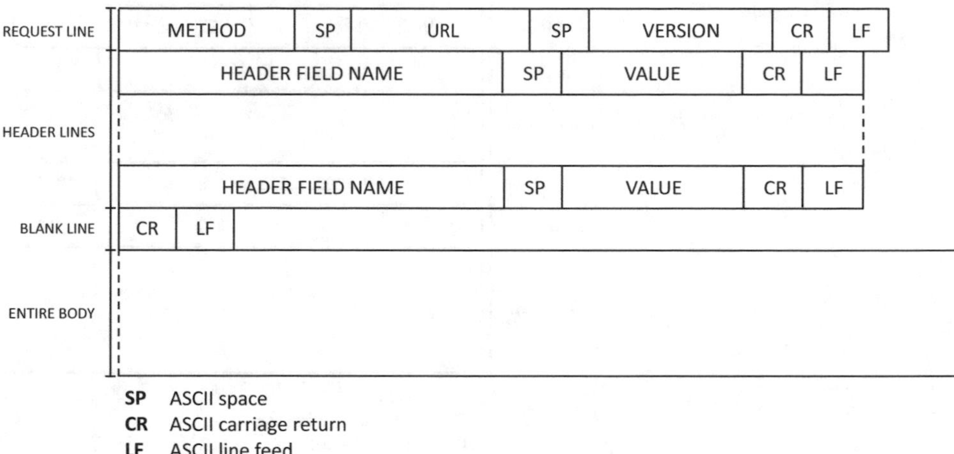

Fig. 5.8 HTTP request message

the client talks to the server through an intermediate proxy server. In this case, the header tells the proxy server who the final destination of the message is. The second header is the *Connection* header that indicates whether the connection is persistent or not. The third and last header is the *User-Agent* header that identifies the application that is making the request to the sensor. This field lets the server make presentation decisions based on the type of application that is originating the request. Other HTTP requests like POST requests include a message body that follows the carriage return and line feed after the last header in the message. The message body is used to include large amounts of information that can be transferred from the client to the server. This is particularly important in certain actuation scenarios. The message body is also used in responses to HTTP GET requests to carry object representations. The HEAD and GET methods are similar with the difference that the former is used to signal that the client does not want to receive the message body if present. Essentially, the HEAD method enables the client to examine the object before requesting it in order to minimize network utilization. On the other hand, the PUT method is usually used to create and upload objects, while the DELETE method enables a client to delete an object from the server. Figure 5.8 shows the format of a generic HTTP request message.

The message below is an example of an HTTP response transmitted by a sensor including a temperature readout:

```
HTTP/1.1 200 OK
Connection: close
Date: Tue, 23 July 2019 15:44:04 GMT
Server: www.l7tr.com
Last-Modified: Tue, 23 July 2019 15:44:04 GMT
Content-Length: 5
Content-Type: text/plain

23.1C
```

In a similar way to a request, a response has three sections: a status line, several headers, and a message body. The status line, in turns, includes a version field *HTTP/1.1*, a status code *200*, and a status message *OK* that maps the status code. A response with a *200 OK* status indicates that the request was received and reply information is returned as part of the message. Other status codes are (1) *301 Moved Permanently* to indicate that the object representation has been permanently moved and a new URL is specified in the location, (2) *400 Bad Request* that is a generic error code used to indicate that the request was not understood by the server, (3) *404 Not Found* that indicates that the object does not exist in the server, and (4) *505 HTTP Version Not Supported* that tells the client that the HTTP protocol version is not supported by the server. Table 5.2 illustrates the numerical range of possible response codes and their meaning. The first header is the *Connection* header that tells the client that the connection must be closed after the message is sent. The second header is the *Date* header that specifies when the HTTP response was created. The third header is the *Server* header that identifies the entity that created the response. The fourth header is the *Last-Modified* header that indicates when the object representation was last changed. The fifth header is the *Content-Length* header that indicates the length of the message body. Finally, the sixth header is the *Content-Type* header that identifies the encoding of the body. In this case, the message body is a plain ASCII text that encodes in 5 bytes the string *23.1C* representing a temperature of 23.1° degrees. Figure 5.9 shows the format of a generic HTTP response message.

Table 5.2 HTTP response codecs

Range	meaning
100–199	Informational
200–299	Success
300–399	Redirect
400–499	Client errors
500–599	Server errors

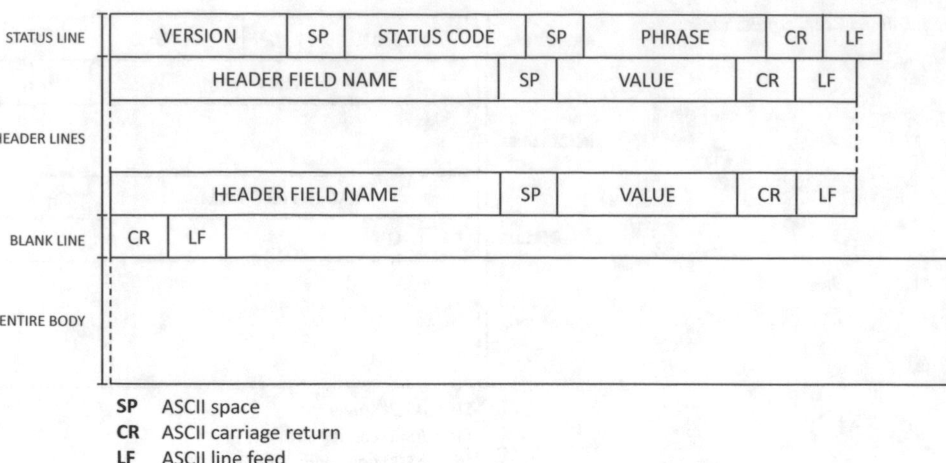

Fig. 5.9 HTTP response message

5.2.2.2 Cookies

As presented in Sect. 5.2.1, one of the main characteristics of REST architectures is that they are stateless. HTTP, fitting the REST model, is stateless, but for many applications, it is convenient to have a mechanism to treat certain transactions as stateful. For example, if an application user is interested in temperature readouts presented as Centigrade degrees, it may be interesting for the device to be able to implicitly identify these requests by keeping track of a user state. Cookies are an add-on mechanism that can be used to introduce a state in the interaction between clients and servers. Cookies are associated with users, and although they violate the REST principle, they are useful under certain circumstances. Essentially cookies are short strings that enable IoT devices to identify multiple applications. The cookie infrastructure relies on two databases: one at the client and another at the server. It also relies on two cookie headers, *Cookie* and *Set-Cookie*, that are added to HTTP requests and responses, respectively.

Figure 5.10 shows the interaction between a client application and a sensor server. The client has previously interacted with another server that has populated a cookie identified by *1912* in its own database. The figure describes the process in several steps: (1) the application sends a request to retrieve a readout of temperature in Centigrade degrees from the sensor. This information is specified as part of the URL, but since this is the first time that this interaction occurs, the client does not include a Cookie header in the request; (2) the server then responds to the incoming request by transmitting the temperature readout and associating cookie *2105* to the client. The cookie is transmitted as a *Set-Cookie* header to the application; (3) the client stores the cookie and associates it with the sensor; then it requests a new readout, but this time it does not specify a full URL but the *2105* cookie instead; (4) the device associates the request with the user by means of the received *2105* cookie and responds with a new temperature readout in Centigrade degrees; (5) any later time, even after the client has been powered down, the requests from the client include the *2105* cookie identifier that is used, in turn, by the sensor to identify the application and respond in accordance with pre-specified preferences and states.

5.2.2.3 Proxy Servers

A proxy server, also known as a web cache, is a device that sits between client applications and sensors as well as devices acting as HTTP servers in order to respond to client requests on behalf of these servers. Proxy servers can reply on behalf of HTTP servers by keeping copies of the information stored in them. Since proxy servers are close to client applications, they are useful in lowering core traffic throughput and thus accomplishing better channel utilization, reducing data access latency, and providing redundancy. Proxy servers attempt to keep in their own storage the most recent copies of any server object that may be requested by applications. Proxy server interaction with clients is shown in the example of Fig. 5.11. The following steps describe the flow of messages shown in that figure: (1) the client connects to its pre-configured proxy server to request, via HTTP, a specific temperature readout; (2) the proxy server checks whether it has the most recent readout of the temperature (if so, jump to step (6)); (3) if the proxy server does not have a recent copy, it connects to the sensor and requests, via HTTP, the temperature readout; (4) the sensor sends an HTTP response including an object representation of the temperature readout; (5) the proxy server stores the copy of the temperature object; and (6) the proxy server sends the object representation of the temperature readout to the client. These steps are summarized as a flow in Fig. 5.12.

The main question is how a proxy server remains synchronized with a web server; HTTP provides a mechanism known as conditional GET that allows the proxy server to verify that a specific object is up to date. The message below is an example of a conditional GET request:

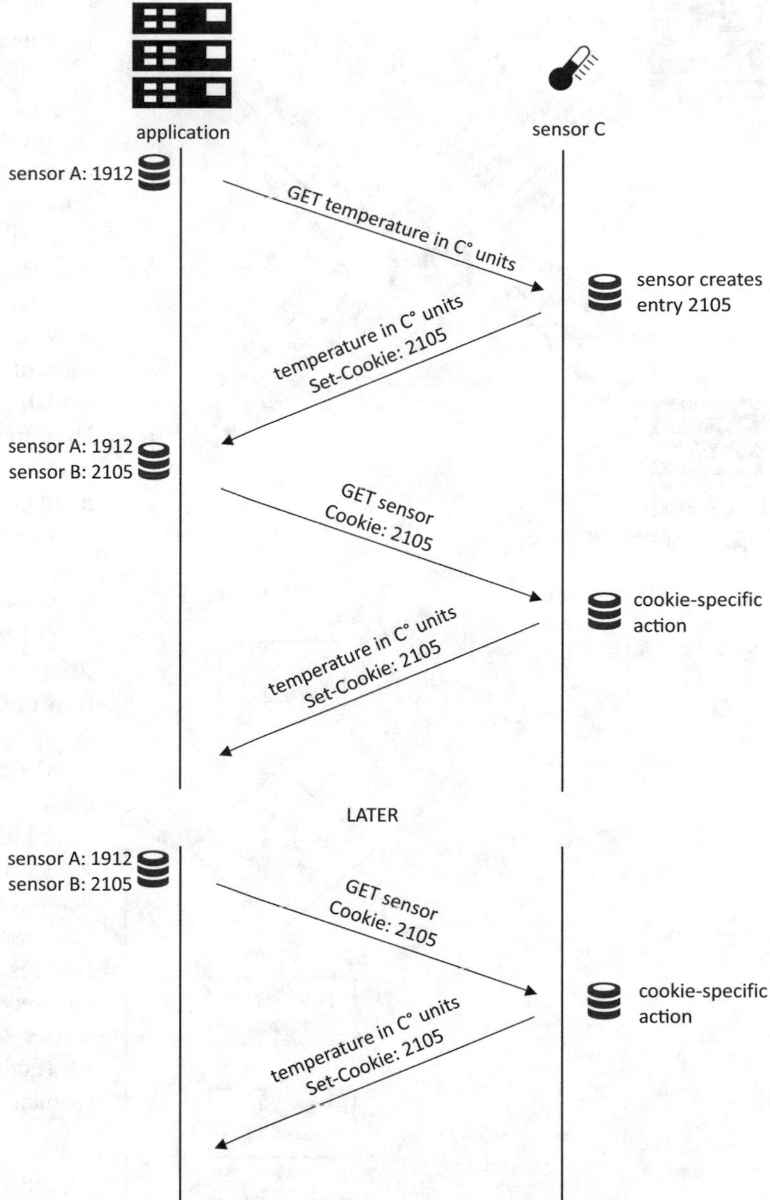

Fig. 5.10 HTTP cookies

GET /sensor102/temperature HTTP/1.1
Host: www.l7tr.com
Connection: close
User-Agent: PXO Sensor
If-modified-since: Tue, 23 July 2019 15:44:04 GMT

The request includes a *If-Modified-Since* header that is used by the proxy server to obtain a new object representation only if it has changed since it was last requested on *Tuesday, July 23 2019 at 3:44:04 PM GMT*. If there have been no more new readouts since the last request, the sensor replies by transmitting a *304 Not Modified* response:

HTTP/1.1 304 Not Modified
Date: Sat, 15 Oct 2011 15:39:29
Server: Apache/1.3.0 (Unix)

On the other hand, if there have been more readouts since the last request, the sensor replies with a regular *200 OK* response:

HTTP/1.1 200 OK
Connection: close
Date: Tue, 23 July 2019 15:44:04 GMT
Server: www.l7tr.com
Last-Modified: Tue, 23 July 2019 16:04:04 GMT
Content-Length: 5
Content-Type: text/plain

25.2C

In general, since the *304 Not Modified* response is much smaller than the *200 OK* response, this mechanism provides a way to improve channel utilization efficiency.

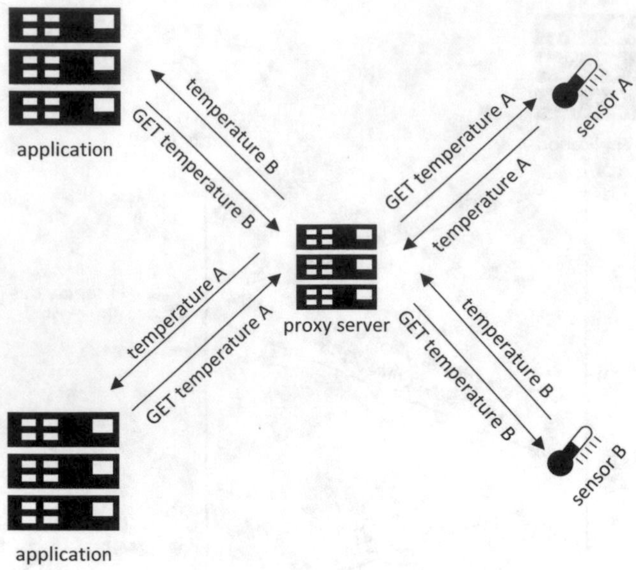

Fig. 5.11 HTTP caching

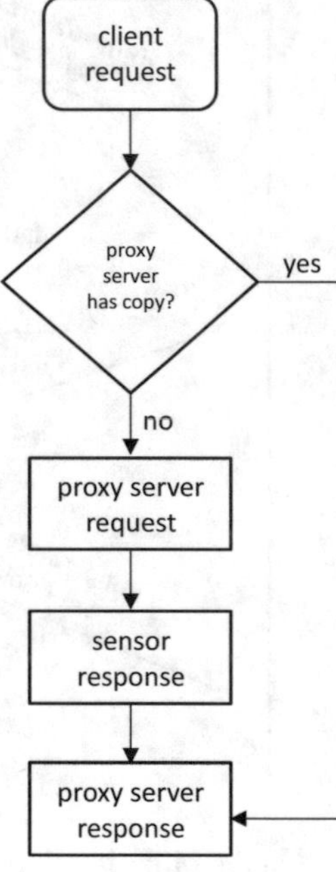

Fig. 5.12 HTTP caching flow

5.2.3 XMPP

Another protocol that loosely follows the *REST* paradigm is the *eXtensible Messaging and Presence Protocol* (XMPP). XMPP, standardized as IETF RFC 6120, is an open session layer protocol that relies on XML as the main format for information exchange. XMPP is a highly decentralized client-server architecture supporting an unlimited number of servers without a single point of failure [16]. XMPP is also extensive, and it has been upgraded to support several applications ranging from video games to IIoT. When compared to HTTP, XMPP supports a built-in REST resource observation that relies on a push model of information that minimizes latency by enabling sensors to push their readouts whenever they become available.

XMPP provides several basic services: (1) channel encryption and authentication to support end-to-end security, (2) presence to enable entities to share information about their availability within the network, (3) contact lists that provide a mechanism for servers to store lists of devices and applications, (4) one-to-one messaging to support the infrastructure of communication that enables users to send XML messages, (4) multi-party messaging and notifications that provide a mechanism to send a single message to multiple subscribers in a similar way to that of the publish/subscribe model, (5) service discovery to let entities (i.e. applications) know the capabilities of other entities (i.e. devices), (6) structured data forms that enable endpoints to exchange structured but flexible forms (i.e. IoT configuration information) with other endpoints, and (7) peer-to-peer media sessions to support, negotiate, and manage media and other real-time sessions between devices and applications.

XMPP decentralization relies on messages being sent from one source device to its home server that, in turn, forwards the message directly to the server that serves the destination endpoint. This scenario is illustrated in Fig. 5.13. In XMPP networks, servers are directly connected to each other to form a composite mesh network. This structure is the basic building block of the so-called XMPP federation. The XMPP federation additionally includes automated clients, known as bots, that enable a wide range of services ranging from assistance in chat rooms to interfaces to non-XMPP services. As with any other IP based protocol, XMPP relies on every device being identified by an address known as *Jabber Identifier* (JID). JIDs follow the same format as e-mail addresses that combine the username associated with an account with a domain as *username@domain*. The domain

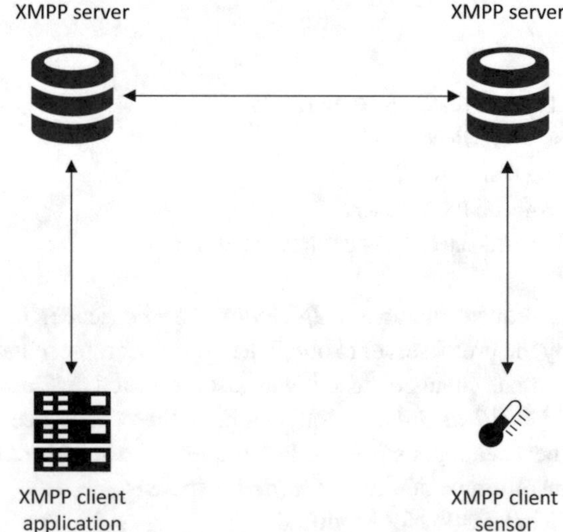

Fig. 5.13 XMPP network

is a *fully qualified domain name* (FQDN) that can be mapped to one or more IP addresses in accordance with *domain name system* (DNS) mechanisms.

When a client connects to an XMPP server, a resource identifier is assigned to the connection. This identifier is added to the end of the JID as *username@domain/resource*. Since a device is typically mapped to a resource and there is a single connection per device, this functionality enables other entities to query or exchange messages with specific devices associated with a given account. Each device is a specific point of presence of a given user that includes different states and capabilities. Note that although the *username@domain* is not case sensitive, the resource identifier section */resource* is. Moreover, JIDs follow URI schemes with full representation as *xmpp:username@domain/resource?action* with *xmpp* being the service identifier and *action* a particular command that can be used, for example, to send a message.

5.2.3.1 XMPP Messages

XMPP is in essence a mechanism to signal and stream XML over TCP connections. Specifically, if a device wants to start an XMPP session, it must first create a persistent TCP connection to the server, and, when security is also required, a TLS session must be negotiated over the connection. The XMPP stream consists of the sequential build-up interaction between a client and its home server made of three main possible XML stanzas: (1) <message>, (2) <presence>, and (3) <iq>. An XML stanza is a discrete semantic unit of structured information that is sent from one entity to another.

Figure 5.14 shows an XMPP XML session where an application acting as client/resource (*client@example.org/amr*) requests a sensor data readout from a device (*device@example.org*) and, later, the same client/resource performs actuation on another device (*digital.output@example.org*). The client establishes a persistent TCP connection to its local server, and all bidirectional traffic including XML stanzas between the client (shown prefaced as *C*) and the server (shown prefaced as *S*) are transmitted inside a *stream/stream* element. If encryption and authentication are needed, a TLS session is established on top of TCP to encrypt traffic [17]. Because under XMPP TCP connections are persistent, devices can push readouts down the client as soon as they become available. This asynchronous behavior, which minimizes latency and lowers throughput, is conceptually like that of the observation introduced for IoT by ETSI. Obviously, the price to pay for persistent connections is the use of additional computational and memory resources. A client can simultaneously send several requests to a device via its home server. These requests, in turn, are responded by the device in the order in which they were transmitted. Note that the client is never blocked as it can transmit new requests while waiting for previous responses.

Stanzas are the messages generated and processed by the XMPP application layer. They are defined by (1) their element name, that is, *message*, *presence*, or *iq*; (2) their type attribute, for example, *get*, *set*, or *result*; and (3) their child elements known as payloads. The message stanza is used to push data from, for example, a device to a client. Since the transport layer is reliable (TCP), these messages are transmitted through a fire-and-forget mechanism that does not rely on acknowledgments and retransmissions. In the context of IoT, messages are used to propagate sensor readouts. Message types are diverse, and they range from normal to chat and groupchat to address unicast and multicast real-time communications. Messages of type error are sent in response to incoming messages to indicate a problem processing them. Other attributes included are *from* and *to* attributes that are used to specify sender and receiver, respectively, and the *id* attribute that provides a mechanism to track messages. Note that to prevent address spoofing, the *from* attribute is filled in by the home server of the sender. Message payloads vary depending on the type and nature of message ranging from body and subject for basic chat to fields for IoT sensor traffic. Some message payloads are specified as part of extensions to the core XMPP specification.

The presence stanza is used to advertise the availability of XMPP clients and resources in order to know whether they are online and available for communication. Presence is based on a subscriber/notification scheme where entities issue presence subscriptions in order to obtain presence information about other entities. When these subscriptions are approved, any change of the presence status, for example, online or offline, is transmitted to each subscriber. The iq stanza, also known as info/query stanza, enables request/response interaction between entities by providing CRUD support as illustrated in Fig. 5.15. Each request is replied by transmitting a response that relies on an *id* attribute for tracking. The iq stanza, as the message stanza, also includes *from* and *to* attributes that are used to specify sender and receiver, respectively. The representational state of an object can be read if the iq stanza type attribute is *get*, or, on the other hand, it can be created, updated, or deleted, if it is *set*. The type attribute *result* is used in responses to specify a readout value or to acknowledge a set request. The *xmlns* attribute, included in iq stanza payloads, specifies the particular *namespace* or XMPP protocol extension associated with the requests and responses.

As previously indicated, because TCP transport guarantees delivery, there is no need for XMPP message, presence, and iq stanzas to be acknowledged. If for some reason a stanza is delivered but an error occurs due to, for example, lack of resources, a response with type attribute *error* including a payload of type error is transmitted to the original sender. Figure 5.16 shows an error response that indicates that the end device failed to process an iq

Fig. 5.14 XMPP XML sequence (C = client, S = server)

```
C: <stream:stream>

C: <iq type=''get''
      from=''client@example.org/amr''
      to=''device@example.org''
      id=''50001''>
        <req xmlns=''urn:xmpp:iot:sensordata''
            seqn=''1''
            momentary=''true'' />
   </iq>

S: <iq type=''result''
      from=''device@example.org''
      to=''client@example.org/amr''
      id=''50001''>
        <accepted xmlns=''urn:xmpp:iot:sensordata''
        seqn=''1'' />
   </iq>

S: <message from=''device@example.org''
            to=''client@example.org/amr''>
      <fields xmlns=''urn:xmpp:iot:sensordata''
            seqn=''1''
            done=''true''>
         <node nodeId=''Device01''>
            <timestamp value=''2018-03-07T16:24:30''>
               <numeric name=''Temperature''
                        momentary=''true''
                        automaticReadout=''true''
                        value=''23.4''
                        unit=''C''/>
            </timestamp>
         </node>
      </fields>
   </message>

C: <iq type=''set''
        from=''client@example.org/amr''
        to=''digital.output@example.org''
        id=''50002''>
      <set xmlns=''urn:xmpp:iot:sensordata''
          xml:lang=''en''>
         <boolean name=''Output''
                  value=''true''/>
      </set>
   </iq>

S: <iq type=''result''
        from=''digital.output@example.org''
        to=''client@example.org/amr''
        id=''50002''>
      <setResponse xmlns=''urn:xmpp:iot:sensordata''/>
   </iq>

C: </stream:stream>
```

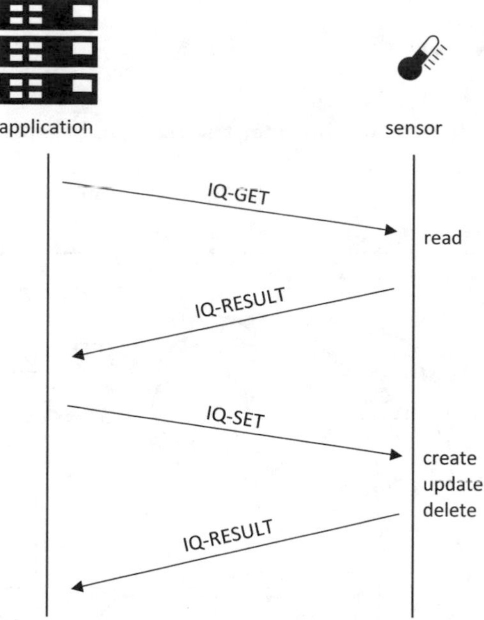

Fig. 5.15 IQ interaction

request. These types of errors are recoverable, and they just indicate that there was a problem processing the request. Transport layer problems that affect the integrity of the XMPP stream are not recoverable and typically result in the termination of the stream. Note that when compared to HTTP, XMPP only relies on persistent TCP connections that are always present. This improves latency and response times but increases the computational requirements of the devices. Encryption and authentication problems associated with TLS follow under the umbrella of transport problems.

5.2.3.2 XMPP IoT Extensions

The core XMPP specification is functionally upgraded by means of XMPP extensions that are formally called *XMPP Extension Protocols* (XEPs) [23]. In the context of IoT, there are multiple extensions including (1) XEP-0000-IoT-BatteryPoweredSensors that defines how devices affected by power duty cycles must be handled; (2) XEP-0000-IoT-Events that deals with events and subscriptions; (3) XEP-0000-IoT-Interoperability that describes interoperability mechanisms between different types of devices; (4) XEP-0000-IoT-Multicast that presents how sensor data can be multicast in efficient ways; (5) XEP-0000-IoT-PubSub that indicates how sensor data can be published; (6) XEP-0000-IoT-Chat that describes *human-to-machine* (H2M) interfaces relying on chat messages; (7) XEP-0323 that introduces the architecture, operations, and data structures for sensor data transmission; (8) XEP-0325 that describes how to provide actuation and control; (9) XEP-0326 that defines how to handle architectures containing servers handling multiple devices; and (10) XEP-0347 that introduces a device discovery mechanism under XMPP.

The stream shown in Fig. 5.14 presents a stream composed of the interaction between a client and a device in accordance with, first, XEP-0323 and then XEP-0325. The client starts by transmitting a get type iq stanza to the device identified by *50001* with a request payload with sequence number *1* in order to obtain a sensor readout. The device responds right away by transmitting a result type iq stanza also identified by *50001* with an accepted payload indicated by sequence number *1* that acknowledges the request. As soon as the device gets a readout, it transmits a message stanza to the client that includes a fields payload identified by the sequence number *1* and indicating that is the only message to be transmitted by means of the done attribute. The payload includes a node element that indicates the device source of the readout that, in turn, includes a timestamp element that provides the actual sensor data readout. This readout is carried inside of a numeric element that provides value and unit attributes. Next, actuation is again started by the client that sends a set type iq stanza to the device identified by *50002* with a set payload that includes a boolean element. The element name attribute identifies the port under which actuation is performed, and the value attribute indicates the actual value to be set. The device sets the value and replies with a result type iq stanza also identified by *50002* that carries an empty setResponse payload.

5.2.4 CoAP

As opposed to HTTP and XMPP that were not designed with IoT in mind, CoAP is a lightweight session layer protocol that has been intended from the very beginning to support sensor and actuation data transmission in LLNs. CoAP was standardized by IETF as RFC 7252 *The Constrained Application Protocol (CoAP)* in 2014. It is expected that CoAP will become the default protocol for access IoT in the coming years with support of billions of deployed devices all over the world. CoAP applications range from smart energy, smart grid, building automation and control to intelligent lighting control, industrial control systems, asset tracking, and environment monitoring [21]. CoAP also provides its own proprietary mechanism for resource discovery that is presented in detail in Sect. 6.4.

One of the most important differences between CoAP and competing technologies like HTTP and XMPP is that the former relies on UDP as a transport layer. Since UDP, as opposed to TCP, is connectionless, it can be used not only in unicast scenarios but also with multicast and broadcast M2P applications that are representative of many IoT scenarios. UDP also improves latency since it does not include any built-

Fig. 5.16 XMPP XML error attribute

```
<iq type=''error''
      from=''device@example.org''
      to=''client@example.org/amr''
      id=''50001''>
   <error type=''cancel''>
      <service-unavailable xmlns=''url:ietf:params:xml:ns:xmpp-stanzas''/>
   </error>
</iq>
```

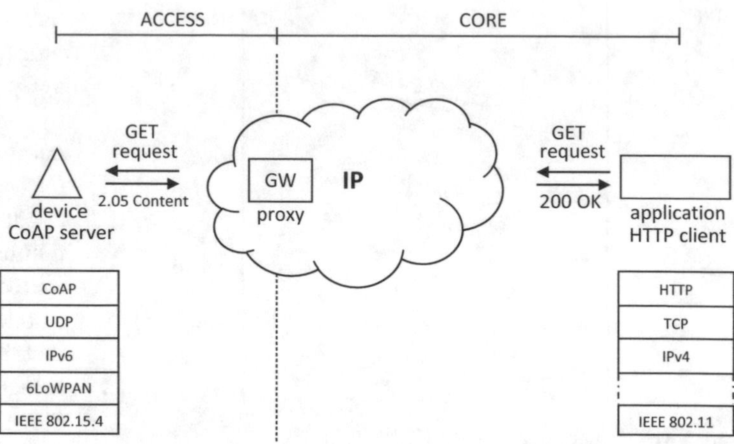

Fig. 5.17 CoAP/HTTP proxy

in retransmission mechanism that might introduce additional delays due to packet loss. This is in opposition to TCP transport as described in Sect. 4.3.7. In addition, because UDP is not a stream protocol like TCP, it provides a technology for the delivery of independent messages, enabling CoAP for asynchronous message exchanges that eliminate HOL blocking exhibited by certain versions of HTTP. Note that although it is not recommended, CoAP TCP transport has been introduced by IETF RFC 8323 as an alternative for certain scenarios where the complex TCP congestion control mechanisms provide a viable solution in certain LLNs [7]. To improve latency, CoAP has incorporated block data transmission as part of IETF RFC 7959 *Block-Wise Transfers in the Constrained Application Protocol (CoAP)* [6]. This mechanism enables the efficient transmission of streams and other structured data.

As a REST protocol, CoAP was designed to map some of the functionality provided by HTTP such that in an IoT infrastructure, CoAP is used in the access side, while HTTP is used in the core of the network with the gateway acting as a proxy that translates traffic from one protocol to another. This situation is shown in Fig. 5.17. The application, typically performing analytics, issues an HTTP GET request to retrieve sensor readouts. This message is transmitted over both TCP and IPv4 and encapsulated in IEEE 802.11. The gateway maps the HTTP request into a CoAP GET message that is transported over both UDP and IPv6. IPv6 is adapted by means of 6LoWPAN for encapsulation under IEEE 802.15.4. At the sensor, the request triggers the transmission of a *readout* as a CoAP *2.05 Content* response that is translated into an HTTP *200 OK* response at the gateway. Because both CoAP and HTTP are stateless protocols, the mapping occurring at the gateway is also stateless.

Besides providing a session layer mechanism like HTTP but relying on UDP to enable multicast transmissions and low latency, CoAP uses binary encoding to lower its transmission rate. Since CoAP traffic is intended to be encapsulated by technologies with small MTU sizes like IEEE 802.15.4, ASCII encoding like that of HTTP and XMPP results in too much overhead that makes it impractical in LLNs. The inherent lack of reliability of UDP is compensated by an optional retransmission mechanism embedded in CoAP known as confirmable mode. The alternative mode of operation, known as non-confirmable, is based on a fire-and-forget approach where messages are sent without expecting any acknowledgment. Note that under CoAP, transport layer functionality responsible for providing network reliability is moved to the session/application layer whenever confirmable mode is enabled.

5.2.4.1 CoAP Basic Flows

CoAP relies on a two-sublayer structure, shown in Fig. 5.18, with the lower message sublayer providing the interface with UDP transport and enabling retransmissions when confirmable mode is configured and the higher request/response sublayer that is responsible for building and parsing the binary messages. In addition to confirmable and non-confirmable messages, the message sublayer also includes acknowledgment and reset messages. The acknowledgment messages are sent in response to incoming confirmable requests or responses, while the reset messages are used to indicate a far end failure in processing an

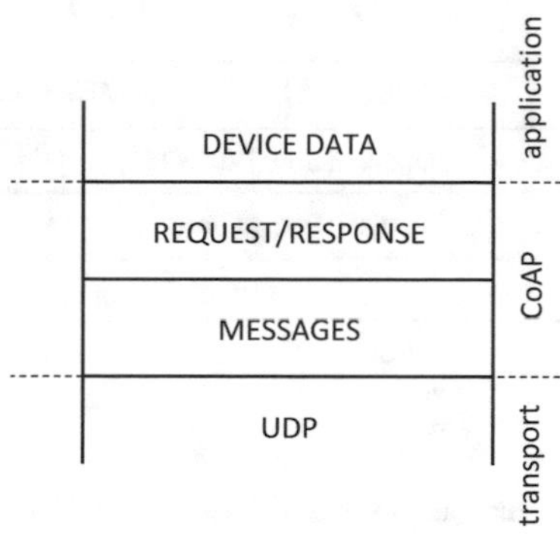

Fig. 5.18 CoAP two-layer structure

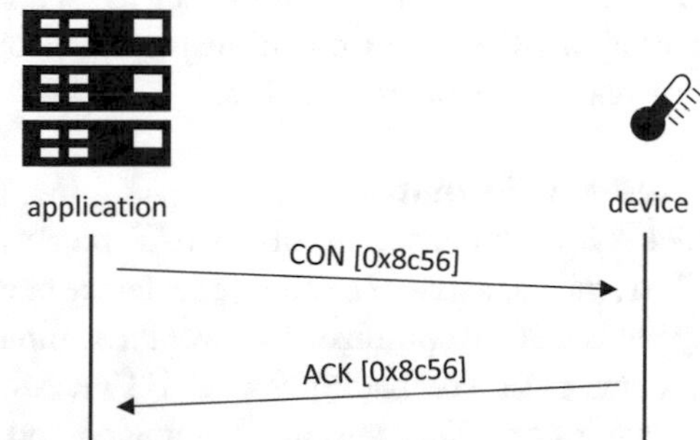

Fig. 5.19 Confirmable CoAP transaction

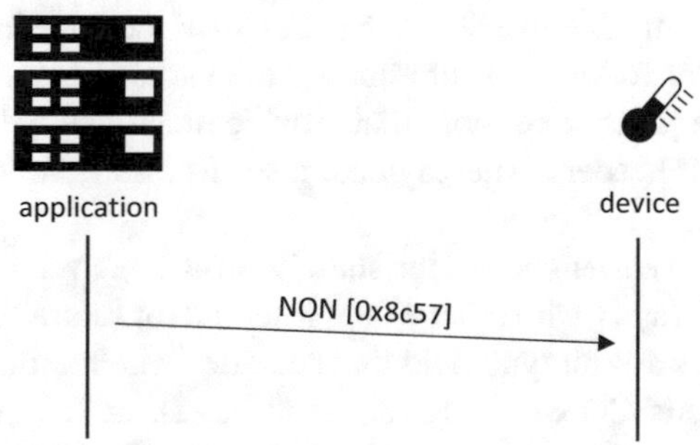

Fig. 5.20 Unreliable CoAP

incoming request or response. In all scenarios, confirmable, non-confirmable, acknowledgment, and reset messages are signaled by means of CON, NON, ACK, and RST message types, respectively.

Reliable message transport, shown in Fig. 5.19, relies on the client transmitting a confirmable (CON) request identified by a 16-bit 0x8c56 message identifier. The device responds to the request by transmitting an acknowledgment (ACK) also identified by the same 16-bit message identifier. A single request/response interaction indicated by a common message identifier is a CoAP transaction. If, for whatever reason, the confirmable message or its acknowledgment is not received, the sender attempts to retransmit the message three more times, exponentially incrementing the timeout each time. Similarly, Fig. 5.20 shows unreliable message transport that relies on the client transmitting a non-confirmable

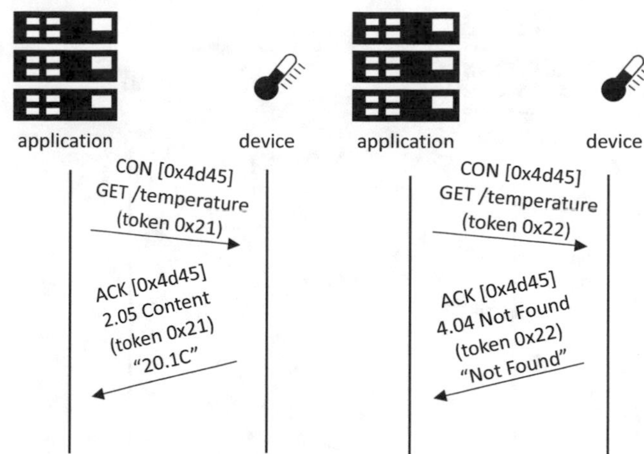

Fig. 5.21 Piggybacked request/response

(NON) request identified by a 16-bit 0x8c57 message identifier. In this case, the device does not acknowledge the request. For both cases, confirmable and non-confirmable requests, if the receiver fails to process the message, it replies by transmitting a reset (RST) response.

Independent of reliability, the request/response sublayer is responsible for generating and parsing requests and responses. Figure 5.21 shows an example of a client requesting a *temperature* readout from a sensor acting as a server. Since the request is a CON CoAP GET, an ACK is transmitted back by the server. Assuming the sensor has access to the readout value, it piggybacks this value to the acknowledgment as a CoAP 2.05 Content response. On the other hand, if the sensor does not support temperature readouts or it does not have a valid value, it responds back with a CoAP 4.04 Not Found. Note that all CoAP responses include a numerical response code in a similar fashion to the mapping of HTTP shown in Fig. 5.2. Response codes follow a $a.bb$ format where a is associated with the class of message: $a = 2$ means success, $a = 3$ means redirection, $a = 4$ means client errors, and $a = 5$ means server errors. Besides a message identifier that identifies the transactions, Fig. 5.21 also shows the presence of a field called token that is common throughout the transactions associated with a single session. In other words, the token uniquely identifies the session through its lifetime. A session, in turn, is given by the interaction between the client and the device to retrieve different readouts of sensor data as time progresses.

Alternatively, if the sensor does not have a temperature readout value available by the time it transmits the ACK message, it delays its transmission until a readout is obtained either by polling or by means of hardware interrupts. This is illustrated in Fig. 5.22. The session is identified by token 0x21. First, the client initiates a transaction identified by message identifier 0x4d45 by issuing a CON CoAP GET request to retrieve a *temperature* readout. The sensor acknowledges

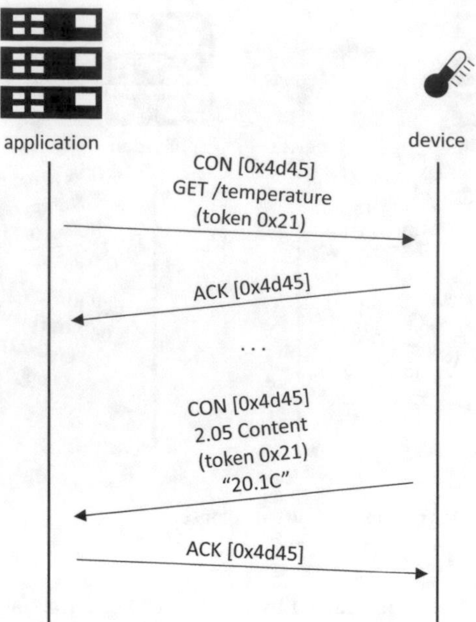

Fig. 5.22 Separate response model

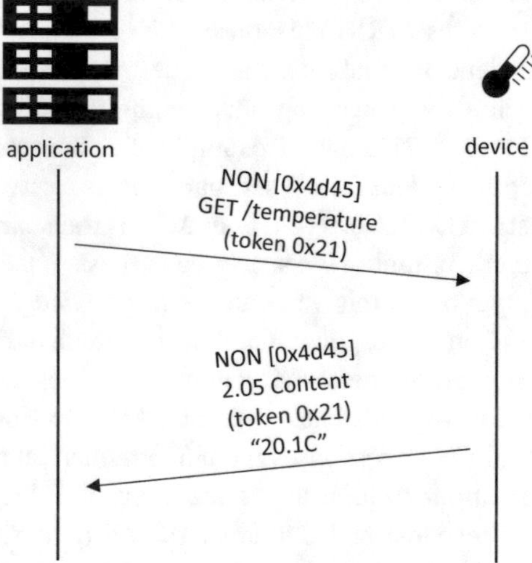

Fig. 5.23 Non-confirmable request/response model

the request by sending an empty ACK response. When the temperature readout becomes available, the sensor transmits the response to the original request by transmitting a CON 2.05 Content message. Since this transmission is related to the original CoAP request, it belongs to the same transaction and, therefore, shares the same message identifier. Finally, the content message by being confirmable requires the client to acknowledge it by transmitting an empty ACK response.

Figure 5.23 shows a non-confirmable request and response interaction. Since it is unreliable, there is no need for acknowledgment. The client first sends a NON CoAP GET request to retrieve a *temperature* readout. The message identifier 0x4d45 identifies the transaction, while the token id

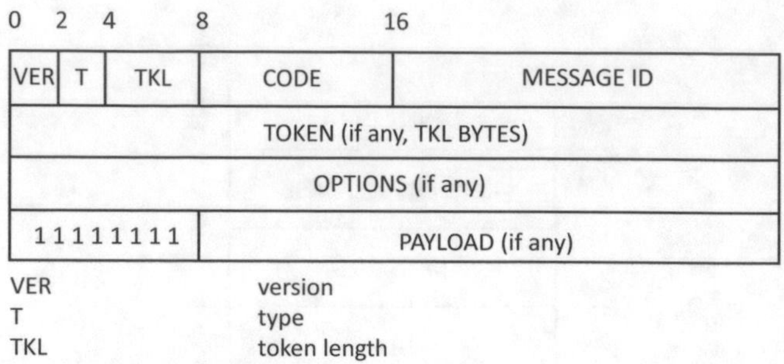

Fig. 5.24 CoAP message format

0x21 identifies the session. The sensor acting as server replies by transmitting a NON 2.05 Content response as soon as a temperature readout becomes available.

5.2.4.2 Message Format

Figure 5.24 shows the encoding of a CoAP request or response. The format consists of a variable sequence of extensible binary fields that are optimized to lower throughput while providing some basic compatibility with HTTP. Moreover, since CoAP relies on UDP transport (running on port 5683), and because of its compact format, it is ideal for transmission on top of 6LoWPAN and wireless IoT link and physical layer mechanisms like IEEE 802.15.4. A CoAP message has a 32-bit header followed, if present, by the variable length token plus a sequence of options that provide functionality like that of HTTP headers. The payload, also if present, follows the options.

The fixed-length header starts with a 2-bit version field that is always *1* with other values reserved for future versions; it follows a 2-bit type field that specifies whether the CoAP message is CON (0), NON (1), ACK (2), or RST (3) and continues with a 4-bit token length that specifies the length in bytes of the session identifier token (0 through 8 with lengths larger than 9 being reserved). The next field includes an 8-bit code that is encoded, as previously mentioned, as *a.bb* where *a* is a 3-bit class digit between 0 and 7 and *bb* is a 5-bit detail number between 0 and 31. If *a* is 0, the message is a request, and otherwise it is a response encoded as specified in the previous paragraph. If it is a request, the value of *bb* identifies its nature (GET if *bb=01*, POST if *bb=02*, PUT if *bb=03*, and DELETE if *bb=04*). The fixed-length header ends with a 16-bit message identifier used for transaction identification.

The options are encoded, as shown in Fig. 5.25, as a sequence of TLVs used for both requests and responses. The original option types, as defined by IETF RFC 7252, are shown in Table 5.3. The option type is differentially encoded in order to use fewer bytes to represent it. The option length is encoded as an absolute value. Options are encoded in ascending order of their type such that for a given option,

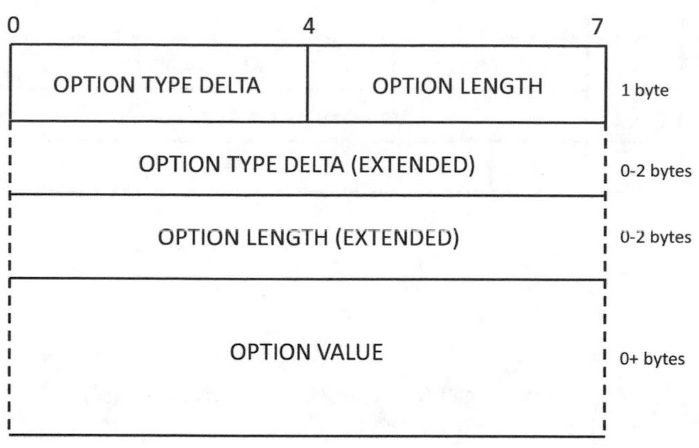

Fig. 5.25 Options encoding

Table 5.3 CoAP option types

Option name	Option type
If-Match	1
Uri-Host	3
ETag	4
If-None-Match	5
Location-Path	8
Uri-Path	11
Content-Format	12
Max-Age	14
Uri-Query	15
Accept	17
Location-Query	20
Proxy-Uri	35
Proxy-Scheme	39
Size1	60

its type value is represented as a delta with respect to the previously encoded option type. The very first option of the sequence is encoded assuming that the previous option type is zero. The option type delta is encoded as a 4-bit number if it smaller than *13*; otherwise, if it is smaller than *256*, it is encoded as *13* with an extension byte representing the actual type delta. Similarly, if the delta is larger than *255* but smaller than *65536*, it is encoded as *14* with a 2-byte extension representing the actual type delta. Multiple instances of the same option type are encoded using a delta of zero. The option length is encoded as an absolute value following the same mechanism, attempting to use the fewer number of bits to represent its value by relying on extension bytes whenever necessary. In all cases, the option value follows the encoded type and length. Figure 5.26 illustrates an example where the Uri-Path is an 11-byte string *Temperature* and Max-Age is a 2-byte integer *3600*. Since the Uri-Path option type is encoded as *11*, the Max-Age option type, which has a value *14*, is encoded as a *3*.

Example 5.2 Consider an end-to-end non-confirmable CoAP scenario where an application requests a readout from a sensor:

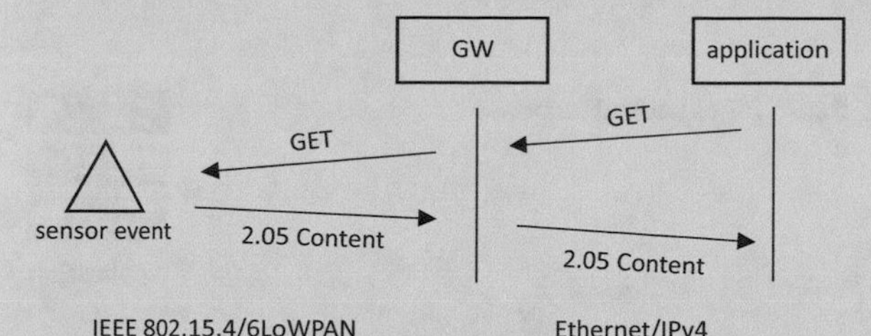

Assume the following: (a) the IEEE 802.15.4 transmission rate is 250 KBps, (b) the Ethernet transmission rate is 1 Gbps, (c) the only delay is the transmission delay, (d) the UDP header is $L_{\text{UDP}} = 8$ bytes long, (e) the CoAP GET request is $L_{\text{GET}} = 20$ bytes long, (f) the CoAP 2.05 Content response is $L_{2.05\ \text{Content}} = 10$ bytes long, (g) the IPv4 header is $L_{\text{IPv4}} = 20$ bytes long, (h) the 6LoWPAN header is $L_{\text{6LoWPAN}} = 10$ bytes long, (i) the Ethernet header is $L_{\text{eth}} = 16$ bytes long, and (f) the IEEE 802.15.4 header is $L_{\text{IEEE 802.15.4}} = 10$ bytes long. How long does it take for the application to receive the readout? How does it compare to the results obtained in Example 5.1?

Solution: The overall delay is given by $T = T_1 + T_2$ where T_1 is given by $T_1 = \frac{8\times(L_{\text{eth}}+L_{\text{IPv4}}+L_{\text{UDP}}+L_{\text{GET}})}{10^9} + \frac{8\times(L_{\text{IEEE 802.15.4}}+L_{\text{6LoWPAN}}+L_{\text{UDP}}+L_{\text{GET}})}{250000} = 1.54$ ms. T_2 is given by $T_2 = \frac{8\times(L_{\text{eth}}+L_{\text{IPv4}}+L_{\text{UDP}}+L_{2.05\ \text{Content}})}{10^9} + \frac{8\times(L_{\text{IEEE 802.15.4}}+L_{\text{6LoWPAN}}+L_{\text{UDP}}+L_{2.05\ \text{Content}})}{250000} = 1.22$ ms. The total delay is, therefore, $T = 2.76$ ms. As expected, a CoAP session exhibits a latency that is a lot smaller than that of an HTTP session.

The Uri-Host, Uri-Port, Uri-Path, and Uri-Query options are used to specify the target resource of a request to a CoAP origin server. The Proxy-Uri option is used to make a request to a forward-proxy. The Content-Format option indicates the representation format of the message body. The Accept option can be used to indicate which Content-Format is accepted from the server. The Max-Age option specifies the maximum time a response may be cached before it is considered not fresh. The *Entity Tag* (ETag) is used as a resource local identifier for differentiating between representations of the same resource that vary over time. The Location-Path and Location-Query options together indicate a relative URI that consists of an absolute path, a query string, or both. The If-Match option is used to make a request conditional on the current existence or value of an ETag for

Fig. 5.26 Uri-path and max-age encoding example

0	2	4	8	16
1	0	0	GET = 1	MID = 0x7d34
11		11	"temperature" (11 BYTES)	...
...		3	2	0xe10 (3600)

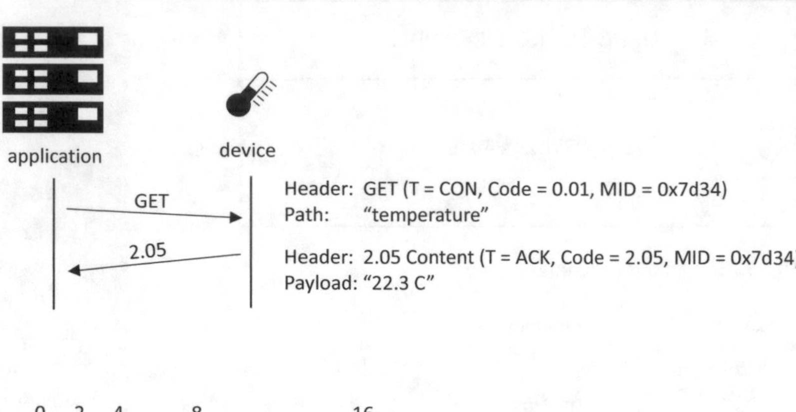

Fig. 5.27 Piggybacked response

0	2	4	8	16
1	0	0	GET = 1	MID = 0x7d34
11		11	"temperature" (11 BYTES)	

0	2	4	8	16
1	2	0	2.05 = 69	MID = 0x7d34
11111111			"22.3 C" (6 BYTES)	

one or more representations of the target resource. The If-None-Match option is used to make a request conditional on the nonexistence of the target resource. The Size option provides size information about the resource representation in a request.

Figure 5.27 shows a single transaction between a client and device acting as a server where the request is a CON GET that includes a single Uri-Path *temperature* option. The response is an ACK that piggybacks a 2.05 Content that carries a *22.3C* temperature payload. Figure 5.28 shows another transaction for the same scenario that belongs to a larger session identified by the token number *0x20*.

Figure 5.29 shows a client sending a CON GET request to a device that responds back with an ACK that piggybacks a 2.05 Content. The response never arrives to the client as it is dropped due to network packet loss. Since it is a confirmable transaction, the request timeouts, and it is retransmitted a few seconds later. When it arrives at destination, the sensor retransmits the ACK with the 2.05 Content response that is then received by the client. This finalizes the transaction. The CoAP standard identifies three parameters that control transmissions, MAX_RETRANSMIT, ACK_TIMEOUT, and ACK_RANDOM_FACTOR, with default values 4, 2, and 1.5, respectively. For a new CON message, the initial timeout is set to a random duration between ACK_TIMEOUT and ACK_TIMEOUT * ACK_RANDOM_FACTOR seconds, and the retransmission counter is set to zero. When the timeout is triggered and the retransmission counter is less than MAX_RETRANSMIT, the message is retransmitted, the retransmission counter is incremented, and the timeout is doubled. If the retransmission counter reaches MAX_RETRANSMIT on a timeout, or if the endpoint receives a RST message, then the attempt to transmit the message is canceled and the application process informed of failure.

An alternative scenario is shown in Fig. 5.30; in this case, the client issues a CON GET request. When the request arrives at the destination, if the device does not have access to a recent readout, it just transmits a plain ACK. When, later, the device, either by polling or by means of hardware interrupts, obtains a valid readout, it transmits this value in a new CON 2.05 Content response. Of course, this response is acknowledged after it is received.

In Fig. 5.31, the client transmits a CON GET request, but its application suffers an exception and crashes. The device replies with an ACK that arrives to the client by the time it has already rebooted. The client ignores this acknowledgment as it cannot figure out what originated it. Moreover, since it is

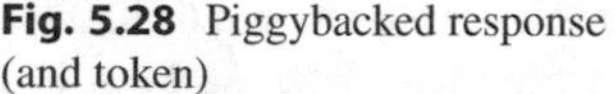

Fig. 5.28 Piggybacked response (and token)

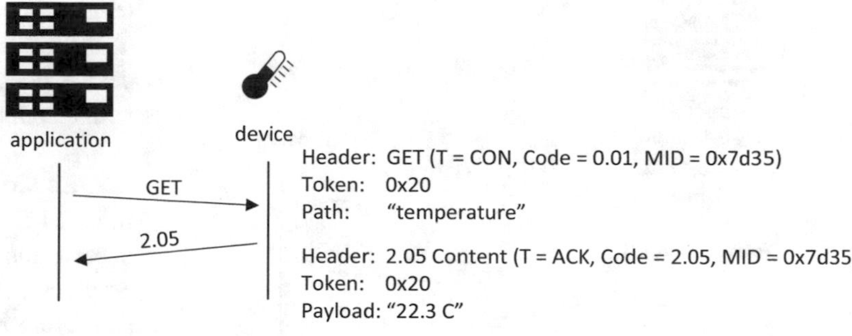

0	2	4	8	16
1	0	0	GET = 1	MID = 0x7d35
0x20				
11		11	"temperature" (11 BYTES)	

0	2	4	8	16
1	2	0	2.05 = 69	MID = 0x7d35
0x20				
11111111			"22.3 C" (6 BYTES)	

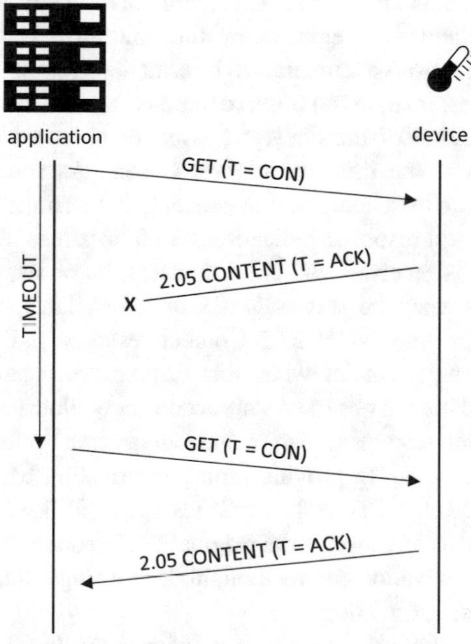

Fig. 5.29 Confirmable request with piggybacked response

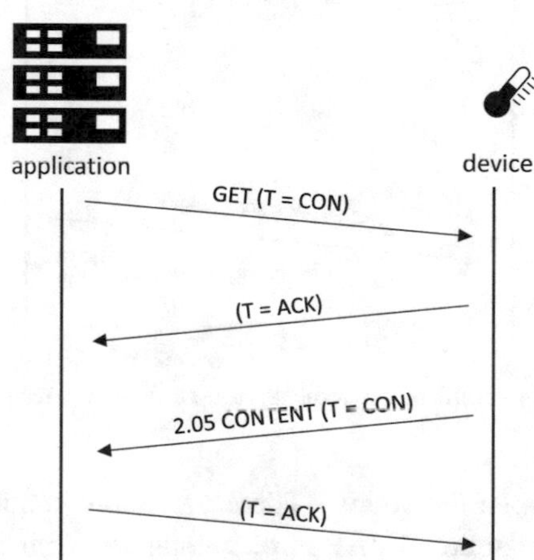

Fig. 5.30 Confirmable request with separate response

just an acknowledgment and there is no action associated with the message, it makes sense to ignore it. When the sensor readout becomes available, the device transmits a CON 2.05 Content message that when, arrives at destination, causes the device to transmit a RST message type to indicate that the session is no longer available.

If network reliability is not a concern, the client can transmit, as indicated in Fig. 5.32, a NON request that is not acknowledged when it arrives at the device. If the message is dropped, there is no standard mechanism for the sender to know that it has not arrived at the receiver. In either case, if it does arrive, whenever a readout is available, the device transmits a NON 2.05 Content response.

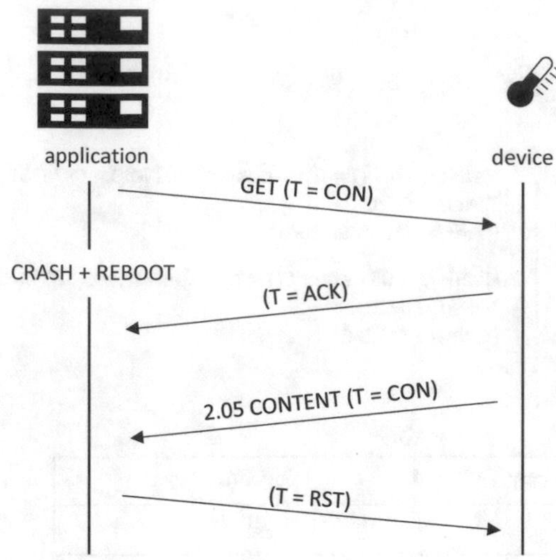

Fig. 5.31 Confirmable request; separate response (unexpected)

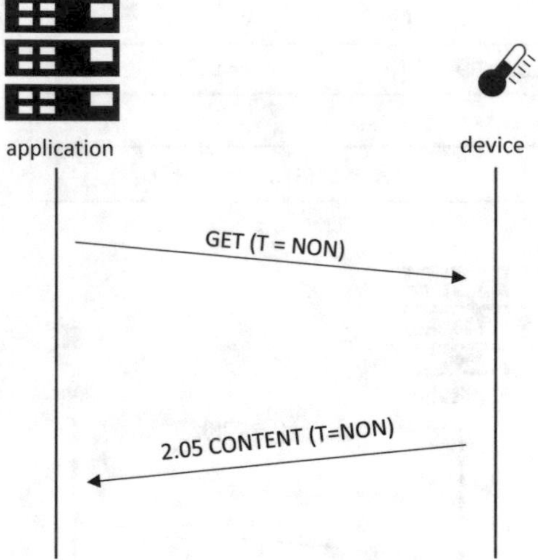

Fig. 5.32 Non-confirmable request; non-confirmable response

► **CoAP Implementations** There are many commercial and open-source CoAP implementations with support ranging from common constrained RTOSs to general-purpose Linux distributions. The list below shows some of the most popular CoAP implementations.

cancoap	BSD-licensed C/C++ implementation of client and server.
eCoAp	MIT-licensed C implementation of client and server.
Erbium	BSD-licensed C implementation (for Contiki) of client and server.
libcoap	GPL-licensed C implementation of client and server.
nanocoap	LGPL-licensed C lightweight implementation of client and server.

One last scenario to consider is shown in Fig. 5.33. The client transmits a multicast NON GET request to obtain temperature readouts from all three devices in the link. Two of them reply with NON 2.05 Content responses. One last device does not provide temperature readouts and replies with a NON 4.04 Not Found response.

5.2.4.3 CoAP Observation

IETF RFC 7641 extends CoAP to natively support resource observation [10]. Specifically, a new observe CoAP option provides a mechanism for GET requests to retrieve current and future representations of an object. Under this scheme, each server keeps a list of observers that include entries identifying clients and tokens associated with different CoAP sessions. To this end, as soon as the GET request arrives at the destination, the server adds an entry that identifies the client as part of its list of observers. Consequently, this CoAP GET request is an observation registration where the value of the observe option is 0. From that point on, whenever there is a new sensor readout, either by polling or by means of a hardware interrupts, the server transmits a 2.05 Content to the client. The server sends the 2.05 Content response to each one of the clients and sessions in its list of observers. When a client does not wish to receive sensor readouts anymore, it must deregister by transmitting a new CoAP GET with an observe CoAP option set to 1.

Figure 5.34 shows a basic example of CoAP observation with a client that registers by transmitting a NON GET request to observe temperature information from the sensor. The request includes an observe option set to 0. The resource representation of the observed asset (or observed state) is unknown at the time the request is sent. As soon as the sensor obtains a temperature readout, it transmits a NON 2.05 Content response indicating a 18.5 °C value. When this response is received, the client updates the observed state. Similarly, when the next readout becomes available, the client transmits a new NON 2.05 Content response indicating a 19.2 °C value. Again, when this response is received, the client updates the observed state accordingly. Note that CoAP observation responses also include an observe option that is set, in this case, to provide timing information based on a centralized clock. The token and message identification fields are the same as those of the original GET requests since the overall observation can be thought of as a single transaction within a single session.

The session continues as it is indicated in Fig. 5.35. The sensor sends a 19.2 °C readout and then crashes. When it reboots, it has no context of the observation session, so it stops transmitting updates even when new readouts become available. The client eventually timeouts and issues a new NON GET request to reregister and observe temperature information from the sensor. The request includes new token and message identification fields as it starts a new transaction in a new session. The server proceeds by transmitting NON 2.05 Content responses carrying temperature readouts.

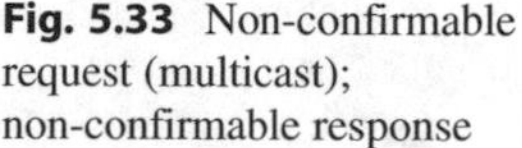

Fig. 5.33 Non-confirmable request (multicast); non-confirmable response

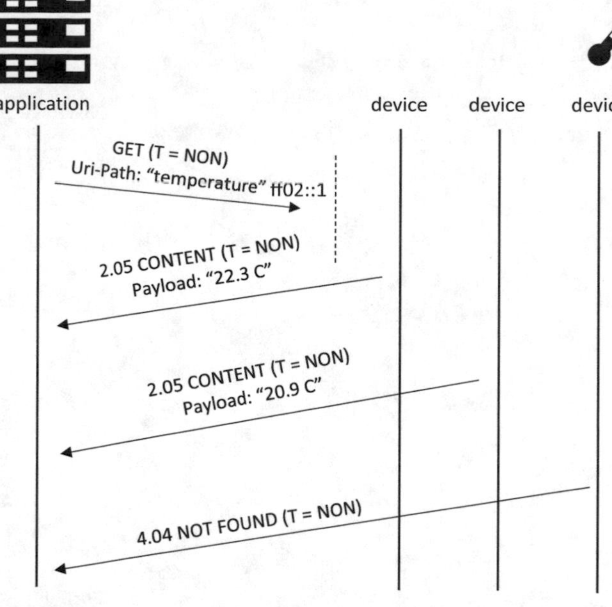

Fig. 5.34 CoAP observation

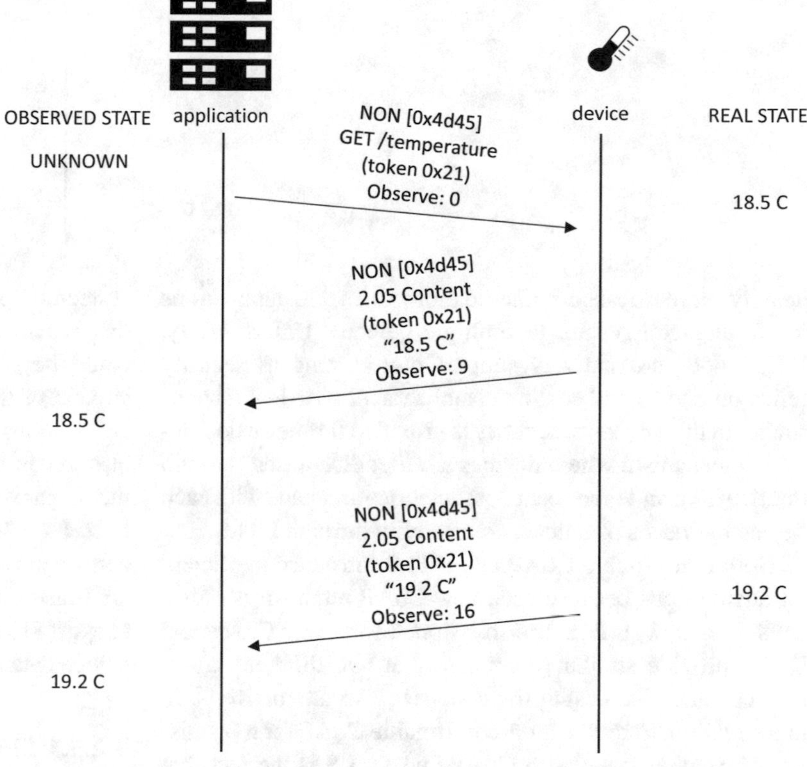

5.2.4.4 CoAP and DTLS

CoAP security is provided by means of the DTLS protocol presented in Sect. 4.3.6.3. The main motivation for using DTLS is the fact that CoAP uses UDP transport and DTLS is intended to work on top of UDP. IETF RFC 7925 addresses some of the requirements needed for DTLS to run in the context of CoAP [22]. Note that although DTLS is simpler than IPSec/IKE and HIP, some features of the protocol are too complex to be supported with lightweight CoAP. One of these issues, already introduced in Sect. 4.3.6.3, is the fact that the DTLS handshake exchanges large certificates that lead to fragmentation and, thus, datagram loss in LLNs. Moreover, alternatives like *elliptic-curve cryptography* (ECC) that provides smaller keys and certificates, for comparatively same levels of security as traditional RSA and DSA security, are too complex to be supported in embedded devices. Another important issue is that DTLS is not prepared to work well with CoAP proxies. Moreover, CoAP relies

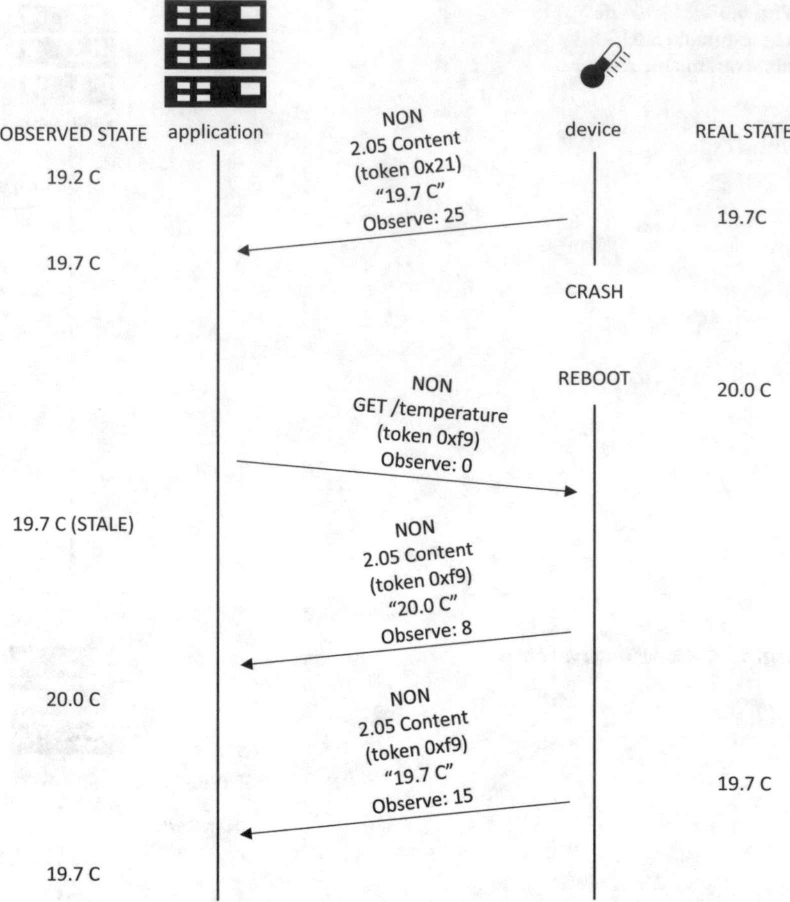

Fig. 5.35 Client reregistration

heavily on multicast communications to enable applications to simultaneously talk to multiple devices. Unfortunately, DTLS does not natively support multicasting as security relies on end-to-end session establishment based on a handshake. In this scenario, security is provided through a double-stage mechanism where devices are first discovered through multicasting and then secure associations are made with each device by means of unicast connectivity through DTLS.

Both confirmable CoAP and DTLS introduce inefficient retransmissions because each new CoAP retransmission results in a new DTLS transmission. Moreover, CoAP and DTLS provide similar functionality at two different layers unnecessarily increasing the code size. An alternative is, in this situation, to rely on non-confirmable CoAP for transmissions. Another issue with CoAP and DTLS is the fact that the latter does not support fragmentation of application data. It is therefore recommended for applications to attempt to fit every single CoAP message and associated headers into a single IP datagram to avoid any fragmentation.

5.2.5 SIP and RTP

SIP, briefly described in Sect. 1.4, provides a mechanism to create, manage, and finish sessions. From a layered architecture perspective, it is an application layer protocol that provides session layer functionality. Although sessions could be of any type, SIP has been mostly used in the context of RTC with special focus on *Media over IP* (MoP) applications like bidirectional speech, audio, and video. IETF also recommends a specific protocol known as RTP for media packetization. Note that RTP was briefly described in Sect. 1.4. Media *packetization* is the process that enables the conversion of a stream of media into messages or packets that are transmitted over IP. There have been several attempts to support SIP and RTP in the context of IoT for both media and device data transmission.

5.2.5.1 SIP

SIP is a client/server protocol that is structurally like HTTP as it includes ASCII encoded requests and responses with headers and methods [18]. SIP, as opposed to HTTP, however, was designed for UDP transport in order to lower the latency usually associated with TCP connection setup and retransmissions. Later when security was introduced to SIP by means of TLS, TCP became an additional transport option. Regardless of the transport, SIP is highly inefficient in IoT scenarios because it relies on plain text messages to establish and manage sessions. Moreover, because SIP does not really comply with the RESTful model either, it is not really useful for IoT device management.

A SIP client is called *User Agent Client* (UAC), while a SIP server is called *User Agent Server* (UAS). UACs generate requests, while UASs generate responses for incoming requests. A SIP transaction is a request/response interaction between a UAC and a USC. Similarly, a SIP dialog is several transactions associated with a given session. A UAS can be (1) a *media server* that generates audio, speech, and/or video traffic to UACs; (2) a *proxy server* that, similar to HTTP proxies, acts on behalf of a client to interact with other UASs; (3) a *redirect server* that redirects UAC requests to other UAS; or (4) a *registrar* that keeps track of the location of UACs to support incoming sessions.

By not being RESTful, SIP relies on different request types or methods that are transmitted by a UAC: INVITE to start sessions, ACK to provide reliability since SIP is primary supported over unreliable UDP transport, BYE to terminate sessions, CANCEL to cancel a pending request, OPTIONS to request UAS capability information, REGISTER for a UAC to register with a registrar, and INFO for out-of-band signaling of, for example, audio tones.

SIP messages are ASCII encoded, and therefore they can be read by humans on the wire by means of network sniffers and analyzers. The message below is an example of a SIP request transmitted by a UAC attempting to start a new session:

```
INVITE sip:deviceB@server2.local SIP/2.0
Via: SIP/2.0/UDP server1.local;branch=z9hG4bK2105Gbc12
Max-Forwards: 70
From: <sip:deviceA@server1.local>; tag=1921052105
To: <sip:deviceB@server2.local>
Call-ID: 53a19e23b12afe01@server1.local
CSeq: 1021 INVITE
Contact: <sip:deviceA@server1.local>
Content-Type: application/sdp
Content-Length: 112

v=0
o=session 5439349 2394349324 IN IP6 deviceA.local
s=SDP exchange
c=IN IP6 2001::21:10
t=0 0
m=audio 5100 RTP/AVP 0
a=sendrecv
```

The format of SIP requests follows that of HTTP requests illustrated in Fig. 5.8. It includes a request line that specifies the method, in this case INVITE, the request URI, and the SIP version that is typically 2.0. The *Via* header indicates the local address that is used to receive the response. The Via header includes a *Max-Forward* field that specifies the maximum number of hops the request must support before being dropped. This field is used to prevent SIP routing loops. The *From* and *To* headers specify the source and destination endpoints including tags that are random numbers assigned to the entities for identification. The *Call-ID* header value, in this case *53a19e23b12afe01@server1.local*, combined with the From and To header values is used to identify the dialog. The *CSeq* header value, in this case *1021 INVITE*, contains an integer and the method. When a transaction starts, the first message is given a random CSeq value that is incremented upon transmission of new messages. Again, this enables endpoints to detect out of order and lost messages as UDP is the preferred SIP transport method. The *Contact* header specifies a direct route to the UAC used by certain responses. The *Content-Type* header indicates the type of content the request includes. In this example, the content is *application/sdp* that specifies session characteristics by means of the SDP protocol briefly described in Sect. 1.4. The *Content-Length* indicates how big the body is.

The SDP is a whole different protocol that is used to describe media sessions in a way that can be understood by all endpoints in a network [14]. As such, SDP provides a mechanism for negotiation, where an endpoint can decide to accept an incoming session based on whether it supports specific media types. The SDP content is ASCII and, as SIP messages themselves, is not efficient in the context of IoT communications. In either case, an SDP typically indicates the list of specific media encoding technologies or codecs that are offered by a given endpoint, along with connectivity parameters like IP addresses and transport ports. Information is presented as lines formatted as *parameter=value*, where the parameter is a single character. In the example above, for example, the *v=0* line specifies the SDP version, *c=IN IP6 2001::21:10* indicates the endpoint IP address, and *m=audio 5100 RTP/AVP 0* signals the type of media (audio), the media protocol (RTP), the UDP port (5100), and the codec code (0) that specifies ITU G.711 μ-Law coding. The *a=sendrecv* indicates that media traffic is intended to be full duplex.

Like requests, the format of SIP responses follow that of HTTP responses shown in Fig. 5.9. The message below is an example of a SIP response transmitted by a UAS as a reply to an INVITE request:

```
SIP/2.0 200 OK
Via: SIP/2.0/UDP server2.local;branch=z9hG4bK1chab10;
received=2001::21:20
From: <sip:deviceA@server1.local>; tag=1921052105
To: <sip:deviceB@server2.local>; tag= 42194562
Call-ID: 53a19e23b12afe01@server1.local
CSeq: 1021 INVITE
Contact: <sip:deviceB@server2.local>
Content-Type: application/sdp
Content-Length: 110
```

Table 5.4 SIP response codecs

Range	Meaning
100–199	Provisional
200–299	Success
300–399	Redirect
400–499	Client errors
500–599	Server errors
600–699	Global failure

```
v=0
o=session 2841303 91021765 IN IP6 deviceB.local
s=SDP exchange
c=IN IP6 2001::21:20
t=0 0
m=audio 5102 RTP/AVP 0
a=sendrecv
```

The first line of the response is the status line that includes the SIP version, the numerical status code, and the corresponding status message or reason phrase. Although SIP response codes follow the HTTP response codes, there are some slight differences as shown in Table 5.4. Responses between 100 and 199 are provisional, in the sense that they indicate that a final response 200 or above will follow at some point. For the example above, the SIP version is 2.0, the numerical status code is 200, and the reason phrase is OK. The *Via* header indicates the local address that is used to receive the response including the specific IP address of the interface that received the message. The *From*, *To*, *Call-ID*, and *CSeq* headers are copied over from the INVITE request. The UAS adds a tag to the *To* header. As with the UAC, the UAS *Call-ID* header value, in this case *53a19e23b12afe01@server1.local*, combined with the *From* and *To* header values is used to identify the dialog. The *Contact* header specifies a direct route to the UAS used by subsequent requests. Because the response to an INVITE has media negotiation information, the *Content-Type* and *Content-Length* header values indicate that the message body has an SDP content that signals the negotiated codec.

Figure 5.36 shows a dialog between a UAC and a media server UAS. This dialog is made of two transactions: one to establish a media session and another one to tear it down. The session establishment transaction starts when the client transmits an INVITE request to the server. The INVITE carries an SDP that provides the client side media offering. The UAS transmits a provisional 100 Trying response to let the client know that the INVITE has been received and it is trying to respond. Right after, the UAS transmits another provisional 180 Ringing response to tell the client to generate a ring tone, while the INVITE request is being processed. Both provisional responses do not carry a body and therefore do not include any media information. Next, the UAS transmits a 200 OK that contains negotiated media. Because of the unreliable UDP transport, the client responds to the 200 OK with an ACK on what it is called a three-way handshake. Once media endpoints have been identified, a bidirectional RTP session starts. At this point, packetized media datagrams flow between endpoints. After a while, the client transmits a BYE request to indicate it wants to terminate the media session. Upon reception, the server transmits another 200 OK to confirm. This latter 200 OK does not carry a body as it does not need to specify media information.

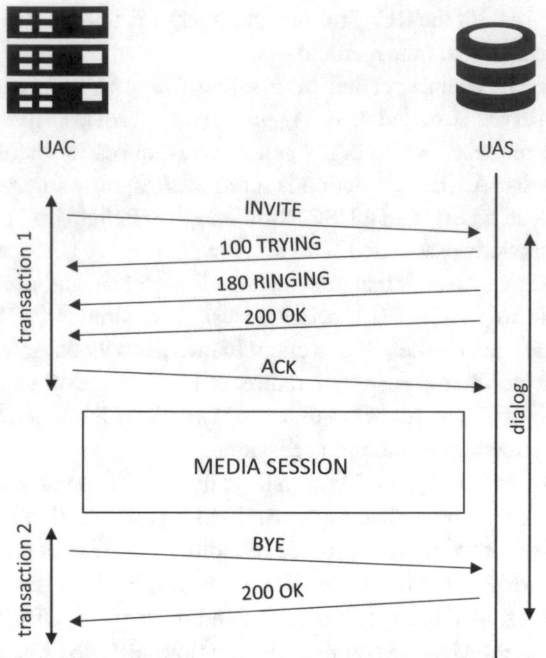

Fig. 5.36 SIP call

SIP, as HTTP, is a protocol that was not designed to be used in the context of LLNs. For example, SIP messages are unnecessarily too big to fit in a single frame of link layer technologies like IEEE 802.15.4. This situation leads to fragmentation, and, thus, it increases the chances of network loss. Although there have been several proposals intended to adapt SIP to IoT networks, none of them have been really standardized. In this regard, it is important to mention *constrained SIP* (CoSIP). CoSIP attempts to compress SIP header information by relying on a message format identical to that of CoAP shown in Fig. 5.24. Under CoSIP, message types do not encode REST methods like GET and POST but encode session establishment, maintenance, and termination messages like INVITE and BYE. CoSIP follows the same convention of CoAP for the inclusion of headers in messages by means of differentially encoded options. Also, in order to support the efficient transmission of messages in LLNs, CoSIP relies on UDP and DTLS transport as opposed to the

TCP and TLS transport of traditional SIP. Fortunately, there is standard mechanism to compress SIP headers that despite the fact that it was not developed with IoT in mind, it can be used in LLNs. Specifically, IETF RFC 5049 *Applying Signaling Compression to the Session Initiation Protocol* merges traditional SIP with *signal compression* (SigComp). SigComp is a standard mechanism for the compression of signaling information associated with call establishment, termination, and network registration.

5.2.5.2 RTP and RTCP

RTP provides the end-to-end real-time delivery of audio, speech, and video sessions [19]. An RTP session is set up through SIP negotiation by means of SDP information exchange. RTP is transported over UDP to guarantee that latency is minimized, as in the context of media transmissions any media packet that arrives too late can be considered lost. In fact, most RTP applications rely on playout buffers that queue packets during a typically short interval of time before they are sequentially played. The idea behind this mechanism is to reduce the choppiness that results from playing packets that arrive at an irregular pace. Additionally, since RTP relies on UDP for transport, it is ideal for multicast transmissions. Specifically, in multicast conferencing, a single transmitter can reach multiple receivers with a minimal amount of traffic. RTP supports multiple simultaneous media types, so a single SIP dialog can set up multiple synchronized audio and video streams. Other applications of RTP include transcoding where a device re-encodes a media streaming using a codec that is different to the one originally used. Note that transcoding is performed by means of signal processing techniques that are computationally complex.

Media streams are packetized and encoded with an RTP header shown in Fig. 5.37. The header starts with a 2-bit version field that is always set to 2, and it continues with a padding bit to signal whether additional padding bytes, needed by certain codecs, follow the payload. The last byte of the padding indicates how long the padding is. The header then includes an extension bit to signal that an extension header follows the RTP header; a 4-bit *CSRC count* field that specifies the number of *contributing source* (CSRC) fields, described below, included in the header; a marker that is used for multiple purposes; a 7-bit payload type that identifies that codec that encodes the media packet payload; a 16-bit sequence number used by receivers for reordering and packet loss detection; and a 32-bit timestamp that specifies the sampling instant of the first byte in the payload. The timestamp is critical for synchronization and jitter estimations. The headers also include a 32-bit *synchronization source* (SSRC) random number that identifies the stream and a list of 32-bit CSRCs for the payload contained in the packet. This is typically used in the context of mixers that are applications that combine multiple media streams into one.

For the purpose of synchronization, a receiver relies on SSRCs to identify streams and the sequence number to determine what media packets to play next. Specifically, since the sequence number is incremented every time a packet is transmitted, the receiver can estimate packet loss and attempt to restore the packet sequence. Along with the sequence number, the timestamp field provides another key piece of information. This field tells the receiver when to play media packets and provides information about synchronization. There are certain applications like video, where several sequential RTP packets carry the same timestamp as they represent a single video frame. In most audio and speech scenario, however, timestamps change with the sequence numbers.

RTP is a generic protocol that provides functionality that has different meanings depending on the codec under consideration. For example, the extension bit is used to indicate that an extension header follows the RTP header. The extension header, in turn, may carry additional fields that are relevant to the codec. Similarly, the marker bit is typically used, depending on the codec, to specify when a stream starts. This is important when a codec supports *discontinuous transmissions* (DTX). Under DTX, *voice activity detection* (VAD) may trigger a period of silence under which the codec does not produce any frames. Once speech is detected again, the media stream restarts, and this is signaled by setting the marker bit in the RTP header.

The RTP suite provides an additional protocol called *Real Time Control Protocol* (RTCP) that is used to provide quality control over media streams. Essentially, for each RTP stream, there is an optional associated RTCP stream that enables receivers to provide feedback about the quality of the incoming media packets. The RTCP stream, as the RTP stream, is transported over UDP typically using contiguous ports. As such RTCP is particularly important in the context of conferencing multimedia applications. RTCP is associated with an RTP stream through its *canonical name* (CNAME) and not through its SSRC, since the SSRC may change during a session. RTCP can include extra information for reference like an e-mail address or a phone number. Figure 5.38 shows the format of a generic RTCP header. It starts with a 2-bit version field that is always set to 2 and continues with a padding bit that indicates that additional bytes follow the format specific information payload. The header also includes a 5-bit item count that indicates the list of items carried in the payload, an 8-bit packet type that specifies the RTCP message type, and a 16-bit length that specifies the size of the payload that follows the RTCP header.

There are many types of RTCP messages that are used to indicate different types of information: (1) a *sender report* (SR) for transmission and reception statistics from active senders, (2) a *receiver report* (RR) for reception statistics from passive senders, (3) a *source description* (SDES)

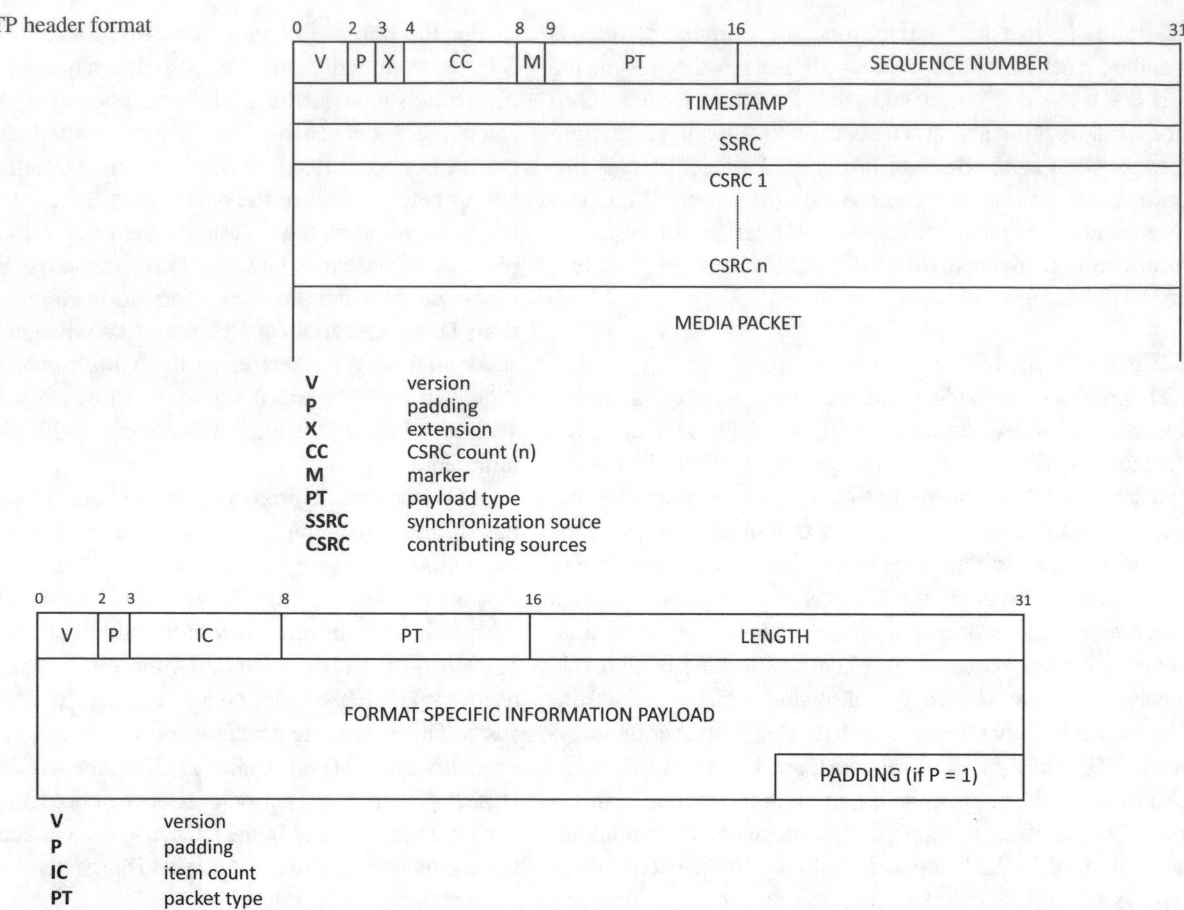

Fig. 5.37 RTP header format

Fig. 5.38 RTCP header format

for a description of a sender including the CNAME, (4) a *BYE* message to indicate the end of participation in a session, and (5) an *APP* message for application-specific functions.

One good thing about both, RTP and RTCP, is that they exhibit a highly optimized structure with small headers that are ideal for LLNs. Moreover, since RTP packets typically carry codec compressed media frames, the RTP payload is also optimal for IoT. In all cases, chances of fragmentation when transmitted over link layer technologies like IEEE 802.15.4 are greatly reduced. When compared to SIP, HTTP, and other mainstream protocols, and because of its IoT readiness, there have not been that many attempts to further optimize the RTP protocol suite. Moreover, there have been some proposals, but none of them have led to standards. One of these proposals introduces IoT-RTP and IoT-RTCP that, respectively, optimize RTP and RTCP for support in LLN networks. IoT-RTP adds a few new fields to the standard RTP header to divide large media sessions into simpler sessions that can adapt better to network changes while keeping track of energy consumption. On the other hand, IoT-RTCP is also upgraded to carry additional information about the transmitter including energy levels. The price to pay for the functionality provided by IoT-RTP and IoT-RTCP is slightly larger packets due to the overhead.

5.2.6 OPC UA

The *Open Platform Communications United Architecture* (OPC UA) provides a protocol stack that supports the SOA approach [20]. This protocol stack, shown in Fig. 5.39, complies with the request/response scheme of session protocols like CoAP and HTTP, but as a stack, it specifies additional functionality. Moreover, OPC UA not only addresses communication, security, and session management needs, but it also deals with data modeling that enables the interaction of heterogeneous devices in the context of IIoT.

With constrained networks in mind, OPC UA introduces two different mechanisms for data encoding: (1) XML and (2) binary. The generic UA XML encoding provides messages that are human readable, but it is expensive from a networking and processing perspective. UA binary encoding, on the other hand, is more efficient and designed to work in constrained environments. Unfortunately, both mechanisms are transmitted over traditional TCP and subjected to all the

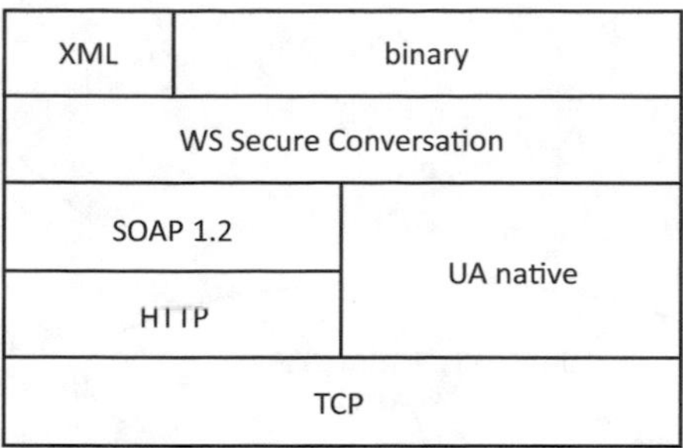

Fig. 5.39 OPC UA stack

limitations associated with TCP in LLNs. On the other hand, the selection of TCP as transport is because TCP traffic is a lot less likely to be rejected by backbone firewalls than UDP traffic.

UA XML payloads are transmitted over HTTP and optionally transported over TLS. Rather than transmitted directly over HTTP, XML messages are first encoded by means of the standard *Simple Object Access Protocol* (SOAP). SOAP introduces both a structure for the encoding of XML data and a mechanism for the exchange of messages. OPC UA introduces an additional level of security by means of the Web Services Security Conversation WS Security Conversation. WS Security Conversation, as opposed to TLS that supports transport security, provides session level security that optimizes the exchange of long messages. UA binary data can also be transmitted over HTTP but, more importantly, over a UA native protocol that provides simple mapping directly over TCP. Note that there is an increasing tendency to support OPC UA data and message models over other IoT session mechanisms including not only request/response ones like CoAP but also publish/subscribe protocols like MQTT and AMQP.

5.3 Publish/Subscribe

The publish/subscribe model, also known as subscribe/publish model, consists of an event-based architecture that provides built-in observation in compliance with IoT requirements. As opposed to request/response, the publish/subscribe paradigm relies on one or many entities known as brokers that provide services by collecting, storing, and forwarding events to and from endpoints. Each broker has a number of endpoints associated with it such that a single endpoint only communicates with a single broker. In addition, endpoints can behave like a publisher and/or like a subscriber. In the context of IoT, an analytics application is typically a subscriber, while a device plays the role of publisher. The publisher advertises the type of event, known as topic, that it can generate, and the brokers broadcast this information to all other endpoints. Those endpoints interested in each topic respond by issuing a subscription. The broker binds the topic to the endpoint in its internal database, and the endpoint then becomes a subscriber. Whenever the publisher generates a sensor readout or event, the broker forwards it, as a notification, to all the corresponding subscribers.

Publish/subscribe systems have been traditionally used in several application domains that involve the distribution of large-scale events, including (1) financial information systems, (2) streaming of live feeds of real-time traffic, (3) support for cooperative working, (4) support for ubiquitous computing, and (5) monitoring applications.

Publish/subscribe systems have two main characteristics: The first is (1) heterogeneity where brokers can handle subscriptions and notifications from entities that are not designed for interoperability. Essentially, multiple topics and events are present in a single publish/subscribe scenario where the broker just forwards the messages without processing their content. Only requirement is a common interface for brokers to understand how to process subscriptions and notifications. The other characteristic is (2) asynchronicity where both subscribers and publishers do not need to be synchronized as the broker provides the infrastructure necessary to keep track of events by means of buffering and storing. Specifically, if an endpoint publishes an event, under the right conditions, it is guaranteed that a subscriber will receive that event at some point.

Figure 5.40 summarizes the small operations involved in the publish/subscribe model: (1) *advertise(t)* is sent by a publisher to declare a topic *t* that represents future events to be published, (2) *unadvertise(t)* is sent by a publisher to revoke the previous advertisement of a topic *t*, (3) *subscribe(t)* is sent by an endpoint to subscribe to a specific topic *t*, (4) *unsubscribe(t)* is sent by a subscriber to revoke an active subscription on topic *t*, (5) *publish(e)* is sent by a publisher to forward an event *e* to the broker, and (6) *notify(e)* is sent by the broker, as a result of an incoming published event *t*, to broadcast it to all the subscribers associated with the topic that the event belongs to.

The broker is the main component responsible for dispatching events from publishers to interested subscribers in the publish/subscribe system. The publish/subscribe model can be either centralized or distributed. The centralized model relies on a single broker with P2P connections from the publishers to the broker and from the broker to the subscribers. The broker has one or more internal queues where incoming events are stored before they can be distributed to subscribers. If the event arrival rate is larger than the queue processing rate, the queues can become full, and newly arrived events get dropped causing application packet loss. Therefore, in this centralized scenario, the broker can end up being a bottleneck due to congestion. On the other hand, the decentralized model attempts to prevent this and other problems by replacing the broker by a network of

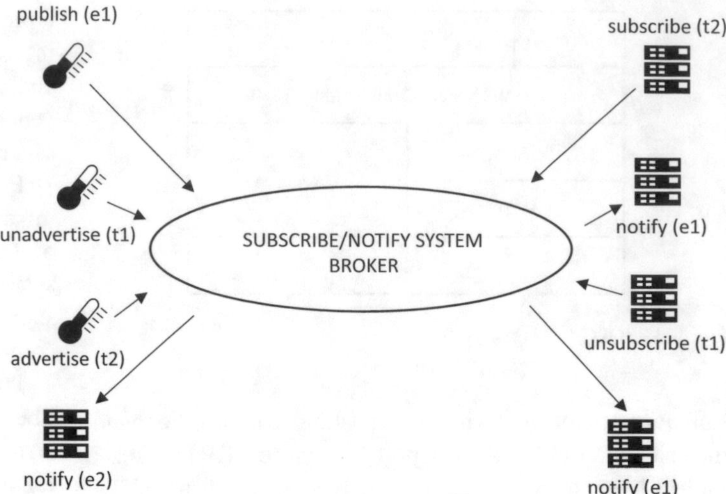

Fig. 5.40 Publish/subscribe model

brokers that cooperate to provide redundancy and enhance functionality and service levels.

5.3.1 MQTT

MQTT is a lightweight broker-based publish/subscribe application protocol that provides session layer functionality. MQTT was developed and designed by IBM in 1999 for low transmission rate constrained devices [2]. Because of its simplicity and low overhead, MQTT is one of the preferred mechanisms for communications in IoT networks. MQTT incorporates and improves some of the publish/subscribe functionality described in Sect. 5.3, while some other features like the advertising of topics are assumed to be accomplished by means of other out-of-band mechanisms. In general, MQTT targets scenarios that are characterized by LLNs where low transmission rates translate into high latency and devices with low computational complexity and limited memory resources.

MQTT, like any other publish/subscribe protocol, enables the wide distribution of device events to multiple subscribers and defines a generic session management mechanism that is agnostic of the payload content. MQTT was designed to rely on TCP transport (on port 1883), including TLS support for security, with three QoS levels that provide additional reliability. These three levels are (1) QoS 0, at most once, that provides a fire-and-forget delivery where the publisher sends events relying on the best effort of the TCP transport; (2) QoS 1, at least once, where any event sent by a publisher is guaranteed to arrive at least once to the broker; and (3) QoS 2, exactly once, that enables the publisher to transmit exactly one event to the broker. Note that although TCP provides, by design, reliable transmission of data, in the context of IoT, network packet loss and overall infrastructure stability cause connections to go down. To prevent application traffic from being lost when TCP fails, MQTT introduces an additional level of reliability by means of the QoS levels. Unfortunately, TCP transport, as indicated in Sect. 4.3.7, when subjected to network packet loss results in retransmissions that increase latency and prevent applications from successfully transmitting real time events. In 2013 MQTT was extended as *MQTT for Sensor Networks* (MQTT-SN) [11] to deal in IoT scenarios by introducing a new QoS level 3 (also known as -1) that provides simple event delivery over UDP. MQTT-SN requires an MQTT-SN gateway that translates the events into traditional MQTT messages for interaction with a broker. The MQTT-SN gateway, as indicated in Fig. 5.41, can be configured as a (1) transparent gateway where each MQTT-SN event is translated into individual MQTT events and as a (2) aggregating gateway where multiple MQTT-SN events are aggregated into a single MQTT event. In 2018 the newest version of MQTT, MQTT v5, was standardized [1]. It includes several new features and improvements that include the support of request/response scenarios in compliance with the mechanisms discussed in Sect. 5.2.

5.3.1.1 Message Format

Figure 5.42 shows the MQTT message format; it consists of a 24-bit fixed size header followed by an optional variable size header and then by the payload. The messages, by being very small, are ideal for processing and transmission of constrained IoT devices. Figure 5.43 shows the format of the fixed header; it includes a 4-bit message type encoded according to Table 5.5, a 1-bit duplicate flag that is set on retransmissions for *at least once* and *exactly once* QoS levels, a 2-bit *QoS* field that indicates the delivery level as indicated in Table 5.6, and a 1-bit retain flag used by a publisher to instruct its broker to hold the published message so it becomes available to future subscribers. Specifically,

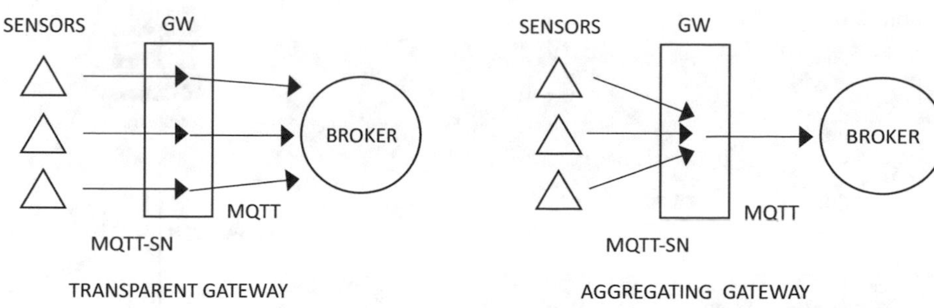

Fig. 5.41 MQTT-SN

when an endpoint subscribes to a topic for which the broker has retained messages, those messages are delivered to the subscriber, with the retrain flag set. The fixed header ends with a 16-bit remaining length field that indicates how big the payload is. The optional variable size header sits in between the fixed size header and the payload, and its presence depends when certain message types are in use. As indicated in Table 5.5, there are many different message types that enable different functionality. Whenever a device connects to a broker, it must first create a connection by transmitting a *connect* (CONNECT) request. The broker replies by transmitting a *connect acknowledgment* (CONNACK) message. Note that this MQTT connection is initiated by the endpoint once the TCP connection to the broker is established. Similarly, the MQTT connection is torn down by means of the *disconnect* (DISCONNECT) message right before the TCP connection to the broker is terminated. If no other messages are to be sent, endpoints periodically transmit keep alive *ping request* (PINGREQ) messages that refresh the MQTT connection to the broker. The broker replies by transmitting a *ping response* (PINGRESP) message. If the endpoint does not receive a PINGRESP or if the broker does not receive a PINGREQ when there are no other transmissions, the connection is deemed terminated. The *subscribe* (SUBSCRIBE) and *unsubscribe* (UNSUBSCRIBE) messages are used by endpoints to, respectively, subscribe and unsubscribe to topics. The broker acknowledges these requests by, respectively, replying with *subscribe acknowledgment* (SUBACK) and *unsubscribe acknowledgment* (UNSUBACK) responses. The *publish* (PUBLISH) message is used by endpoints to transmit events for QoS levels 0 through 2. The PUBLISH message is acknowledged by transmitting a *publish acknowledgment* (PUBACK) for QoS level 1 or by transmitting a *publish received* (PUBREC) for QoS level 2. QoS level 2 includes an additional transaction that follows the PUBREC transmission. In this case, the endpoint transmits a *publish release* (PUBREL) message that is replied by the broker by means of a *publish complete* (PUBCOMP) message.

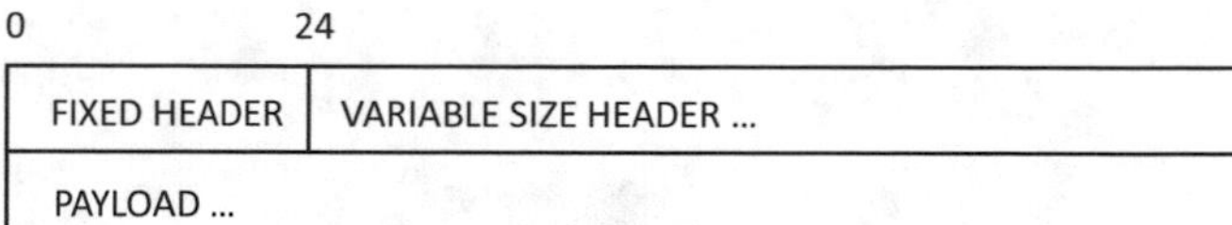

Fig. 5.42 MQTT message format

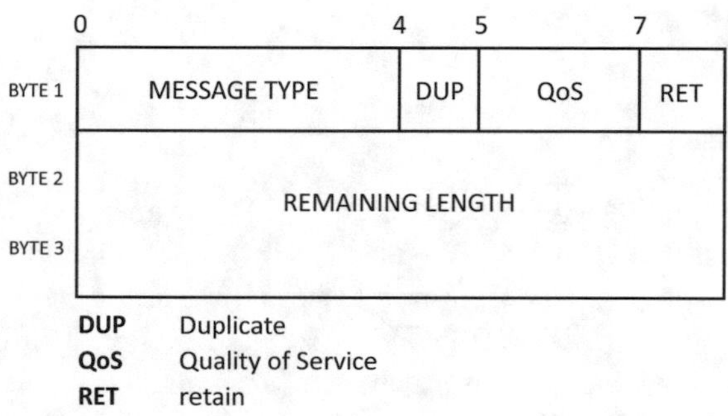

Fig. 5.43 MQTT fixed header

Table 5.5 MQTT message type

Reserved	0	Reserved for future use
CONNECT	1	Client request to connect to broker
CONNACK	2	Connect acknowledgment
PUBLISH	3	Publish message
PUBACK	4	Publish message acknowledgment
PUBREC	5	Publish received (QoS = 2)
PUBREL	6	Publish release (QoS = 2)
PUBCOMP	7	Publish complete (QoS = 2)
SUBSCRIBE	8	Client subscribe request
SUBACK	9	Subscribe acknowledgment
UNSUBSCRIBE	10	Client unsubscribe request
UNSUBACK	11	Unsubscribe acknowledgment
PINGREQ	12	Ping request
PINGRESP	13	Ping response
DISCONNECT	14	Client disconnection request
Reserved	15	Reserved for future use

Table 5.6 QoS level

QoS value	Bit 2	Bit 1	Description
1	0	0	At most once (i.e. fire and forget)
2	0	1	At least once (i.e. acknowledgment delivery)
3	1	0	Exactly once (i.e. assured delivery)
Reserved	1	1	Reserved for future use

Example 5.3 Consider a sensor that is periodically transmitting readouts:

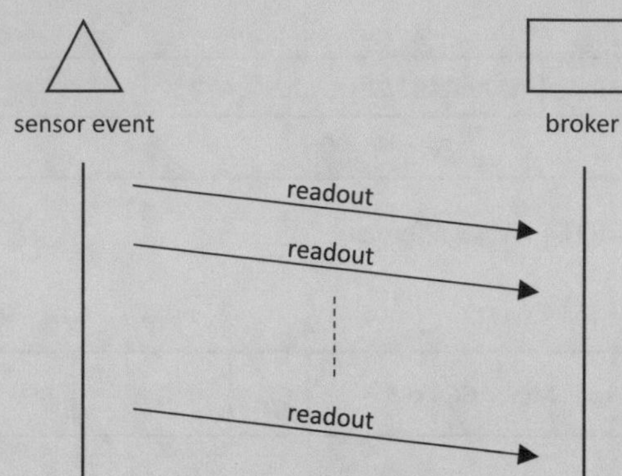

Assuming a 250 Kbps IEEE 802.15.4 transmission rate and an average packet size of 200 bytes and ignoring the propagation delay, how fast can the sensor generate readouts over MQTT QoS 0, 1, and 2 sessions? Also assume that the sensor transmits one readout at a time, making sure that a transaction is terminated before a new one can be initiated.

Solution: For QoS 0, the sensor transmits PUBLISH messages one after another. The sensor readout rate is given by $R_{\text{QoS} \quad 0} = \frac{250000 \text{ bps}}{8 \times 200 \text{ bits}} = 156.25$ readouts per second. Similarly, for QoS 1, the sensor transmit PUBLISH messages that are acknowledged by the broker as it transmits PUBACK responses. The sensor readout rate is given by $R_{\text{QoS} \quad 1} = \frac{250000 \text{ bps}}{2 \times 8 \times 200 \text{ bits}} = 78.125$ readouts per second. Finally, for QoS 2, the sensor transmits PUBLISH messages that are acknowledged by the broker that sends back PUBREC responses. The sensor then replies by transmitting PUBREL messages that are acknowledged by the broker by transmitting PUBCOMP responses. The sensor readout rate is given by $R_{\text{QoS} \quad 2} = \frac{250000 \text{ bps}}{4 \times 8 \times 200 \text{ bits}} = 39.062$ readouts per second. In other words, if TCP can be guaranteed to be reliable, then the best bet is to rely on MQTT QoS 0 sessions to maximize the readout transmission rate.

5.3.1.2 Message Flows

Figure 5.44 shows a PUBLISH message being delivered from a publisher to the broker when QoS is configured as *at most once*. Without support of retransmissions, this level relies on the underlying TCP transport for guaranteed delivery. Conceptually, if TCP fails, the messages will not arrive at the destination and will not be retransmitted either. Once at the broker, the received message is published to all the subscribers.

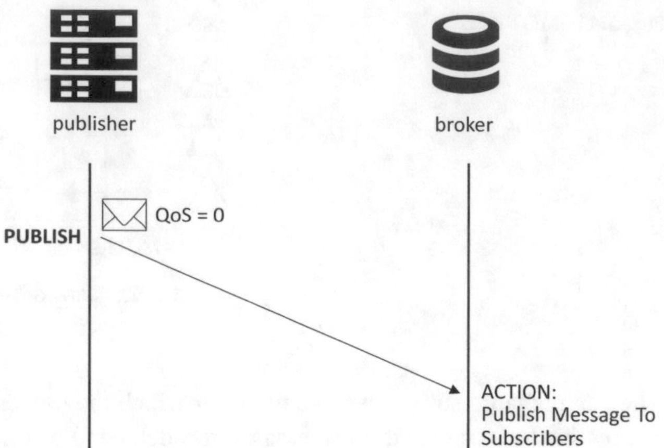

Fig. 5.44 At most once delivery

Figure 5.45 shows a *at least once* delivery scenario. The publisher delivers a PUBLISH message, identified by a message identifier, to the broker. The first transmission has the duplicate flag unset. When received at the broker, an PUBACK message is transmitted back to the publisher. If either of these messages fails to be received at the corresponding far end, a retransmission, with the duplicate flag set, is initiated by the publisher. Once the message is received by the broker, it is forwarded to all the corresponding subscribers. SUBSCRIBE and UNSUBSCRIBE messages are also acknowledged and, therefore, are delivered as *at least once* messages. Note that under this QoS level and depending on what messages are affected by packet loss, multiple delivery of a given message is possible. The message identifier field is used to track the delivery of the different messages involved in this scheme.

This issue is solved by the *exactly once* delivery shown in Fig. 5.46. This scheme guarantees that duplicate messages are not delivered to the receiving application. As in the previous scenario, the publisher delivers a PUBLISH message, identified by a message identifier, to the broker. The first transmission has the duplicate flag unset. When received at the broker, a PUBREC message is transmitted back to the publisher. If either of these messages fails to be received at the corresponding far end, a retransmission, with the duplicate flag set, is initiated by the publisher. When the message is received by the broker, it can be forwarded to all the publishers. Upon reception of the PUBREC message, the publisher transmits a PUBREL message to tell the broker to forward the message to the subscribers (if it has not done it yet) and to delete it. The broker replies by transmitting

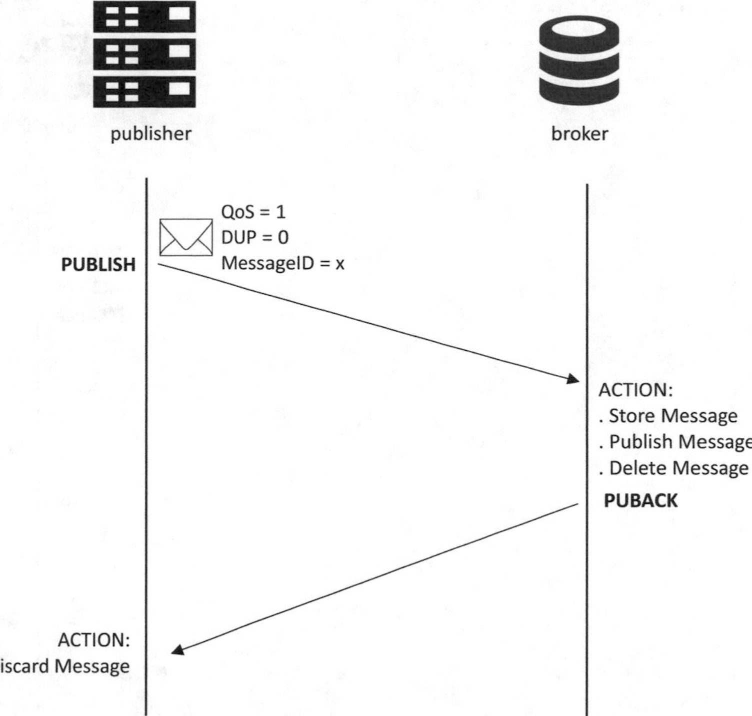

Fig. 5.45 At least once delivery

a PUBCOMP message that triggers the publisher to discard the message. The double transaction mechanism guarantees that messages are delivered only once. The price to pay is added latency and extra throughput. As before, the message identifier field is used to track the delivery of the different messages involved in this scheme.

▶ **MQTT Implementations** There are many commercial and open-source MQTT implementations with support ranging from common constrained RTOSs to general-purpose Linux distributions. The list below shows some of the most popular MQTT implementations.

mosquitto	BSD-licensed C implementation of client and broker.
MQTT-C	MIT-licensed C implementation of client.
Paho MQTT	BSD-licensed C/C++ implementation of client.
wolfMQTT	GNU-licensed C implementation of client.
eMQTT5	MIT-licensed C++ implementation of client.

5.3.1.3 MQTT v5 Request/Response Support

MQTT version 5 provides support for request/response transactions that comply with the REST architectures and protocols introduced in Sect. 5.2.

Rather than natively enabling direct end-to-end transmission of requests and responses, MQTT relies on the broker for the coordination needed for the delivery of messages.

Figure 5.47 shows the transactions associated with this mechanism. Essentially, clients and servers behave as subscribers and publishers in order to support bidirectional communication. The application acting as client subscribes to its own private client topic. In general, all clients must subscribe to their own private topics. The sensor acting as a server, in turn, subscribes to its well-known public server topic. The client transmits the request by publishing to the well-known server topic. At the broker, the request is forwarded to the server. The message includes a response topic field that specifies the client topic that is used by the server to publish the response. Specifically, the server publishes the response to the client topic specified in the request. Again, the response, as the request, is forwarded by the broker to the application. From a performance perspective, MQTT emulates request/response support and does not natively support it. When compared to HTTP, CoAP, and other REST protocols, the support of a request/response scheme under MQTT results in additional latency due to the messages being forwarded by the broker. Moreover, the broker is still a single point of failure that can prevent end-to-end connectivity even when both client and server are fully functional.

5.3.2 AMQP

Advanced Message Queuing Protocol (AMQP) is a session layer protocol, similar to MQTT, that was initially designed for enterprise applications like financial businesses but due to its simplicity and small footprint has become an integral part of many IoT solutions [13]. As MQTT, AMQP relies on TCP transport, with security provided by

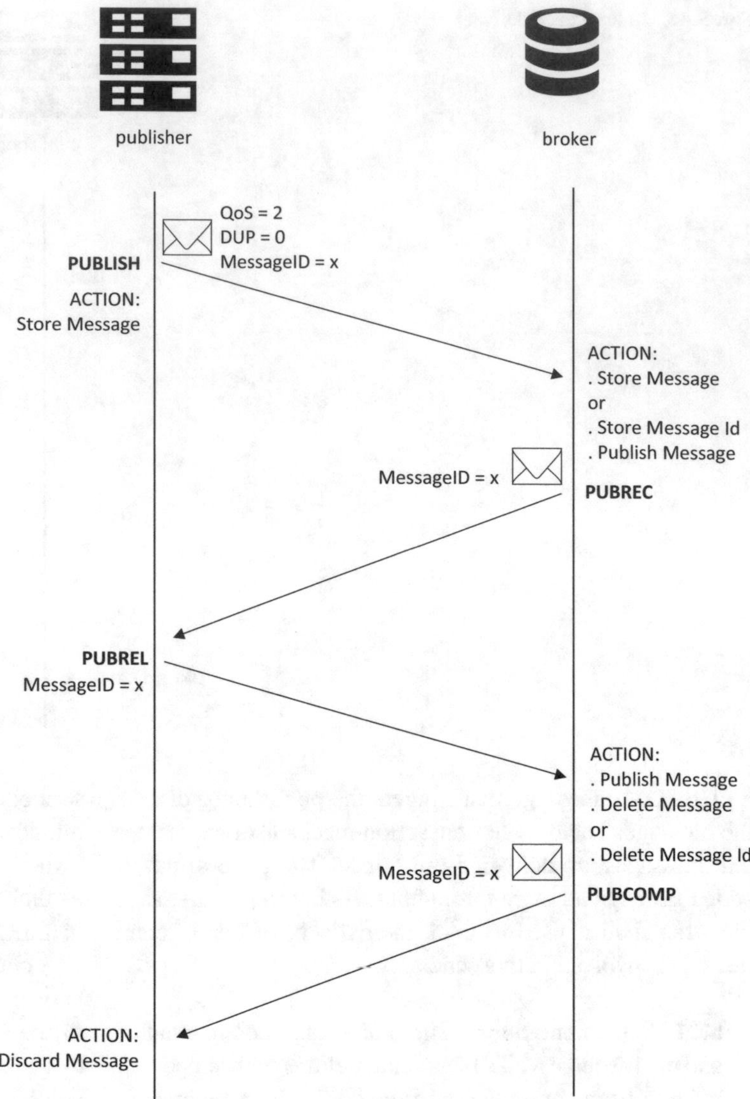

Fig. 5.46 Exactly once delivery

means of TLS, that enables the support of three distinct QoS levels. Several dataflows between peers are transmitted over a single TCP connection. AMQP was first standardized as version 0.8 (formalized as 0–8) in 2006. The latest AMQP version 1.0, released in 2011, has been standardized by *International Organization for Standardization* (ISO) in 2014.

► **CoAP, HTTP, MQTT, and IoT Cloud Services** There are many commercial and non-commercial IoT services that enable devices to push sensor readouts to the cloud for processing by means of analytics or for offline visualization. Some of these technologies include *Amazon Web Services* (AWS), Google Cloud, Kaa, ThingWorx, and ThingSpeak. These solutions typically support HTTP or MQTT protocols that do not natively enable end-to-end CoAP sessions. Specifically, end-to-end CoAP sessions rely on the following conversion:

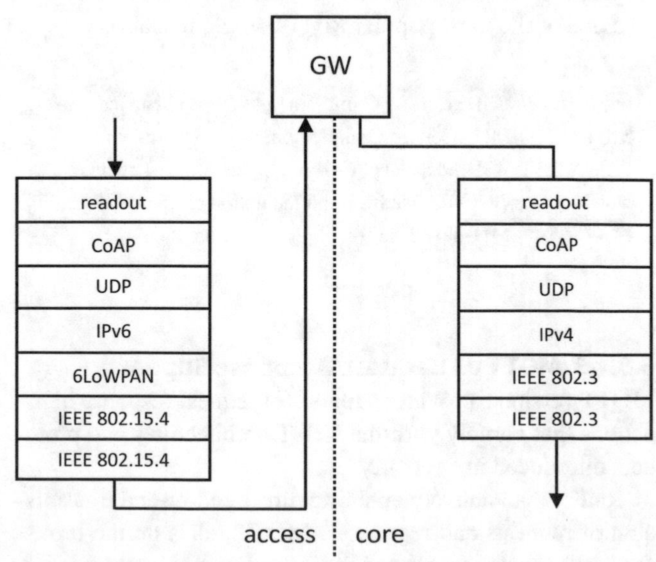

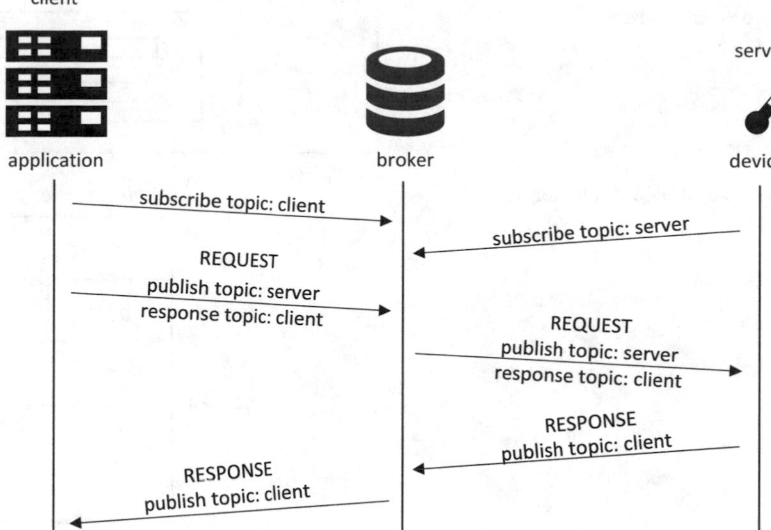

Fig. 5.47 MQTT client/request support

The gateway converts IPv6 datagrams over IEEE 802.15.4 into IPv4 datagrams over Ethernet. All upper layers remain unchanged. The problem with this approach is that the core network includes firewalls that due to security and other performance considerations filter out UDP traffic. Because of this, most core networks only support TCP-based protocols like MQTT and HTTP, and it is the task of the gateway to fully convert the stacks as shown below:

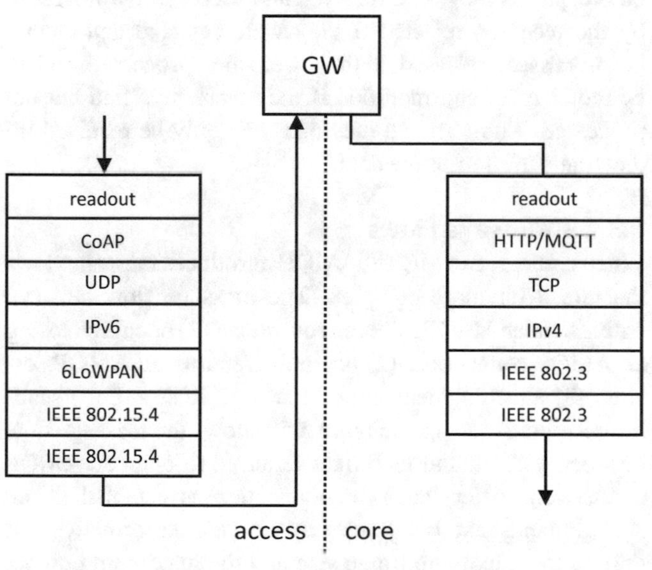

Because IoT access networks do not include firewalls, device connectivity is guaranteed in this scenario.

As CoAP (shown in Fig. 5.18), AMQP also relies on a two-sublayer structure, shown in this case in Fig. 5.48. The transport/framing sublayer defines the connection and

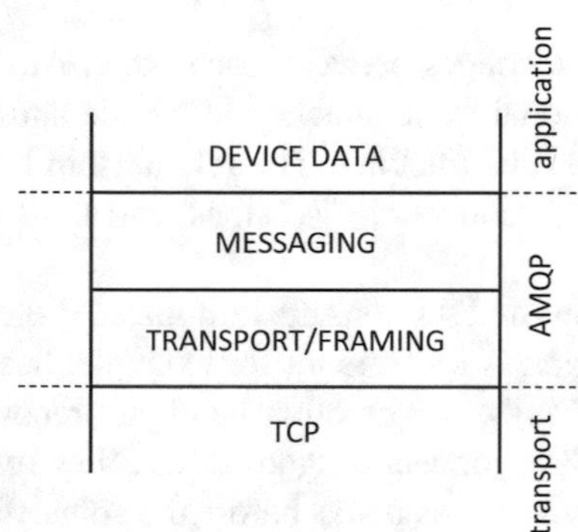

Fig. 5.48 AMQP two-layer structure

security behavior between peers right on top of the TCP transport. It also defines the framing mechanism that rules how fields are formatted and encoded. The messaging sublayer provides the messaging capabilities that relying on the transport/framing sublayer enables peers to exchange messages.

5.3.2.1 Transport/Framing

AQMP transport defines two entities that are used to build peers: (1) nodes that oversee storing and delivering messages and (2) containers that define clients and brokers that contain multiple nodes. In general, nodes can be producers, consumers, or queues. Clients contain producers and/or consumers, while brokers contain queues. An AQMP connection is negotiated on top of TCP connection established between client and broker, with security optionally provided by means of TLS. The AQMP connection supports full-duplex communication by means of an ordered sequence of frames that can be used to multiplex channels. An AQMP session is created when establishing two unidirectional channels that enable communication between nodes. Nodes, in turns, rely on links

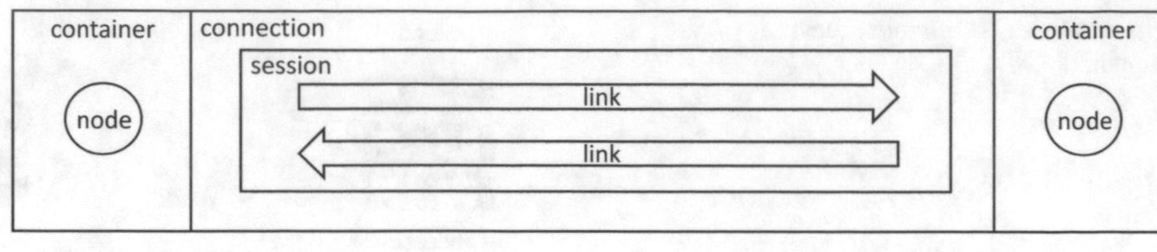

Fig. 5.49 AMQP connections, sessions, and links

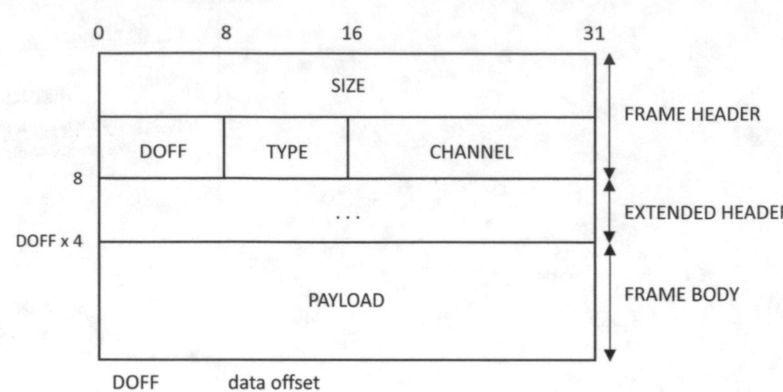

Fig. 5.50 AMQP frame format

to exchange messages between each other. An AMQP link is a unidirectional route attached to a node and responsible for message status tracking. The relationship between containers, nodes, connections, sessions, and links is shown in Fig. 5.49.

Connections and sessions are terminated if the underlying TCP connection is lost due to, for example, heavy network packet loss. Links, on the other hand, are recoverable, and even if the TCP connection goes down, they resume at the same delivery status as it was before the connection loss.

Figure 5.50 shows an AMQP frame divided in three sections: (1) an 8-byte frame header that is included in every single frame, (2) an optional variable length extended header that depends on the frame type, and (3) the variable length frame body. The frame header includes several fields: a 32-bit frame size that accounts for the frame header, extended header and frame body, an 8-bit data offset that points to the start of the payload, an 8-bit frame type that indicates the format and purpose of the frame (frame type 0 indicates an AMQP frame), and a 16-bit channel field that identifies the channel number. The frame body is defined as a performative followed by payload filled by the application with data to send. The performative enables opening/closing an AMQP connection, starting/ending an AMQP session, attaching/detaching a link, transferring content, and handling flow control.

5.3.2.2 Messaging

Messaging capabilities are provided by the messaging sublayer built on top of the framing sublayer. The sublayer defines a well-known format that consists of two main components: bare message and annotated message (Fig. 5.51). The bare message contains the content that is transmitted by the sender to the receiver and remains unchanged by intermediate entities. The bare message includes a body and two properties: system properties that are defined by AMQP and application properties that are defined by the application. Table 5.7 shows some of the possible system properties defined by AMQP. The annotated message is a superset of the bare message that additionally contains a header and annotations that are used by intermediate entities.

The messaging sublayer introduces two message states: they can be terminal or non-terminal. Terminal messages can be accepted if they were received and successfully processed by the receiver, rejected if they were rejected and cannot be processed, released if they were not processed and to be redelivered, and modified if they were modified but not processed. Non-terminal messages can only be received indicating partial message data.

5.3.2.3 Message Flows

When compared to MQTT, AMQP introduces message flows that are a bit more complex. The message flow between entities under AMQP relies on four steps: (1) opening/closing an AMQP connection, (2) beginning/ending an AMQP session, (3) attaching/detaching an AMQP link, and (4) sending/receiving messages. Figure 5.52 shows the message flow between a client and its broker required to open an AMQP connection. After the TCP connection is established, an AMQP handshake is followed by an *open* performative that defines the maximum frame size and the maximum number of channels. The maximum frame size defines the maximum size of each frame that is exchanged. A session is then started by means of the *begin* performative that specifies the window size. Each endpoint has an incoming and an outgoing window with a size defined as frame count that is decremented on each transmission. The frame exchange stops when the sender does not have window space to transmit and the receiver

Fig. 5.51 AMQP messaging format

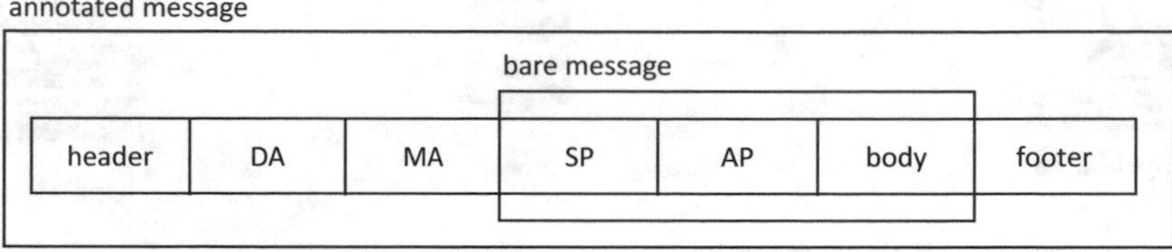

Table 5.7 CoAP option types

System property	Meaning
Message id	Message identifier
To	Message destination
Subject	Information about message content
Reply to	Address to send replies to
Correlation id	Identifier to link different messages
Content type	Message content type
Absolute expiry time	Expiration time of message

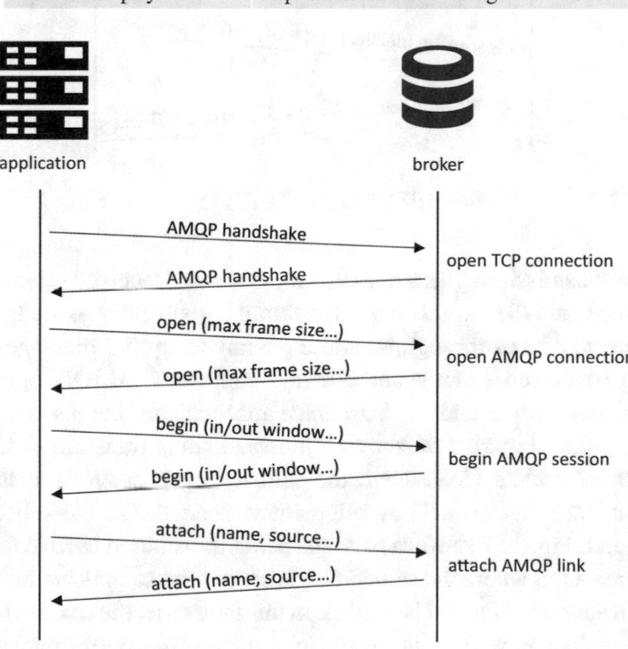

Fig. 5.52 AMQP open

does not have window space to receive. Once a session is established, a link is attached by transmitting an *attach* performative. Each link relies on a link credit that indicates the number of messages a receiver can receive.

Figure 5.53 shows the message flow where a client sends a message to its broker. After an AMQP link is attached, the receiver transmits a *flow* performative, and the client sends a message by means of a *transfer* performative. If the QoS level is configured as *at least once*, the broker replies by transmitting a *disposition* performative. A single *disposition* performative can serve as a reply to multiple

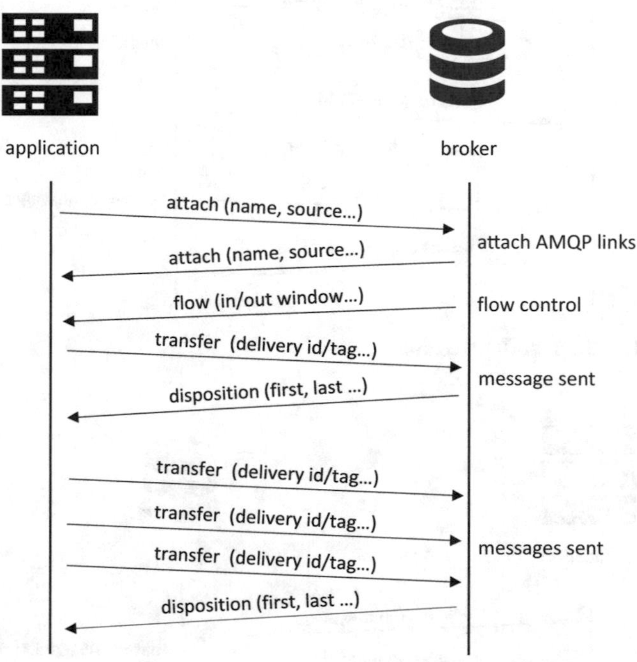

Fig. 5.53 AMQP send

transfer performatives associated with multiple transmitted messages.

Similarly, Fig. 5.54 shows the message flow corresponding to a client receiving messages from its broker. Roles are reversed in this case; the client transmits a *flow* performative to tell the broker that it is ready to receive data. The broker then sends a *transfer* performative to send data, which is responded, if the QoS level is *at least once*, by a *disposition* performative. Again, if multiple messages are transmitted, a single *disposition* performative can be used as response.

Figure 5.56 shows the message flow that enables the closing of the AMQP connection. First, the client issues a *detach* performative to detach the link which is followed by an *end* performative the terminate the session. Finally, as shown in Fig. 5.55, the client issues a *close* performative to finish the AMQP connection. The TCP connection is then torn down.

AMQP supports that same three QoS levels that MQTT supports: (1) *at most once* that as a *fire-and-forget* mechanism transmits a message only once, (2) *at least once* that through

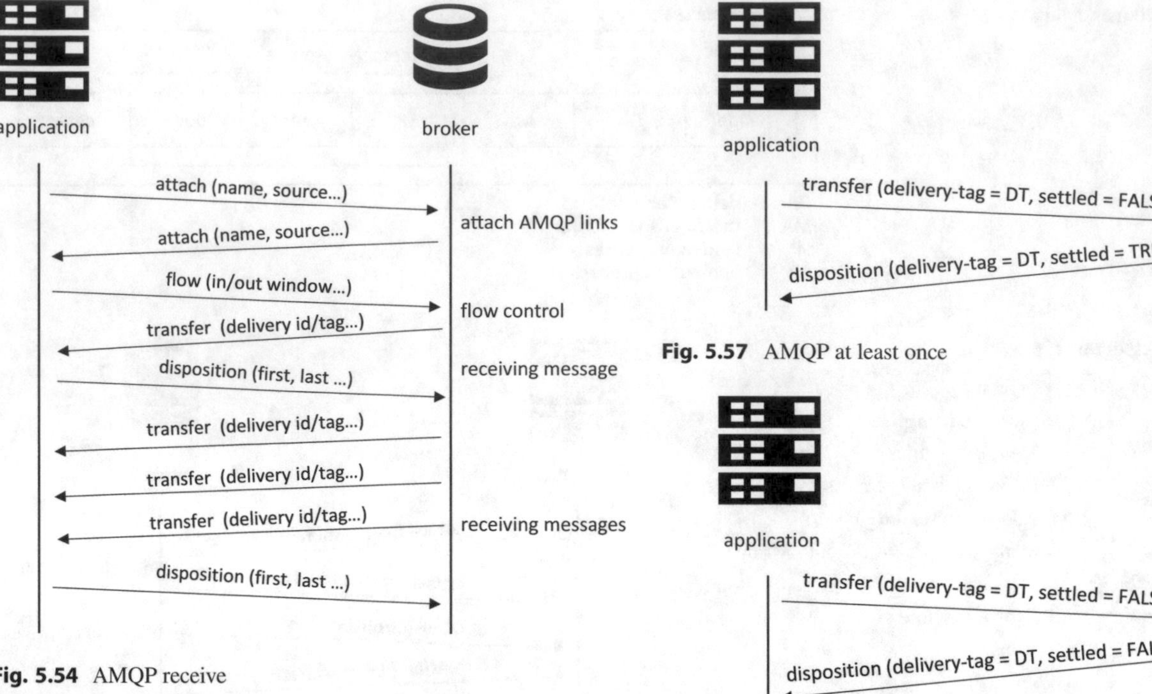

Fig. 5.54 AMQP receive

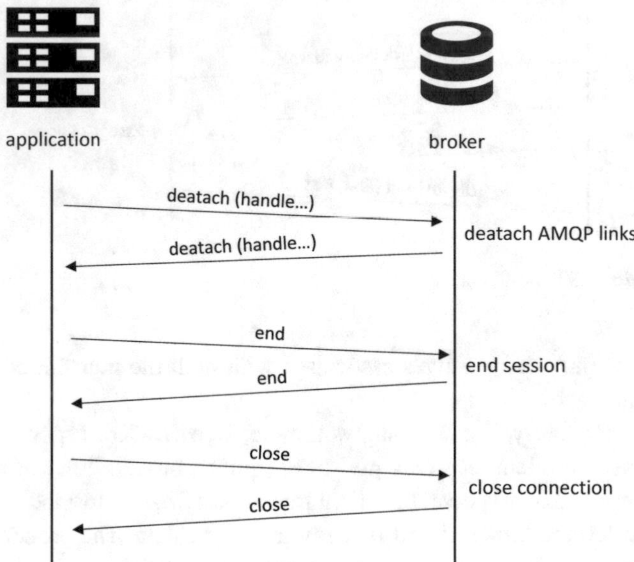

Fig. 5.55 AMQP close

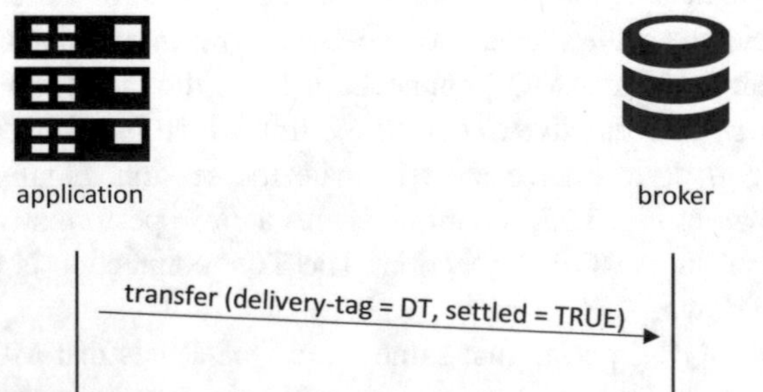

Fig. 5.56 AMQP at most once

Fig. 5.57 AMQP at least once

Fig. 5.58 AMQP exactly once

retransmissions guarantees that messages are received at least once, and (3) *exactly once* that through double acknowledgment reduces throughput while guaranteeing that messages are received exactly once. In the context of AMQP, only *at most once* and *at least once* mechanisms are used in practice. Figure 5.56 shows a message being transmitted as *at most once* QoS where the settled flag is set to true to indicate that no acknowledgment is needed. On the other hand, Fig. 5.57 shows a message being transmitted as *at least once* QoS where the settled flag is set to false to indicate that an acknowledgment is needed. In this later case, the acknowledgment is sent in the form of a *disposition* performative with the settled flag set to true. If an acknowledgment is not received, the sender retransmits the message. Similarly, Fig. 5.58 shows a message being transmitted as *exactly once* QoS where double acknowledgment is performed.

Summary

Interaction between devices and applications rely on two main architectures that are common under IoT: request/response and publish/subscribe. The chapter started by describing the characteristics of these mechanisms with initial focus on request/response architectures that have REST as key component. REST, in turn, is extended by

Table 5.8 Application layer protocols

Protocol	Client/req or pub/subs	Transport	Overhead	Security or QoS
HTTP	Client/req	TCP	High	Both
XMPP	Both	TCP	High	Security
CoAP	Client/req	UDP	Low	Both
SIP	Client/req	UDP/SIP	High	Security
RTP	Client/req	UDP	Low	Both
OPC UA	Client/req	TCP	High	Security
MQTT	Pub/subs	TCP	Low	Both
AMQP	Pub/subs	TCP	Medium	Both

supporting observation that improves both latency and QoS. The chapter explored HTTP detailing its main characteristics like message formats, cookies, and proxy support. The chapter then introduced XMPP and its IoT extensions and continued by presenting CoAP as a lightweight alternative to HTTP. Both SIP and RTP were introduced as mechanisms for the establishment and management of media sessions in the context of IoT. The chapter ended by detailing publish/subscribe protocols like MQTT and AMQP. Table 5.8 provides a comparison between the different application layer protocols.

Homework Problems and Questions

5.1 When comparing request/response and publish/subscribe architectures, what are their differences from energy, packet loss, and latency perspectives?

5.2 Consider the following publish/subscribe scheme:

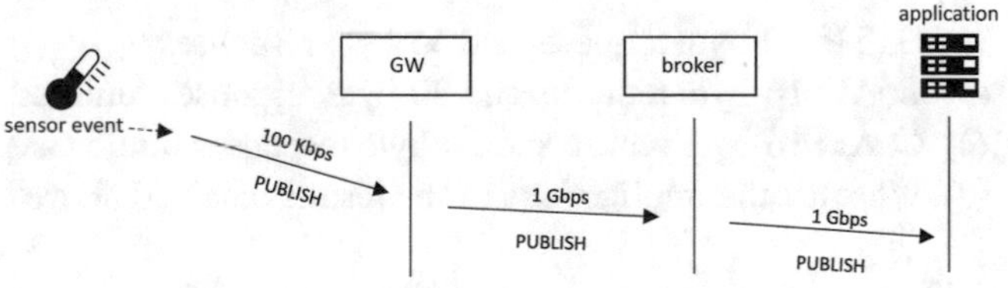

Assuming a lossless network where the only delay is the transmission delay, how long does it take for a 200-byte packet with the sensor event data to arrive to the application? What happens if the transmission rate between the gateway, the broker, and the application is also 100 Kbps?

5.3 Consider the following request/response scheme: Assuming a lossless network where the only delay is the transmission delay, how long does it take for 200-byte response packet with the sensor event data to arrive to the application? Also assume a 100-byte request packet and a half a second delay between the event generation and the actual application event polling. What happens if the trans-

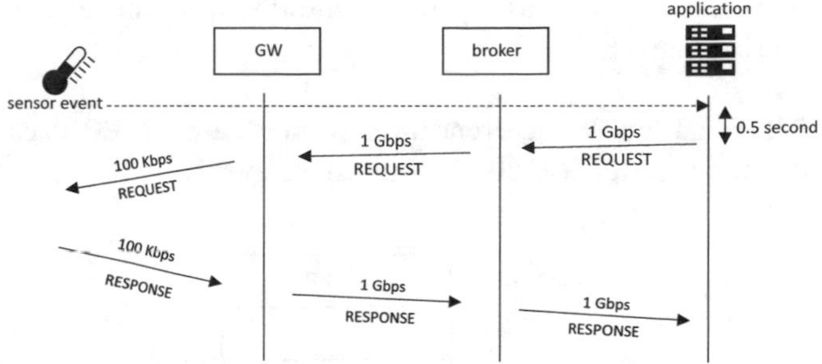

mission rate between the gateway and the application is also 100 Kbps? How do these results compare to those of the Problem 5.2?

5.4 Describe a scenario where HTTP is better suited than CoAP for sensor data transmission.

5.5 As indicated in this chapter, IoT cloud services do not typically support end-to-end CoAP sessions. Why is not possible to support end-to-end HTTP or MQTT sessions instead? What are some of the reasons for this?

5.6 Consider an HTTP application requesting three readouts from a sensor:

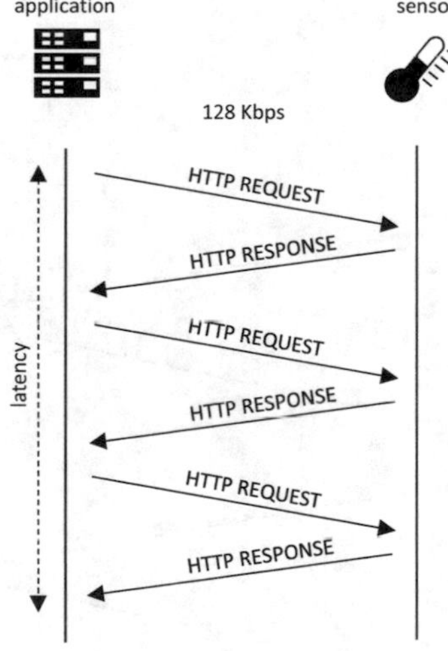

Assuming a lossless network where the only delay is the transmission delay, if the transmission rate is 128 Kbps (bidirectional), what is the overall transmission latency for both persistent and non-persistent connection scenarios? Also assume 150-byte request and response packet sizes as well as 30-byte connection establishment request and response packet sizes.

5.7 XMPP is many times considered a hybrid application layer protocol because it can be seen either as a request/response or as a publish/subscribe mechanism. Can you explain why?

5.8 What are the different layers of a secure XMPP stack that relies on a IEEE 802.15.4 physical layer?

application	XMPP
security	
transport	
network	
adaptation	
link	
physical	IEEE 802.15.4

5.9 The following flow shows a device transmitting three sensor readouts to an application. The assumption is that there has been a CoAP GET request from the application to the device to enable CoAP observation. Add the missing information: (1) type of message (examples: CON, NON, ACK, RST), (2) message code (when needed) (examples: 2.05 Content, 4.04 Not Found), (3) message id, and (4) token id to each message.

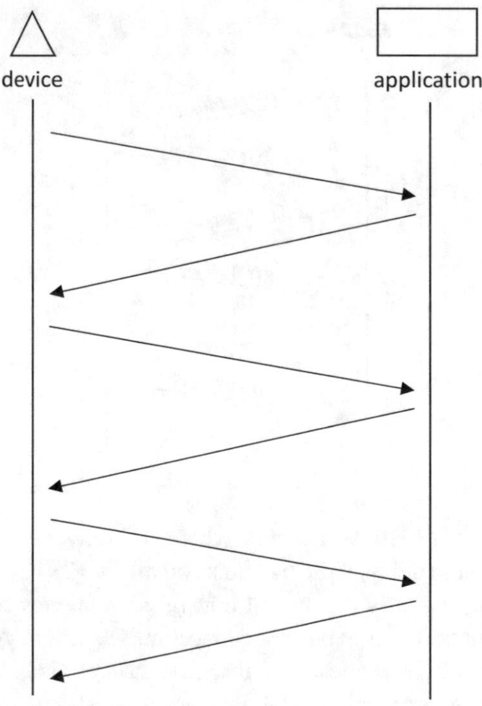

5.10 Show the packet flow between a CoAP client and a sensor when the client attempts to observe a temperature sensor on the sensor. Assume non-confirmable operations where ten server updates are received with a 10% packet loss.

5.11 Under an observation scenario, *12.31 C* temperature readouts are transmitted every half a second by a sensor to an application as a payload of *2.05 Content* CoAP messages. The messages are transported over UDP and IPv6. Assuming the shortest possible 6LoWPAN and IEEE 802.15.4 headers as well as a 1-byte token and no options as part of the CoAP message, what is the overall transmission rate of the scheme?

5.12 Consider an application simultaneously polling eight sensors to request readouts at a rate of twice a second in a lossless topology:

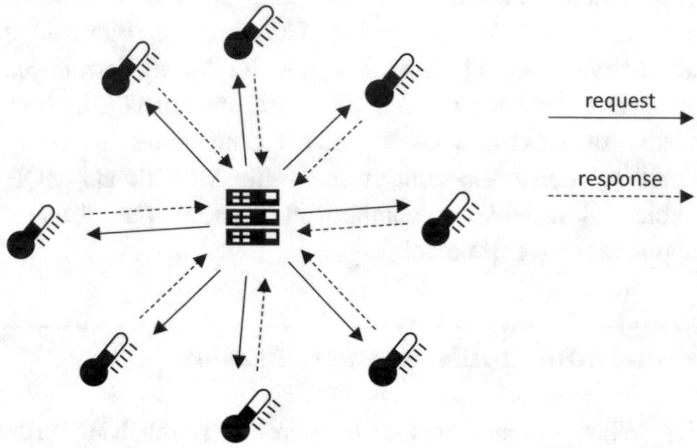

For maximum size IEEE 802.15.4 and 6LoWPAN headers with no security, what is the application transmission rate and throughput for the following scenarios?

(a) HTTP: 30-byte requests and 35-byte responses.
(b) CoAP: 16-byte requests and 20-byte responses (unicast)
(c) CoAP: 16-byte recuse and 20-byte responses (multicast)
(d) What are the implications of the results obtained above?

5.13 A microphone generates 8000 samples of audio per second. Each sample is, in turn, compressed as an 8-bit value. These samples are buffered into 160-byte frames that are transmitted as insecure RTP packets over UDP and IPv6. Assume the physical layer is Ethernet. What is the transmission rate of the microphone?

5.14 In an MQTT scheme, a sensor transmits readouts by means of blocking transactions. In other words, if a transaction is initiated to transmit a readout, the different messages associated with this transaction must be processed before the following readout can be sent. Assuming 100-byte packets, 128 Kbps transmission rates, no network impairments, and no delays other than the transmission delay, what is the maximum readout transmission rate of the scenario for each QoS level?

5.15 Describe a scenario where MQTT is better suited than CoAP for sensor data transmission.

References

1. Banks, A., Gupta, R.: Mqtt version 5.0 oasis committee specification (2019). https://docs.oasis-open.org/mqtt/mqtt/v5.0/mqtt-v5.0.html
2. Banks, A., Briggs, E., Gupta, R.: Mqtt version 3.1.1 oasis committee specification (2014). http://docs.oasis-open.org/mqtt/mqtt/v3.1.1/mqtt-v3.1.1.html
3. Bellavista, P., Corradi, A., Reale, A.: Quality of service in wide scale publish-subscribe systems. IEEE Commun. Surv. Tuts. **16**(3), 1591–1616 (2014)
4. Belshe, M., Peon, R., Thomson, M.: Hypertext Transfer Protocol Version 2 (HTTP/2). RFC 7540 (2015). https://doi.org/10.17487/RFC7540. https://rfc-editor.org/rfc/rfc7540.txt
5. Bishop, M.: Hypertext Transfer Protocol Version 3 (HTTP/3). Internet-Draft draft-ietf-quic-http-29, Internet Engineering Task Force (2020). https://datatracker.ietf.org/doc/html/draft-ietf-quic-http-29. Work in Progress
6. Bormann, C., Shelby, Z.: Block-wise transfers in the constrained application protocol (CoAP). RFC 7959 (2016). https://doi.org/10.17487/RFC7959. https://rfc-editor.org/rfc/rfc7959.txt
7. Bormann, C., Lemay, S., Tschofenig, H., Hartke, K., Silverajan, B., Raymor, B.: CoAP (Constrained application protocol) over TCP, TLS, and WebSockets. RFC 8323 (2018). https://doi.org/10.17487/RFC8323. https://rfc-editor.org/rfc/rfc8323.txt
8. Fielding, R.T.: REST: architectural styles and the design of network-based software architectures. Doctoral Dissertation, University of California, Irvine (2000). http://www.ics.uci.edu/~fielding/pubs/dissertation/top.htm
9. Fielding, R., Gettys, J., Mogul, J., Frystyk, H., Masinter, L., Leach, P., Berners-Lee, T.: Rfc 2616, hypertext transfer protocol – http/1.1 (1999). http://www.rfc.net/rfc2616.html
10. Hartke, K.: Observing resources in the constrained application protocol (CoAP). RFC 7641 (2015). https://doi.org/10.17487/RFC7641. https://rfc-editor.org/rfc/rfc7641.txt
11. Hunkeler, U., Truong, H.L., Stanford-Clark, A.: Mqtt-sn - a publish/subscribe protocol for wireless sensor networks. In: 2008 3rd International Conference on Communication Systems Software and Middleware and Workshops (COMSWARE'08), pp. 791–798 (2008)
12. Nepomuceno, K., d. Oliveira, I.N., Aschoff, R.R., Bezerra, D., Ito, M.S., Melo, W., Sadok, D., Szabo, G.: Quic and tcp: a performance evaluation. In: 2018 IEEE Symposium on Computers and Communications (ISCC), pp. 00045–00051 (2018)
13. OASIS: advanced message queuing protocol (amqp) version 1.0 part 2 (2012). http://docs.oasis-open.org/amqp/core/v1.0/os/amqp-core-transport-v1.0-os.html
14. Perkins, C., Handley, M.J., Jacobson, V.: SDP: session description protocol. RFC 4566 (2006). https://doi.org/10.17487/RFC4566. https://rfc-editor.org/rfc/rfc4566.txt
15. Rodríguez-Domínguez, C., Benghazi, K., Noguera, M., Garrido, J.L., Rodríguez, M.L., Ruiz-López, T.: A communication model to integrate the request-response and the publish-subscribe paradigms into ubiquitous systems. Sensors **12**(6), 7648–7668 (2012). https://doi.org/10.3390/s120607648. https://pubmed.ncbi.nlm.nih.gov/22969366.22969366[pmid]
16. Saint-Andre, P.: Extensible messaging and presence protocol (XMPP): core. RFC 6120 (2011). https://doi.org/10.17487/RFC6120. https://rfc-editor.org/rfc/rfc6120.txt
17. Saint-Andre, P., Alkemade, T.: Use of transport layer security (TLS) in the extensible messaging and presence protocol (XMPP). RFC 7590 (2015). https://doi.org/10.17487/RFC7590. https://rfc-editor.org/rfc/rfc7590.txt
18. Schooler, E., Rosenberg, J., Schulzrinne, H., Johnston, A., Camarillo, G., Peterson, J., Sparks, R., Handley, M.J.: SIP: session initiation protocol. RFC 3261 (2002). https://doi.org/10.17487/RFC3261. https://rfc-editor.org/rfc/rfc3261.txt
19. Schulzrinne, H., Casner, S.L., Frederick, R., Jacobson, V.: RTP: a transport protocol for real-time applications. RFC 3550 (2003). https://doi.org/10.17487/RFC3550. https://rfc-editor.org/rfc/rfc3550.txt
20. Schwarz, M.H., Borcsok, J.: A survey on opc and opc-ua: about the standard, developments and investigations. In: 2013 XXIV International Conference on Information, Communication and Automation Technologies (ICAT), pp. 1–6 (2013)
21. Shelby, Z., Hartke, K., Bormann, C.: The constrained application protocol (CoAP). RFC 7252 (2014). https://doi.org/10.17487/RFC7252. https://rfc-editor.org/rfc/rfc7252.txt
22. Tschofenig, H., Fossati, T.: Transport layer security (TLS)/datagram transport layer security (DTLS) profiles for the Internet of Things. RFC 7925 (2016). https://doi.org/10.17487/RFC7925. https://rfc-editor.org/rfc/rfc7925.txt
23. XMPP: Xmpp extensions. https://xmpp.org/extensions/

Part III

Advanced IoT Networking Topics

This part of this book, that includes four chapters, reviews the advanced IoT networking topics that complement the protocols presented in Part II. Chapter 6 explores the different mechanisms used for resource identification and management. Chapter 7 focuses on the technologies that provide traffic routing and message forwarding in the context of IoT. Chapter 8 presents a wide range of full and hybrid IoT LPWAN industry standards. Chapter 9 introduces Thread, a popular home automation WPAN architecture, that provides an alternative approach to traditional IoT stacks.

6 Resource Identification and Management

6.1 IoT Services and Resources

Because IoT devices, like sensors, actuators, and controllers, are becoming smaller and more portable, there is an increasing need for them to operate in networks where deployment and provisioning are carried out without human intervention. This scenario leads to what it is called zero-configuration devices. One key component of zero-configuration is service discovery. There are two classes of service discovery: distributed and centralized. Distributed discovery consists of devices discovering each other without accessing centralized directories. Similarly, centralized discovery relies on one or more service directories that provide a list of services offered by devices. A device accesses these directories to find out what services are provided by devices in the network. The main functions of service discovery include publication, registration, discovery, and resolution. In centralized scenarios, directories result in several maintenance functionalities including updating, removing, and validating directory entries [8, 1, 5].

Publication is the process by which devices publish a list of their supported services. The following information is typically associated with each publication: a service class that specifies the nature of the service provided (i.e. temperature sensing), a service access that specifies the network address of the service (i.e. 2001::21:5), a service name that identifies the specific instance of service supported (i.e. sensor1 vs sensor2), a domain name that identifies the service domain (i.e. building.local), and service properties that provide a list of attributes associated with the service (i.e. units=Fahrenheit).

Registration is the mechanism by which services provided by devices are stored in global directories. Registration can be stateless if the directory listens for broadcast advertisements to fill in its database or if it periodically queries devices for service support updates. In stateful registration, the device knows the address of the directory and explicitly registers directly with it. Stateful registration includes additional functionality including directory discovery that enables devices to find the address of the directories, registration update by which devices update their service support, registration validation used by devices to verify their own service support, and registration removal where devices unregister some their currently supported services.

Some protocols like CoAP [14] provide their own proprietary service discovery mechanism that cannot be extended to other technologies like MQTT [2, 3] or HTTP [10]. Fortunately, most publish/subscribe protocols include basic discovery by design supported through topic advertisement. Request/response deployments, however, require additional functionality to support service discovery. One approach to providing a generic framework for service discovery is by extending the DNS infrastructure.

DNS [9] is a well-established IP suite protocol that is mainly used for address resolution whenever a client wants to find out the IP address of a given hostname. This is shown in Fig. 6.1 where as a client sends an HTTP GET request to the *en.wikipedia.org* HTTP server, it must first translate the hostname into an IP address in order to build up the outgoing datagram. There are several steps: (1) the client starts a recursive query to its configured *local DNS server* to find out what the address of the *en.wikipedia.org* HTTP server is, (2) the *local DNS server* then initiates a sequence of iterative queries by first asking its configured *root DNS server* the *.org DNS server* address, (3) the *root DNS server* responds by transmitting the address of the *.org DNS server*, (4) the *local DNS server* queries the *.org DNS server* about the address of the *wikipedia.org DNS server*, (5) the *.org DNS server* responds by transmitting the address of the *wikipedia.org DNS server*, (6) the *local DNS server* queries the *wikipedia.org DNS server* about the address of the *en.wikipedia.org* HTTP server, (7) the *wikipedia.org DNS server* responds by transmitting the address of the *en.wikipedia.org* HTTP server and finalizing the sequence of iterative queries initiated by *root DNS server*, (8) the *local DNS server* responds by forwarding the address of the

R. Herrero, *Fundamentals of IoT Communication Technologies*, Textbooks in Telecommunication Engineering, https://doi.org/10.1007/978-3-030-70080-5_6

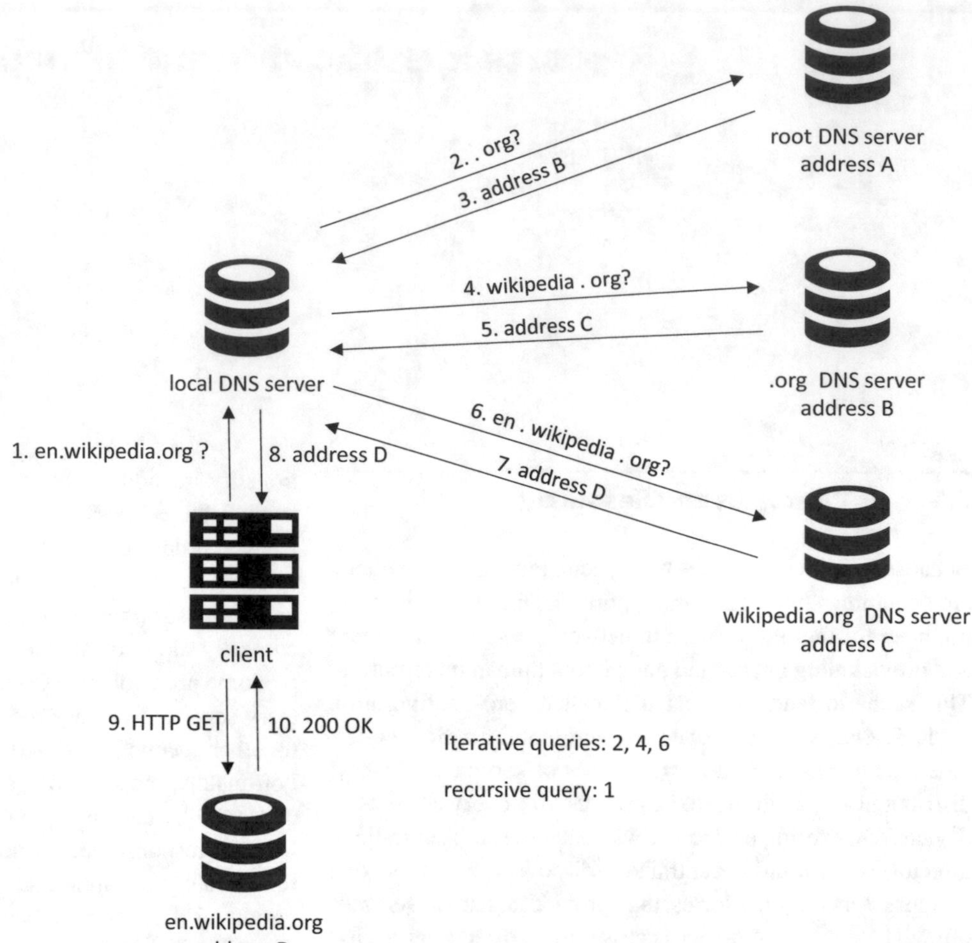

Fig. 6.1 DNS recursive query

en.wikipedia.org HTTP server to the client and terminating the recursive query, (9) the client sends the HTTP GET request to the *en.wikipedia.org* HTTP server, and (10) the *en.wikipedia.org* HTTP server replies by transmitting a 200 OK.

DNS, however, is a lot more than that; it introduces a generic mechanism for querying information and retrieving *resource records* (RRs) that can provide not only network and transport layer information but also configuration parameters. Under IoT and, in order to support zero-configuration, DNS is extended by means of two methodologies that enable distributed service discovery: (1) *multicast DNS* (mDNS) and (2) *Service Discovery DNS* (SD-DNS) [16, 11].

6.2 mDNS

As it can be seen in Fig. 6.1, the DNS infrastructure consists of a network of DNS servers that can be globally reached by any client. In the context of IoT, the deployment of this infrastructure is not always possible due to two main issues: (1) network impairments associated with LLNs may affect the global connectivity to DNS servers and (2) registering RRs for a massive number of devices is too expensive to be practical in most deployments. mDNS provides an alternative to traditional DNS by addressing these two problems. Specifically, mDNS (1) removes the need for DNS servers by moving the resolution capability to each device, and (2) it assigns a portion of the DNS namespace for free local use eliminating the need for global RR registrations and delegations [7]. When compared to DNS, mDNS requires very little or no configuration to function, and it works even when no infrastructure is present, or the current infrastructure is faulty.

From a functional perspective, mDNS relies on the existing DNS message structure and syntax, RR types, as well as DNS operation and response codes. mDNS, however, specifies how devices must coordinate themselves to send and receive multicast requests and responses in a relevant way. DNS/mDNS messages are transmitted over UDP on port 53. Figure 6.2 shows a DNS/mDNS message; it includes a fixed size header and variable number of questions and RRs organized as answer records, authority records, and additional records. Questions include the requested record

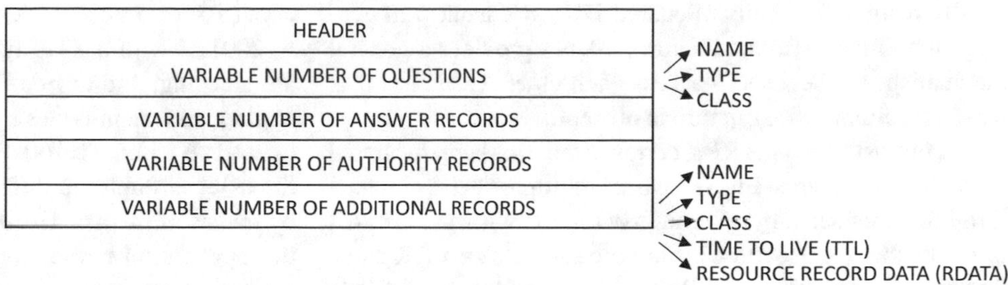

Fig. 6.2 DNS message format

name, type, and class, while RRs include only the record name, type, and class but also the TTL and the corresponding data associated with the record. The name identifies the resource within the device, while the type specified the nature of the resource. Some resource types are A and AAAA to respectively specify IPv4 and IPv6 addresses, PTR for reverse lookups, SRV for service information, and TXT for configuration information. The class is always set to IN for IP resource types. The TTL field identifies the number of seconds for which a resource record is valid. Note that this TTL field is, therefore, a Resource Record TTL (RR TTL) in contrast to the IP TTL field that specifies the hop count limit.

A and AAAA are probably the most relevant RR types since they enable hosts and other entities to convert device hostnames into respective valid IPv4 and IPv6 addresses. For a device to be globally addressable, it must belong to an organization that has control over the DNS namespace such that it can be assigned a name. For example, if the organization owns the DNS namespace *l7tr.com*, a given device can have the subdomain *sensor1.l7tr.com* globally allocated for applications and other hosts to access it. However, RR registration on subdomain DNS servers increments topology complexity and requires global connectivity to work. This is not always available in IoT LLNs where infrastructure limitations and network outages are common. mDNS assigns RRs that are only valid within a given link-local segment of the network and that they are not typically accessible beyond the link-local scope. These mDNS specific RR names take the form *submain.local* where, for example, a device could assign the *sensor.local* name to itself for its own A/AAAA records. It is critical the RR name selection is done in such a way that there are no collisions with names selected by other devices. In this context, RRs are unique RRs. To this end, a device must first verify that no other devices are using the selected RR name. A device must, therefore, support both client and server DNS functionality by being able to generate and process multicast requests and replies. Multicast requests and replies guarantee that they are visible to all devices co-located on the same link. Figure 6.3 shows a state machine that defines how a device must assign an RR name. Specifically, the device must first randomly select a *.local* RR name and acting as a client sends a multicast mDNS question asking all other devices on the same link whether they have

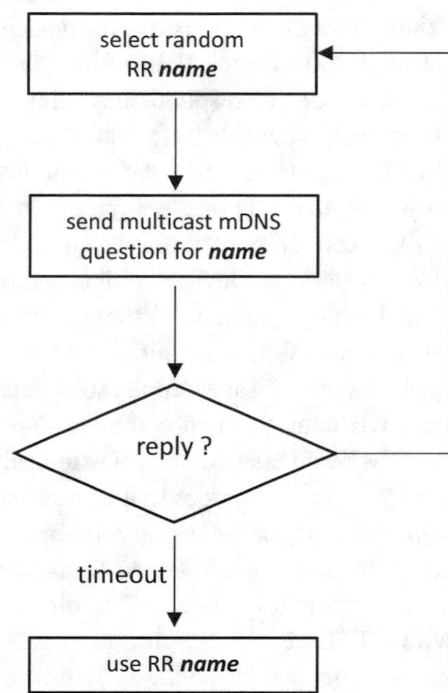

Fig. 6.3 RR name resolution state machine

authority over the selected RR name. If the request timeouts, that is, no other device responds, then the sender can assign the name to the RR, otherwise the device must select a new random *.local* RR name and repeat the process. If a sensor has authority of an RR name like *temperature.local*, it can continue using that name until a conflict occurs. In this case, the device must allocate a new name for the RR in accordance with the procedure specified in Fig. 6.3.

mDNS reserves the top-level domain to *.local* that, as a link-local allocation, implies that all subdomains are only meaningful in the link where they are defined. Requests and responses associated with questions and answers belonging to the .local domain must be sent to the mDNS IPv6 multicast addresses FF02::FB. mDNS supports that other *non .local* questions can be sent multicast to other devices in the link if no global DNS server is configured. This enables any non-authoritative entity to respond to these requests and provide RR name resolution even when connectivity to the traditional DNS infrastructure is not available. For example, if an IoT gateway, due to an outage, cannot access the network core, devices on the same link can still talk to each other using locally resolved globally allocated DNS RR names. In general, due to its distributed nature, mDNS provides a general mechanism for devices to access each other RR names if at least a minimal infrastructure is present.

In contrast to unique RRs, certain scenarios support shared RRs. This is shown in Fig. 6.4 where multiple devices interact through either sensing or through actuation with an asset. All three devices (A, B, and C) share the same AAAA RR name that maps to three different IPv6 addresses: 2001::5, 2001::6, and 2001:7. At initial time, device A is active (devices B and C sleep), at time 100 s device C is active (devices A and B sleep), and at time 200 s device B is active (devices A and C sleep). Each active device responds including a TTL field of 100 s. In essence, the devices have nonoverlapping power duty cycles that attempt to preserve the overall network lifetime. In other words, IoT networks exhibit power limitations that enforce the need of power duty cycles that balance active with sleeping periods. Devices in sleep mode only keep some basic functionality that enables them to become active at some point in time. While sleeping, devices sometimes harvest energy by means of, for example, solar panels. mDNS provides shared RR names to overcome the limitations that result from devices being unavailable for extended periods of time. Specifically, multiple devices that observe an asset can take turns responding to external requests. In the figure, at time 0, device A is active, while B and C are sleeping, any AAAA request querying for *asset.local* results in a mapping to 2001::5 with a TTL of 100. By around time 100, devices A and B are sleeping, while device C is now active. The client issues a new AAAA query that results on a mapping to 2001::7 with a TTL 100. At time 200, devices A and C are sleeping, and device B is active. As the previous query expires, the client issues a new one that results on a mapping to 2001::6 with TTL 100. Essentially, the client interacts with the asset through the different devices in a transparent way by means of the *asset.local* hostname. The client is unaware that application and session layer requests (i.e. CoAP, HTTP) are being processed by three different devices with three different power duty cycles. This mechanism that attempts to address the power limitations by abstracting the different power cycles is known as load balancing.

6.2.1 Queries

Figure 6.5 shows two devices, one acting as a client and another acting as a server. Both client and server consist of

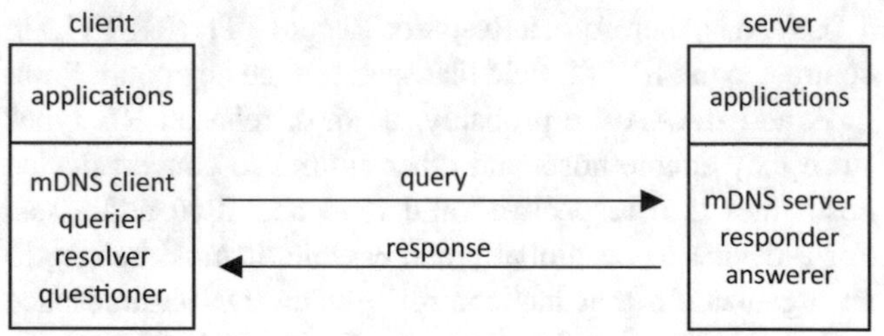

Fig. 6.5 Queries and responses

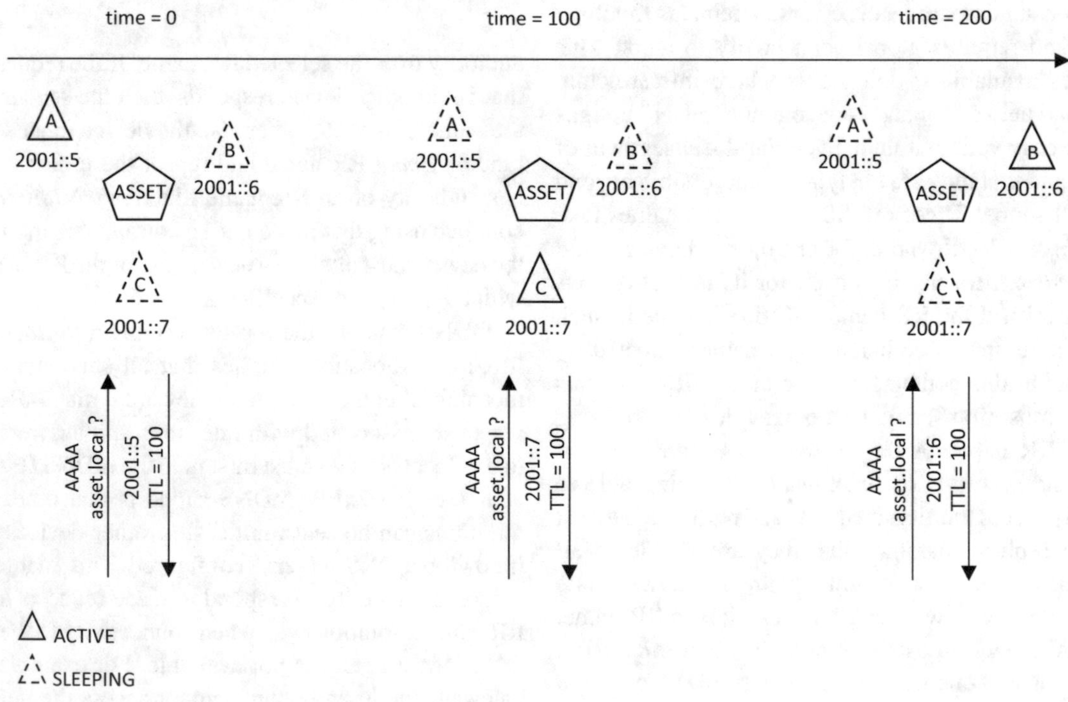

Fig. 6.4 Shared resources

several client and server applications. The client applications rely on an mDNS client, also known as querier, to perform RR name resolution. The querier or resolver (also known as questioner) transmits a query that is answered by the mDNS server, also known as responder (and also known as answerer). mDNS queries can be (1) one-shot queries that are typically associated with legacy DNS queriers and responders and (2) continuous queries that are made by fully compliant mDNS queriers and responders in order to support asynchronous operations like IoT load balancing. In general, because devices advertising unique RRs must be able to perform conflict resolution, they typically include both querier and responder functionality. For devices that only advertise shared RRs, there is no need for them to act as queriers as they are not exposed to conflict resolution.

Under one-shot queries if a device is configured with a local DNS server, it transmits all queries to the aforementioned server, while those falling under the *.local* domain are transmitted multicast to the destination IPv6 address FF02::FB on UDP port 5353. These multicast queries are transmitted using a source port different from 5353 in order to indicate that they are one-shot queries. With one-shot queries the resolver will use the first response it receives assuming the RR is unique, and it will not attempt to request other answers. This behavior may work on certain simple scenarios, but it may not be optimal when dealing with more complex situations where, for example, shared RRs are involved. mDNS introduces continuous queries to address some of the performance issues introduced by one-time queries.

With continuous queries, a single response is not an indication that no more responses follow. For example, in a load balancing scenario where multiple devices sense a single asset, they all may send separate answers to a question about an RR name mapping. Moreover, if one of the devices is sleeping, the one-shot approach would cause the client to forget it. Under continuous queries, operations are asynchronous, and a transaction can be finalized when there is clear indication that no other responses will be received.

If a client application is looking to send and receive datagrams from a device, it must keep processing mDNS responses until there is clear indication that the application is interacting with the device. One common scenario requires a client application to show the list of available services in real time. One naive (and inefficient) approach is to rely on issuing several queries of interest and process them as answers are received. As time progresses the client application must resend the questions to prevent the use of stale information. This can be done either through a timer or by means of the end user clicking on a refresh button. Unfortunately, this compulsive device polling can lead to inefficient network utilization by putting an unreasonable burden on the communication infrastructure. Therefore, a way to prevent this is by introducing a controlled schedule, where the interval between the first two queries is 1 s and any other successive queries are transmitted by multiplying this interval by two until reaching a maximum interval length of 3600 s. Moreover, mDNS makes queriers support a mechanism known as known answer suppression. Figure 6.6 shows the interaction between three devices and a client: first at time 0, device A and B are active, but device C is sleeping. AAAA request querying for *asset.local* results on a mapping to 2001::5 with a TTL of 200 and to 2001::6 with a TTL of 100. At time 100, the client sends a new query for *asset.local*, but since it knows device A is still active, it populates the known answer section including the device A AAAA RR with the updated TTL value of 100. This prevents device A from replying with its own AAAA RR information, thus improving network efficiency by reducing throughput. Obviously, this is why the mechanism is called known answer suppression; known answers are suppressed by not being transmitted. At this point, device B is no longer active, but device C becomes active and replies with its IPv6 address 2001::7 with a TTL of 200. At time 200, the client issues a new query with a known answer section including the device C AAAA RR with the updated TTL value of 100. Since at this point device C is the only one active and the known answer section information is correct, nobody replies to the query. Essentially, the known answer suppression mechanism tells mDNS servers what answers are already known to the querier.

If way too many answers are included in the known answer section of a request, they may not fit in a single mDNS message due to fragmentation restrictions. In this case, the querier sends multiple requests with different sections of the list of known answers with all packets having the *truncation* (TC) bit set other than the last one transmitted. The responder, upon receiving mDNS messages with the TC set, defers transmitting any answer until either receiving the last mDNS message with the TC bit unset or timing out after 500 ms. This mechanism provides a trade-off between latency and wasted network bandwidth, such that when waiting for more known answers, fewer answers are transmitted by the responder.

In continuous queries, the querier retransmits queries to obtain new answers that may not have been available when the questions were first transmitted. To this end, successful implementation of known address suppression is critical to minimize throughput and limit network traffic. Since channel capacity, especially for most low-power IoT technologies like IEEE 802.15.4, is low, instant bursts of traffic result in contention that ultimately leads to latency and packet loss. To prevent this, mDNS enforces that queriers delay the transmission of questions by choosing a random delay in the range of 20–120 ms.

Based on the TTL fields of received replies, clients keep a cache to verify the validity of the different RRs. When an RR expiration time is reached, if no client application has

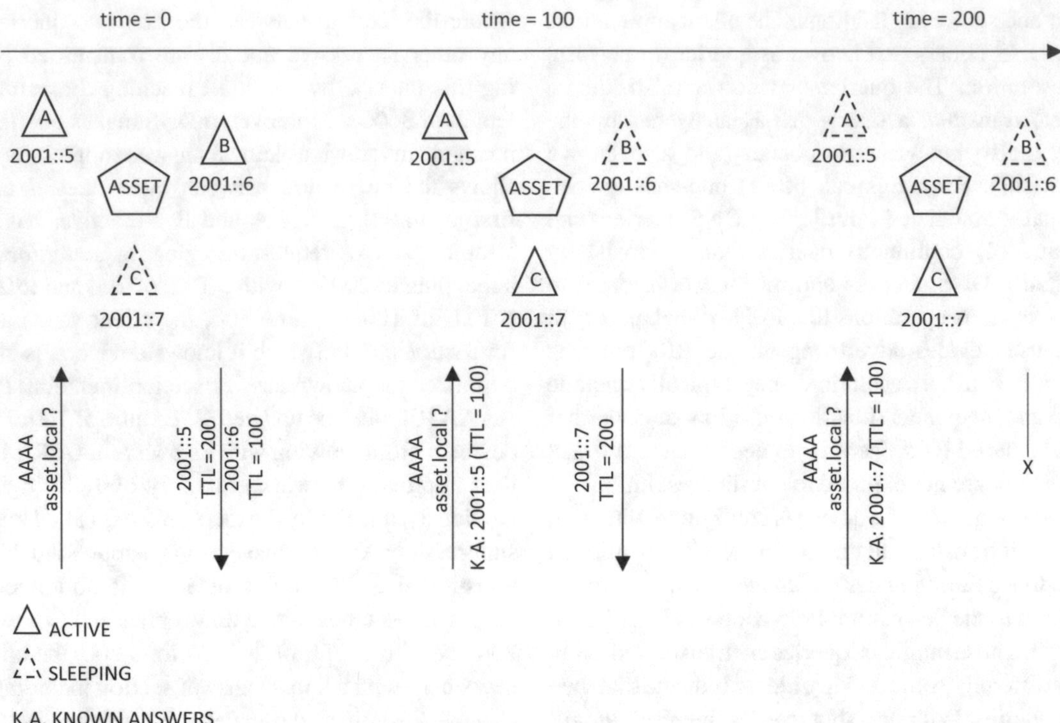

Fig. 6.6 Shared resources with known answer suppression

an active interest in the records, they are removed from the cache. In reality, way before an RR is about to expire, and as long as there is an active interest, the client must transmit a query to make sure that the record is still valid and also to update the cache with a newer TTL value. These query retransmissions must be performed starting after at least half of the RR lifetime has elapsed. Formally, mDNS indicates that a retransmission must first be issued when the lifetime of the RR is at 80%. If no answer is received, the next retransmission is performed when the lifetime of the RR is at 85%. If still no answers are received, two additional attempts are performed when the lifetime of the RR is at 90% and 95%, respectively. In either case, whenever an answer is received, the RR lifetime is updated according to the TTL field. If no answer is received by the end of the RR lifetime, the record is removed from the cache. As in the case of replies, queriers randomly delay the retransmission of questions to minimize the chances of traffic bursts that can lead to packet loss. They introduce a random TTL variation of 2% that implies that retransmissions are performed first between 80 and 82% of the lifetime, second between 85 and 87% of the lifetime, third between 90 and 92% of the lifetime, and finally between 95 and 97% of the lifetime. mDNS clients that comply with the mechanisms of continuous queries must issue requests to the IPv6 multicast address FF02::FB from UDP port 5353.

A single mDNS request can carry multiple queries issued by a unique querier. From a protocol perspective, the effect of multiple questions in a single request is like that of multiple requests carrying a single question. The advantage of multiple queries on a single request is the transmission of fewer lower layer headers that lead to less overhead and more network efficiency. If a device is about to send a question and it notices that the multicast request generated by another device includes the same question and the known answer section does not include a valid answer, then it can avoid sending the query. Specifically, the assumption is that any other device with authority over the RR will answer the query by transmitting a multicast response.

6.2.2 Responses

The transmission of multicast responses enables all link-local devices to benefit from the responses of the different responders. This way devices can keep their caches up-to-date and minimize the transmission of additional queries that may have been resolved by previously received answers. In some situations, however, it is neither practical nor efficient for all devices on the link to receive responses, for example, if a previously inactive interface is activated through a configuration change. In this case, the cache associated with the interface is initially empty so the sequence of queries transmitted by its device can cause a sudden flood of multicast responses that produce traffic bursts that can make the network unusable due to collisions and device contention. Since most other devices on the network are likely to have the answers in their caches, the transmissions of these redundant multicast replies can be avoided. One way to do this is by explicitly indicating that requests need to be replied using unicast responses.

mDNS specifies that the class field of a DNS question must include a unicast response bit that, when set, indicates that the querier wants to receive unicast replies in response to the question. mDNS calls questions that request unicast responses QU questions to distinguish them from the more common QM questions that request multicast responses. When an interface first initializes, the device transmits QU questions only. Question retransmissions are sent as QM questions because they are likely to include a large list of known answers derived from the initial QU question transmission. More known answers mean that fewer answers are transmitted by those devices answering questions, thus, reducing the likelihood of traffic bursts. Sometimes, responders send multicast answers to QU questions; this happens when the mDNS server detects that the time since the last multicast answer transmission is more than quarter of the TTL time. In general, QU responses follow the same timing and packet generation mechanisms than QM responses.

In certain circumstances some client applications require resolvers to send unicast requests to a specific mDNS server even when the queries are associated with *.local* RR names. In these cases, the mDNS server sends answers as if the requests were carrying QU questions. The mDNS server must make sure that the request is being initiated by an mDNS client on the same local link. Specifically, if a request originates outside the local link, it must be dropped by the mDNS server. In some other circumstances, certain mDNS servers are designed to respond to queries initiated outside the local link. In this case, unicast requests associated with *.local* RR names are answered by responders transmitting unicast responses. Since the requests/responses in this scenario become traffic in the Internet backbone, it is convenient to minimize their transmissions by making sure that only those RR for which the mDNS server is authoritative are sent.

In general, an mDNS responder must only answer when it has a non-null positive response to send or if it is authoritative about the nonexistence of a given RR. For the case of unique RRs, where the responder is the sole owner of RRs, the mDNS server must transmit negative answers for queries related to RRs that do not exist. Negative answers are transmitted in terms of *next secure* (NSEC) RRs. As with any records, NSEC RRs include a TTL field that indicates for how long it is nonexistent. For example, IoT devices have IPv6 addresses but typically lack IPv4 addresses. Since addresses are unique, devices are guaranteed to be authoritative about A and AAAA RRs. If a client queries about A/AAAA RRs, the IoT device will send a positive AAAA answer and a negative A answer to indicate that the record does not exist. In general, questions that request information related to the IP addresses of a given interface are answered by means of a single response that includes both IPv4 and IPv6 addresses. The IPv4 address is transmitted as either an A or an NSEC RR, and the IPv6 address is transmitted as either an AAAA or an NSEC RR. The point of including both answers in a single response is to make sure that the resolver receives all the interface information at the same time and minimize the chances of incomplete information because of packet loss. The initial design of mDNS assumed that there was no need for negative answers and just a simple timeout would work as a good indication of the nonexistence of a given RR. The problem with this later approach is that it induces the retransmission of questions that introduce additional latency as well as network packet loss.

mDNS responses are supposed to be true and accurate answers regardless of what questions those answers are addressing. mDNS requests can include, besides questions, answers in the form of the known answers. On the other hand, mDNS responses must only include answers. If an mDNS response has questions in its question section, they must be ignored by the resolver. As in the case of mDNS requests, the transmissions of mDNS responses are randomly delayed minimizing traffic bursts that lead to channel contention and network packet loss. If the responder advertises unique RRs, there is no need to impose any transmission delays as it is guaranteed that only one device will respond to the incoming requests. On the other hand, if the question is regarding shared RRs, then the responders must delay the transmission of the answers by a random amount of time uniformly distributed in the 20–120 ms range. This is also important if multiple devices are attempting to answer the same question, since the random transmission delay enables devices to observe whether the answer has been already transmitted. If so, the device does not need to send the query. This situation is compatible with the case in which mDNS proxy servers respond to queries based on the information previously sent by other devices.

mDNS responses must be transmitted over UDP with source and destination ports 5353 to indicate full compliance with specifications. A resolver typically drops any mDNS responses that do not have 5353 as source port. Responses are transmitted multicast to the link-local IPv6 destination address FF02::FB, unless they are in response to QU questions. To minimize traffic bursts that lead to contention and loss, a responder must wait for at least 1 s before it can resend an answer to a previously transmitted RR. Whenever possible, and in order to be efficient, responders tend to aggregate as many answers as possible into a single request. By doing so, the overhead due to the DNS and lower level headers is minimized. Moreover, a single multicast mDNS message can include answers that address questions from multiple queriers. In general, a responder may delay the transmission of a response by up to 500 ms in order to collect enough questions to guarantee that many answers is included in the packet. Of course, the price to pay for this aggregation is an additional 500 ms latency. In addition, in responding to requests that include known answers, the responder may resend the answers with an updated TTL field if the lifetime of the corresponding RR is less than half the TTL value. If a device is about to send an answer and it notices that the multicast response generated by another device includes the same answer, then it can avoid sending the response if the TTL field of the answer complies with the lifetime of its RR.

Example 6.1 Consider an IEEE 802.15.4 scenario where a querier attempts to discover three sensors that share an asset:

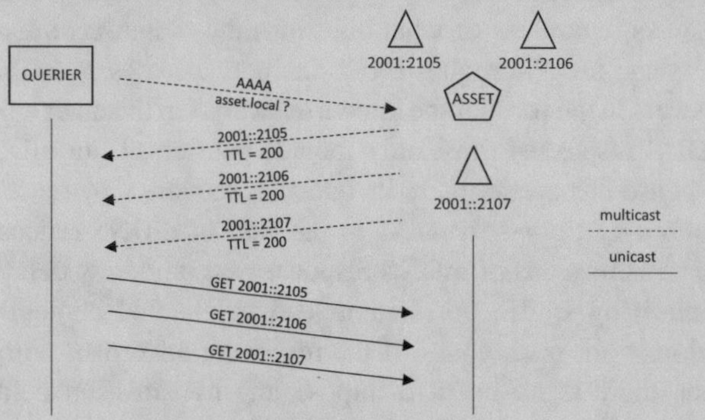

If it takes 11.34 ms for all three unicast requests to arrive at the sensors, what is the average MAC contention delay? Ignore any delays other than the transmission delay. Assume average mDNS and CoAP packet sizes of 30 and 40 bytes long, respectively.

Solution: The contention delay is given by $T_c = T - T_t$ where T is $T = 11.34$ ms and T_t is the transmission delay. T_t is given by $T_t = \frac{L_{\text{mDNS request}} + 3\times L_{\text{mDNS response}} + 3\times L_{\text{CoAP}}}{R} = 7.68$ ms where $R = 250000$ Kbps, $L_{\text{mDNS request}} = L_{\text{mDNS response}} = 8 \times 30$ bits and $L_{\text{CoAP}} = 8 \times 40$ bits. The contention delay is, therefore, $T_c = 3.66$ ms.

6.2.3 mDNS Message Header

Figure 6.7 shows the DNS message header format; it starts with a 16-bit message identifier field that is assigned by the client and identifies the transaction, and it follows the *query/response* (QR) bit that when set indicates that the message is a response, the 4-bit *operation code* (OPCODE) that identifies the request message operation type and specifies whether it is a standard query, an inverse query, or a status request; the *authoritative answer* (AA) bit that, when set, it indicates that the response carries answers owned by the sender; the TC bit that specifies that the request is truncated and more messages are to be expected; the *recursion desired* (RD) bit that it is typically used by the querier to enable recursive queries; the *recursion available* (RA) bit used by the responder to specify support of recursive queries; and the 4-bit *response code* (RCODE) that identifies the response type and it is used to indicate format errors, server failures, RR errors, request refusal, or no errors. Then four 16-bit counters end the message; the *question count* (QDCOUNT) that indicates the number of entries in the question section, the *answer count* (ANCOUNT) that specifies the number of RRs in the answer section, the *authority count* (NSCOUNT) that indicates the number of RRs in the authority section, and the *Additional Count* (ARCOUNT) that accounts for the number of RRs in the additional records section.

In the specific cast of mDNS, all responses including those unsolicited responses coming from IoT devices must be processed. This observation of all incoming responses is typically performed regardless of the content in the message identification field and the question section. In fact, under mDNS, a device can cache the content of all responses even if it does not have a client with an active interest in those RRs. Of course, the mDNS client must respect RR lifetimes when caching those responses.

In the context of mDNS, there are a few expectations on how the fields are used. For example, for multicast mDNS messages, the message identification field is always 0 regardless of whether the message is a request or a response. Only in the case of legacy unicast responses triggered by

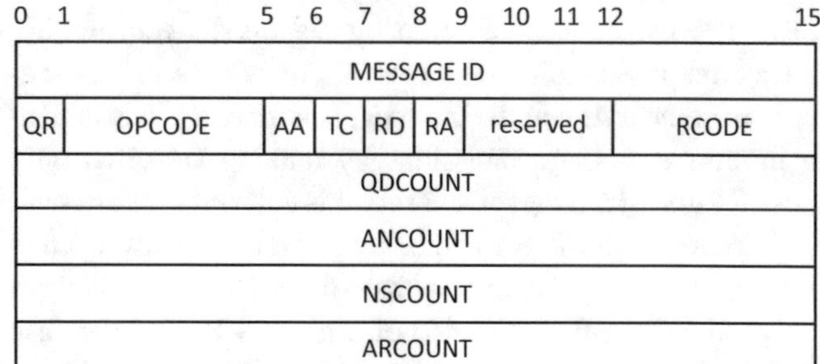

QR	query (0) or reply (1)
OPCODE	operation code
AA	authorative answer
TC	truncation
RD	recursion desired
RA	recursion available
RCODE	response code
QDCOUNT	question record count
ANCOUNT	answer record count
NSCOUNT	authority record count
ARCOUNT	additional record count

Fig. 6.7 DNS message header format

QU requests the message identification field is used to keep track of the transaction. Similarly, the OPCODE is always set to query on transmission, and any other value is ignored by the responder. The AA bit must be unset on transmission and ignored when received. As previously indicated, if the TC bit is set, the responder must wait for at least 500 ms to wait for more messages. This bit only applies to queries, so responses do not set it and queriers ignore the value in incoming responses. Finally, queriers set the RCODE to no error on transmission, and it is ignored by responders upon reception. Another difference between traditional DNS and mDNS is that the former relies not only on UDP but also on TCP transport for unicast transmissions. Since TCP does not support multicast operations, it cannot be used for mDNS transport.

6.3 SD-DNS

mDNS provides the basic infrastructure for the exchange of RRs between devices on the same link [6]. By itself, mDNS does not specify a procedure for the discovery of a newly deployed device in the network. To this end, SD-DNS relies on the infrastructure to enable zero-configuration support. Moreover, SD-DNS specifies how RRs must be named to provide service discovery, but, as mDNS, it makes no changes to the structure of DNS messages, operation, and response codes as well as most DNS protocol values. With SD-DNS a client can specify, by means of DNS queries, the type of service and the domain in which it is looking for that service and obtain a list of the named instances that provide the service.

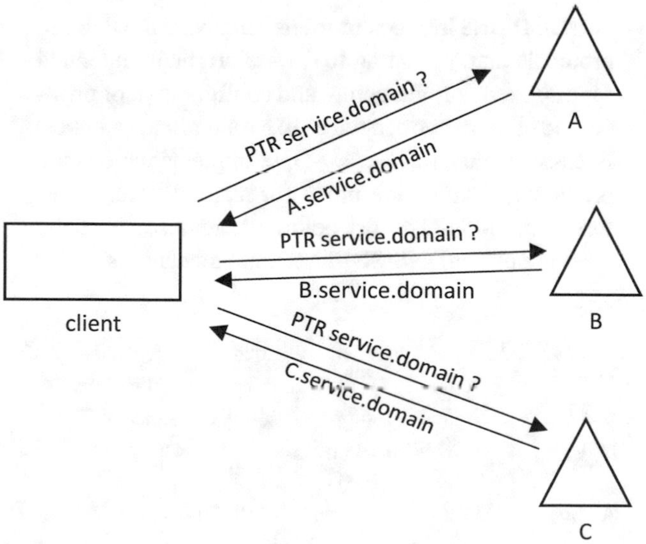

Fig. 6.8 Discovery of instances

Because an instance is related to a service and a domain, they are discovered by querying *service.domain* PTR RRs. Once all instances are identified, then it is possible to retrieve IP addresses, transport ports, and general configuration information by querying *instance.service.domain* A/AAAA, SRV, and TXT RRs respectively. Figure 6.8 shows a client transmitting a multicast *service.domain* PTR request that is answered by three devices (A, B, and C) that transmit their supported instances *A.service.domain*, *B.service.domain*, and *C.service.domain*. In Fig. 6.9 the client sends three queries derived from the list of instances: (1) an *A.service.domain* A/AAAA request to obtain the IP addresses, (2) an *A.service.domain* SRV request to obtain the TCP/UDP ports, and (3) an *A.service.domain* TXT request to retrieve configuration information. Note that SD-DNS is defined to work not only in the context of multicast mDNS transactions but also relying on traditional unicast DNS servers. In this later case a client can send all SD-DNS queries to a local DNS server that through recursion can discover instances and their addresses, transport ports, and configuration parameters.

The *service* component of the *service.domain* element takes the form of the protocols associated with the instances being queried. For example, if a client is interested in HTTP instances, the corresponding service name is *_http._tcp* since HTTP is a session protocol that runs over TCP. Similarly, if the client is interested in CoAP instances, the service name is *_coap._udp* since CoAP is run over UDP. Services can be more specific, for example, *temperature._mqtt._tcp* addresses all instances of temperature sensors that support MQTT over TCP. Similarly, *_coap._dlts._udp* looks for all instances of secure CoAP. Note that protocol names are preceded by an _ symbol. Note that the original service specification supports only two different transport types *_tcp* and *_udp*. *_tcp* is used for TCP transport-based protocols, and *_udp* is used for all other protocols regardless of whether they support UDP or not. In other words, a popular transport

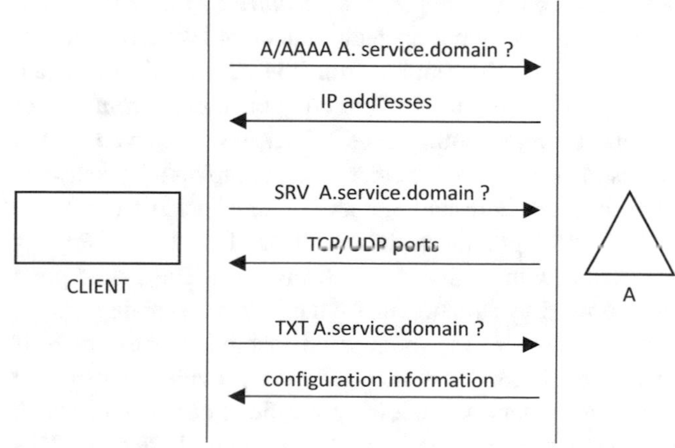

Fig. 6.9 Getting IP addresses, ports, and configuration information

protocol known as *Stream Control Transmission Protocol* (SCTP) that enables efficient transmission of data streams is encoded as *_udp* and not as it would be expected as *_sctp*. Fortunately, most IoT protocols rely on UDP transport so this is never an issue.

Example 6.2 What is the transmission rate overhead due to the transactions shown in Fig. 6.9?

Assume that the TTL fields of the A/AAAA, SRV, and TXT RRs are 100 ms, 250 ms, and 500 ms, respectively. Additionally assume that the size of the mDNS request packets is 30 bytes long and that the sizes of the A/AAAA, SRV, and TXT mDNS responses are $L_{\mathrm{A/AAAA}} = 20$, $L_{\mathrm{SRV}} = 10$ and $L_{\mathrm{TXT}} = 57$ bytes, respectively.

Solution Considering that in 1 s there are ten, four and two A/AAAA, SRV, and TXT transactions, respectively, the rate overhead R is given by $R = 8 \times \left(10 \times L_{\mathrm{A/AAAA}} + 4 \times L_{\mathrm{SRV}} + 2 \times L_{\mathrm{TXT}}\right) =$ 2832 bps.

The *domain* component of the *service.domain* RR name follows the usual *subdomain.local* format for mDNS-based queries or follows the traditional unicast hostname format. Similarly, the *instance* components of the *instance.service.domain* element is a user-friendly name that consists of a string composed of 16-bit unicode characters. In most cases, the instance is a default name that enables a client to access the device without any previous manual configuration. Moreover, to prevent collisions, in most cases these defaults names are generated as a random combination of keywords and numbers. Instance names, as opposed to hostnames, are not constrained by specific rules, and rich text strings are allowed. Spaces and other special characters are possible, for example, *Temperature Sensor 1* is a valid instance name. Some special characters like dots need to be escaped to differentiate from the dots used to separate instance, service, and domain components. Escaping a dot is done by representing it as a backslash followed by the dot itself (i.e. \. On the other hand, escaping a backslash is performed by representing it as a double backlash (i.e. \\).

SRV RRs provide a list of the UDP and/or TCP ports associated with a given service instance. The list of ports can be used to support load balancing by enabling clients to access servers in a sequential fashion. For example, if TCP ports 1883, 1884, and 1885 are provided as response to an SRV query, the client applications can then establish new sessions by sequentially connecting to those ports. This port selection, however, is ruled by two fields; priority and weight associated with each port. Higher priority servers, intended to be accessed more often, are specified by lower priority numbers. Similarly, if several servers have ports with the same priority, the weight specifies what server should be accessed more often. When there is only one RR that specifies the transport port, then the weight and priority fields can be ignored.

TXT RRs provide a list of device configuration parameters that are service and instance dependent. In other words, how their parameters are interpreted depends on the type of service under consideration. For example, the *unit* configuration parameter on a temperature sensor is specified as either *Celsius* or *centigrade* degrees. Similarly, on a barometric pressure sensor, it is specified as either *Pascals* or *pounds per square inch*. What is standardized is the way in which TXT RRs are accessed; SD-DNS compliant devices must be able to provide TXT RRs in addition to SRV RRs with the same name even if they have no relevant configuration information. In this later case, the TXT RR must provide a single 0 byte. In all cases, the lifetime of the record is given based on the TTL field of the RR received.

Because the list of configuration parameters is typically large, it is important to mention the size restrictions that apply to most RRs. As any other record, a TXT RR that includes the list of parameters can be up to 65535 bytes long. This length is part of the length field in the RR header. In the context of mDNS RR, when considering all protocol headers and networking conditioning, the practical upper limit of a TXT RR length is around 9000 bytes. It is always preferable to keep the TXT RR length below the MTU size of a single datagram to prevent, whenever possible, fragmentation. Of course, when relying on 6LoWPAN and low-rate IoT mechanisms like IEEE 802.15.4, fragmentation is, in many cases, unavoidable.

► **mDNS/SD-DNS Implementations** mDNS/SD-DNS are protocols that, predating IoT, were originally intended for discovery, provisioning, and configuration of printers and other network devices like routers and switches. Because of this, mDNS/SD-DNS implementations are pretty stable and widespread for many different platforms and OSs. The list below presents some of the most popular mDNS/SD-DNS implementations.

Bonjour	Apache licensed open source C implementation.
	Available for general purpose OSs like Linux.
	Adapted to support RTOS like Contiki.
lwIP	BSD licensed open source C implementation.
	RTOS specific support.
Avahi	LGPL licensed open source C implementation.
	Available in most Linux distributions included in embedded devices.
OpenMDNS	BSD licensed open source C implementation.
	Ultra lightweight support of mDNS/SD-DNS.

The list of configuration parameters is made of a sequence of key and value pairs. The key represents the parameter name, while the value indicates its associated parameter value. Each pair is encoded as a single string *key=value* where key and value are separated by an equal sign =. For example, a key could be *location* and its value the corresponding GPS coordinates such that the encoding string is *location=41.40338,2.17403*. The key size is between one and nine characters long, while the overall encoded string length size is never larger than 255 bytes. Obviously, the key or parameter name has to be unique within the context of the service being considered. For the most part, the key does not need to be human readable as it is intended to be processed by client applications. Given a service and all its associated parameters, there are four possible scenarios for the parameters in the list included in the TXT RR: (1) if the key is not present, it implies that the parameter takes the default value or that the parameter is unknown; (2) if the key is present with no value, it implies that the parameter represents a boolean condition that is false; (3) if the key is present with an empty value, it implies that the parameter takes the default value; and (4) if the key and its value are both present, it assigns the value to the parameter.

Key/value pairs, encoded as strings, are concatenated in a binary frame and pre-appended by a single byte that specifies the length of the string. This is shown, as example, in Fig. 6.10 where the following strings are encoded: *key=value*, *active=1*, and *units=C*. In general, it is up to the service specifications to purely rely on TXT RRs for configuration or combine TXT RRs with inband protocol mechanisms. For example, under CoAP, configuration information like the units associated with temperature sensing can be transmitted as part of a TXT RR list or as part of a CoAP option.

For efficiency, under SD-DNS, whenever a PTR query is transmitted, the responders transmit not only the list of instances but also all associated A/AAAA, SRV, and TXT RRs. This saves the querier the need of transmitting additional A/AAAA, SRV, and TXT RR queries. Similarly, if a client sends a SRV query, the responder transmits the SRV RRs and the associated A/AAAA RRs. Finally, if a querier transmits a TXT request, the responder sends back just the TXT RR.

Figure 6.11 shows a service discovery example; the client just sends a multicast mDNS PTR RR query for *_temperature._coap._udp.local*. Device *sensor1* answers the query by sending a PTR RR that indicates the *sensor1._temperature._coap._udp.local* session. The response also piggybacks an NSEC RR to indicate that the device does not have an IPv4 address, an AAAA RR that signals a 2001::21:10 IPv6 address, a SRV RR that points to a single UDP port 5683, and a TXT RR that transmits a configuration parameter list *units=C* and *location=41.40338,2.17403*.

6.4 CoAP Service Discovery

CoAP introduces both distributed and centralized approaches to service discovery. The distributed approach is based on simple CoAP resource discovery, while the centralized approach is based on CoAP resource directory [15]. In all cases, as a result of the discovery, URIs that represent the resources and corresponding attributes are obtained. Distributed CoAP service discovery relies on interaction, by means of queries, between applications and devices. On the other hand, under centralized CoAP service discovery, applications and devices do not query each other, and the interaction is by means of a directory. Note that although these mechanisms have not been fully standardized, they are supported by several CoAP implementations [17].

6.4.1 Distributed CoAP Resource Discovery

This method provides the simplest way for devices to support service discovery without the need of directories and other central entities. The scenario is illustrated in Fig. 6.12 where a receiver application discovers the services offered

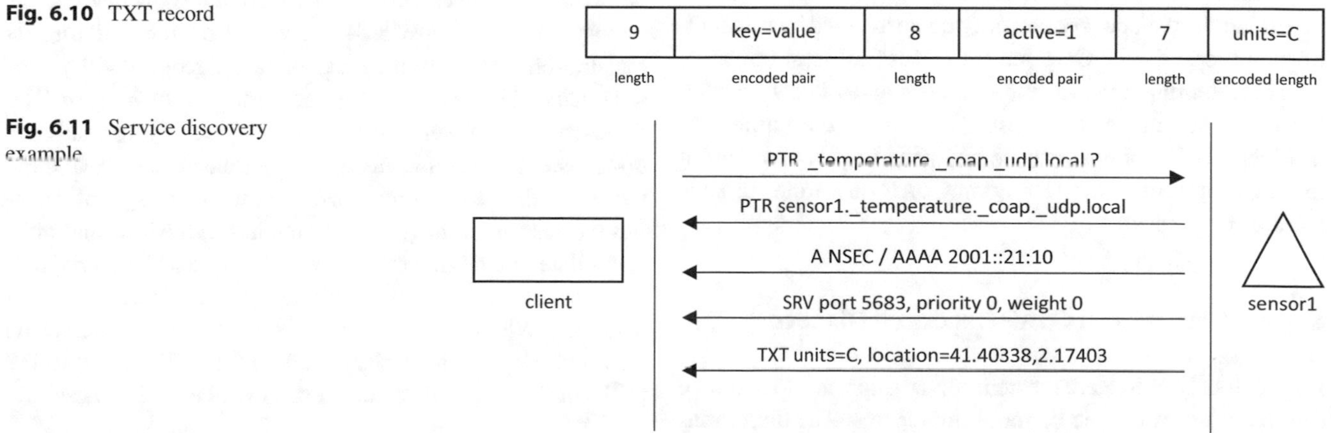

Fig. 6.10 TXT record

Fig. 6.11 Service discovery example

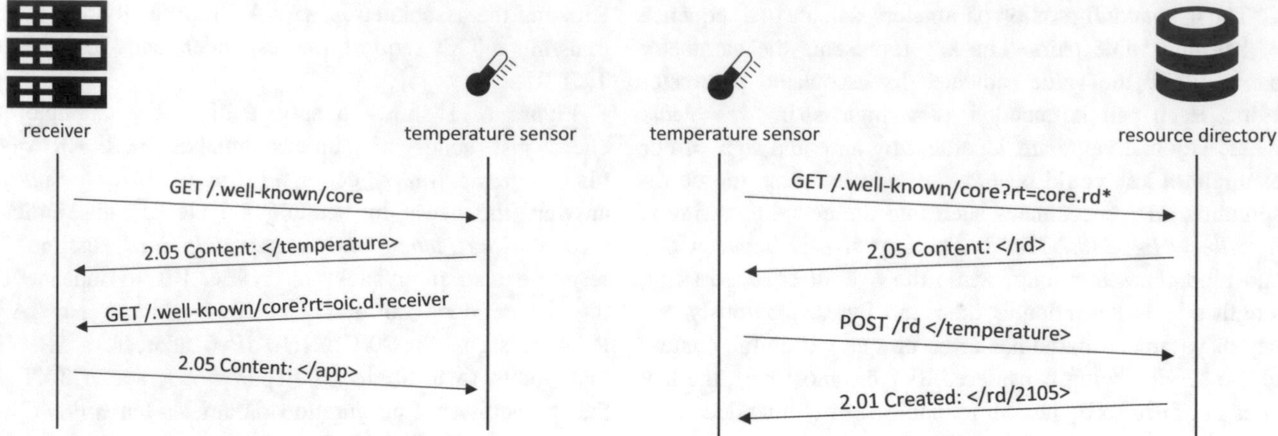

Fig. 6.12 Distributed CoAP resource discovery

Fig. 6.13 Resource discovery registration

by a temperature sensor by transmitting a regular CoAP GET request to the well-known URI */.well-known/core*. The querying application may or may not know the address of the device; for the first case, the application transmits the request directly to the unicast address of the device, while for the second case, the application transmits a multicast request instead. Devices can also discover applications; in Fig. 6.12 the temperature sensor sends a CoAP GET request to the */.well-known/core?rt=oic.d.receiver* URI. Note that the URI includes an attribute *resource type* (rt) that provides additional filtering in order to only reach receiver applications. This is specified by means of the *Internet Assigned Numbers Authority* (IANA) registered attribute value *oic.d.receiver*. Other allowed attributes that can be included in URIs are (1) the *interface description* (if) that is a string that indicates the name or URI used to access the resource, (2) the *maximum size* (sz) that specifies the maximum size of the resource representation returned by performing a GET of the target URI, and (3) the *content type* (ct) that indicates the format of the content returned when accessing the resource. In the context of CoAP, the use of URIs with attributes is specified by IETF RFC 6690 *Core Link Format* as a mechanism for servers to describe hosted resources.

In most cases, however, the device address is not known beforehand, so discovery can only be performed by transmitting multicast CoAP GET requests. But multicast transmissions are unreliable since there is no way to know whether a request has made it to destination since the number of destinations is unknown to begin with. In this situation, it makes a lot more sense to rely on CoAP directories to keep track of resources.

6.4.2 Centralized CoAP Resource Discovery

As previously indicated, the centralized approach to resource discovery relies on one or more directories. The directories store the description of the CoAP services and resources available in the LLN and provide an interface for devices and applications to query them. Devices register with the directories so that any entity can look up for resources by transmitting a single CoAP GET request. Since all queries are submitted to the directory, devices and applications must be able to reach it. There are three possibilities: (1) the device knows what the address of the directory is, (2) the directory address may be the same as that of the border router obtained by means of RAs, and (3) the directory address is discovered through a multicast transmission.

Figure 6.13 shows the device registration process. First, a device sends a CoAP GET request to the */.well-known/core?rt=core.rd** URI in order to determine the specific URI of the directory. This is usually done in all scenarios, including under multicast discovery. The rt attribute points to *core.rd** in association with the resource directory information. Once the directory location and address are known, the temperature sensor can then transmit a CoAP POST to the */rd* URI to add its resource identifier. The directory responds with a *2.01 Created* to indicate that the entry has been inserted into the database and it has been assigned the */rd/2105* URI.

CoAP introduces additional functionality to manage directory entries. Figure 6.14 shows the device updating its registration by transmitting a CoAP PUT request with a new URI */temp*. The request is replied with a *2.04 Changed*. The directory can also verify if the device resource information is up-to-date. In this case, the directory transmits a CoAP GET request to the */.well-known/core* URI with a ETag that specifies the version of the information it has. If the information is up-to-date, the device replies with a *2.03 Valid*; otherwise it transmits the updated information by means of a *2.05 Content* response. When the temperature sensor wants to delete its directory entry, it transmits a CoAP DELETE request to the URI */rd/2105*. The directory replies with a *2.02 Deleted*.

If a device wants to look up for specific resource information in the directory, as illustrated in Fig. 6.15, it can transmit a GET CoAP request to the */rd/res?rt=oic.d.receiver* URI. The lookup type */rd/res* indicates that the device is querying the directory for those registered resources of type *oic.d.receiver*. The directory then replies with a 2.05 Content that includes the *coap://[2001::21:5]:5683/app* URI for direct access to the resource. Similarly, if the device just wants to find out what the domain of the receiver is, it can issue a GET CoAP request to the */rd/ep?rt=oic.d.receiver* URI where the lookup type */rd/ep* specifies that endpoint information of type *oic.d.receiver* is being requested.

6.5 UPnP

The *Universal Plug and Play* (UPnP) architecture provides a framework for the discovery of network elements ranging from computers and printers to gateways and access points [12]. UPnP enabled zero-configuration networking by leveraging IP, TCP, UDP, HTTP, and XML. Each device is associated with a particular XML schema that defines the data structures that are used for interoperability and discovery. Note that most of these protocols and technologies are not efficient in the context of LLN scenarios, so improvements have been proposed to enable UPnP IoT support [13, 4].

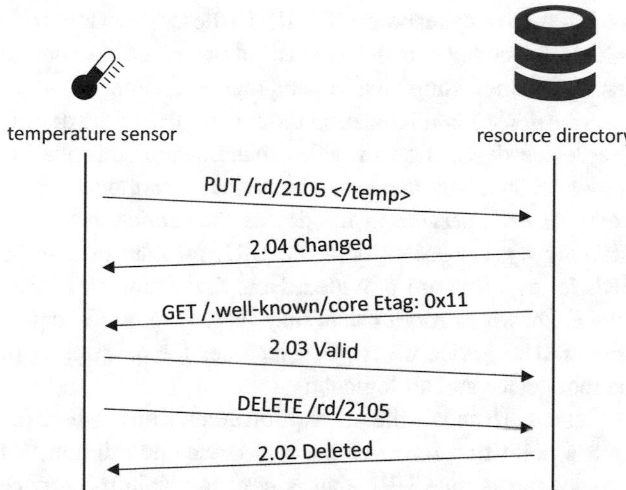

Fig. 6.14 Resource discovery entry maintenance

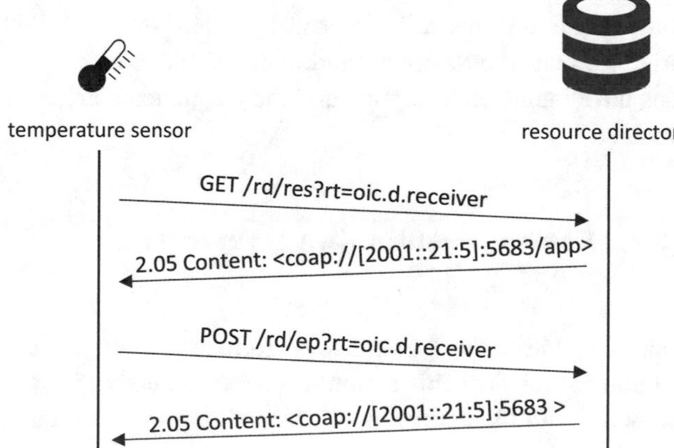

Fig. 6.15 Resource discovery lookup

UPnP entities are classified as either controlled devices or control points. Controlled devices are, from an IoT functional perspective, sensors and actuators that act as UPnP servers, while control points are user applications that run on smartphones and typically act as UPnP clients. UPnP networking is made of five different steps: addressing, discovery, description, control, eventing, and presentation. Addressing is the first step of UPnP networking and allows devices and control points to get network addresses. Once a control point has an address, through discovery, it can find controlled devices and then obtain their capabilities by means of the description step. Through eventing, control points listen for state changes of the devices. The last step presentation enables control points to display a user interface for the devices.

6.5.1 Addressing and Discovery Steps

For IoT applications, as part of the addressing step, a device or a control point acquires an IPv6 address by one of two possible mechanisms: through DHCPv6 or auto-configured as a link-local address. Discovery is then used by devices to advertise their services and discover services provided by other devices in the network. The discovery message includes some basic information about the device including additional pointers to more detailed information. In addition, when a device joins the network, it may multicast a discovery message to search for other devices providing specific capabilities. In general, devices listen for multicast messages and respond accordingly when the search criterion of the query matches the characteristics of the device. The transmission of these messages relies on the *Simple Service Discovery Protocol* (SSDP) for both discovery and advertising. SSDP is an ASCII based protocol that is based on *HTTPU*, that is HTTP over UDP, in order to support multicast transmissions.

There are three types of advertising messages: (1) *alive* messages that enable devices to advertise services that are available, (2) *update* messages that are used by devices to make changes to the available services, and (3) *byebye* messages that allow devices to signal the removal of services. The message below is an example of an SSDP device advertisement message:

```
NOTIFY * HTTP/1.1
Host: [FF02::C]:1900
Cache-ControlL: max-age=3600
NT: uuid:2619d124-5400-2105-af21-90123ae100ff
NTS: ssdp:alive
Location: http://[2001::21:5]:8060/sensor/data.xml
USN: uuid:2619d124-5400-2105-af21-90123ae100ff
```

The message is transmitted by a device as a multicast datagram to destination [FF02::C]:1900 and includes a start line *NOTIFY * HTTP/1.1* that indicates that the request is an advertisement. The message includes a *Host* header that specifies the destination address and port included at the application layer for recovering when NAT is in place; it continues with a *Cache-Control* header that determines how long the message is valid, an *NT* header that indicates the notification type as a service identifier, an *NTS* header that specifies the notification subtype as *ssdp:alive*, a *Location* header that indicates the URL that provides more information about the service provided by the device, and a *USN* header that identifies the service. Note that in this message, the NTS and USN headers carry the same information.

Discovery, as advertising, is carried out by the transmission of a multicast datagram request from a control point to destination [FF02::C]:1900. The message below is an example of an SSDP control point discovery message:

```
M-SEARCH * HTTP/1.1
Host: [FF02::C]:1900
MAN: ssdp:discover
MX: 1
ST: ssdp:all
```

The start line *M-SEARCH * HTTP/1.1* indicates that the message is a discovery request. As for the advertising message, the message includes a *Host* header that specifies the destination address and port included at the application layer for recovering when NAT is in place. The *MAN* header that follows specifies the type of message which for discovery requests it is always *ssdp:discover*. The *MX* header indicates how many seconds can devices wait before responding to the request. The message ends with the *ST* header that specifies the scope of the search; in this case, *ssdp:all* indicates that all devices and services must be searched.

Devices reply by transmitting a unicast response to the device. The message below is an example of an SSDP device discovery response message:

```
HTTP/1.1 200 OK
Cache-Control: max-age=3600
ST: uuid:2619d124-5400-2105-af21-90123ae100ff
USN: uuid:2619d124-5400-2105-af21-90123ae100ff
Server: Raspian/Buster UPnP/1.0 Temperature Sensor/2.4
Location: http://[2001::21:5]:8060/sensor/data.xml
```

The start line *HTTP/1.1 200 OK* follows the convention of the HTTP responses and specifies this is a response to the request transmitted by the control point. The message continues with a *Cache-Control* header that determines how long the message is valid. The message follows with an *ST* header that specifies the identity of the service responding to the discovery request, and a *USN* header also identifies the service. The *Server* header indicates the server information including OS, protocol, and application version. Finally, the *Location* header indicates the URL that provides more information about the service provided by the device.

6.5.2 Description Step

The discovery step gives a control point some basic information about the devices, but it does not provide any specific information about the data structures needed to interact with them. Specifically, to access a device, the control point must be able to download the description of the device and its capabilities that are indicated in the Location header of the 200 OK discovery response. The device description is carried as XML formatted data that consists of two parts: (1) a *device description* indicating physical and logical containers and (2) a *service description* indicating the capabilities of the device. The device description provides manufacturer information like serial number, vendor information, and model name, while for each service in the device the service name, the URL for service description, the URL for control and the URL for eventing are provided. Note that a single physical device, known as *root device*, may include multiple *logical devices*. The device description includes the description of the root device and all logical devices.

Figure 6.16 shows the description architecture, where the control point first retrieves the root device description that includes the service URL that is used to obtain the service description of the logical devices. The service description defines the actions that the control point can perform including the arguments these actions can take. It also includes a list of the possible states a device can take with its associated variables, data types, range, and event characteristics. Actions have inputs and outputs, and their arguments are state variables.

6.5.3 Control, Eventing, and Presentation Steps

Control enables control points to perform actions on devices and poll for results. This action/response interaction is synchronous, with the control point transmitting the action and

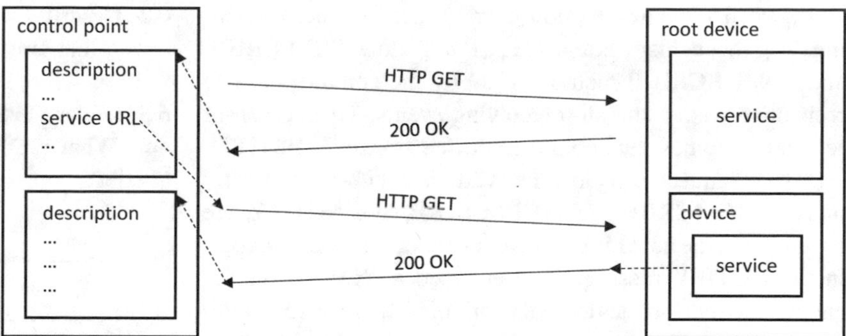

Fig. 6.16 Description architecture

waiting for the response from the device. When the device receives the action, it processes it and replies with either a successful response or an error. The action is sent to the URL associated with the service obtained from the service description. Control messages are encoded as SOAP payloads transmitted over HTTP. As indicated in Sect. 5.2.6, SOAP is an XML data structure that provides a mechanism for exchange of messages. SOAP messages include a mandatory *envelope*, an optional *header*, and a mandatory *body*.

The message below shows an example of an action transmitted from the control point to the device:

```
POST http://[2001::21:5]:8102/upnp/control/Temperature HTTP/1.1
SOAPAction: urn:schemas-upnp-org:service:Temperature:1
Host: [2001::21:5]:8102
Content-Type: text/xml;charset="utf-8"
Content-Length: 324

<?xml version="1.0" encoding="utf-8"?>
  <s:Envelope
  s:encodingStyle="http://schemas.xmlsoap.org/soap/encoding/"
  xmlns:s="http://schemas.xmlsoap.org/soap/envelope/">

    <s:Body>
      <ns0:GetSensorTemperature xmlns:ns0="urn:schemas-upnp
            -org:service:Temperature:1">
      </ns0:GetSensorTemperature>
    </s:Body>
  </s:Envelope>
```

The message is transmitted as a HTTP POST request with a SOAP body. The message not only includes regular HTTP headers like *Content-Type*, *Content-Length*, and *Host*, but it also has a SOAP specific header *SOAPAction* that indicates the intent of the message. The body includes an Envelope that contains a Body that invokes the *GetSensorTemperature* action. The device replies by transmitting the following message:

```
HTTP/1.0 200 OK
Content-Type: text/xml; charset="utf-8"
Server: Raspian/Buster UPnP/1.0 Temperature Sensor/2.4
Content-Length: 328

<?xml version="1.0"?>
  <s:Envelope
  s:encodingStyle="http://schemas.xmlsoap.org/soap/encoding/"
  xmlns:s="http://schemas.xmlsoap.org/soap/envelope/">
    <s:Body>
      <ns0:GetSensorTemperatureResponse xmlns:ns0="urn:
            schemas-upnp-org:
        service:Temperature:1">20C
      </ns0:GetSensorTemperatureResponse>
    </s:Body>
  </s:Envelope>
```

The response is a regular HTTP 200 OK message with a SOAP body that has an Envelope that contains in the Body the response *GetSensorTemperatureResponse* that carries a *20C* readout.

When a device like a temperature sensor generates a new readout, it can trigger an event on the control point. Eventing supports two ways of operation: unicast by relying on HTTP over TCP and multicast by means of HTTPU, that is, HTTP over UDP.

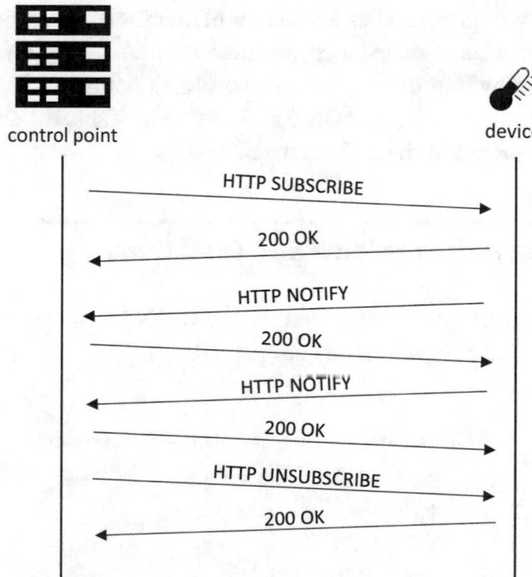

Fig. 6.17 Unicast eventing message flow

Figure 6.17 shows the flow of messages for unicast eventing. The mechanism extends HTTP to include SUBSCRIBE and UNSUBCRIBE methods used by the control points to respectively start and stop receiving events. When a sensor readout becomes available, the device transmits an HTTP NOTIFY request carrying the readout itself. All these requests, SUBSCRIBE, UNSUBSCRIBE, and NOTIFY, are replied with the usual HTTP response codes. Device readouts in the NOTIFY messages are encoded as XML bodies to encode the relevant device state variables. In the case of multicast transmissions, control points do not need to subscribe as the devices transmit the NOTIFY messages directly over HTTPU, and they do not expect a response either.

Lastly, the presentation stage, when signaled through a URL in the device description, provides a web interface for users in control points to control the device and view its status. As such, presentation stage is supported by the device acting as a web server.

Summary

Most IoT solutions rely on a great number of resource and service discovery mechanisms that when combined with SAA enable zero-configuration scenarios. This chapter started by presenting the main concepts and functions of service discovery including publications, registration, discovery, and resolution. It described how traditional DNS can be extended to provide lightweight multicast mDNS support over low-power and constrained devices transmitting in LLNs. The chapter explored the details of mDNS and its use along with DNS-SD to provide a generic mechanism for service discovery. Alternatives to both mDNS and DNS-SD were then presented as part of this chapter. Specifically, CoAP was presented as a provider of an efficient method for centralized and distributed resource discovery. The chapter finished by introducing details and functionality of UPnP as another alternative to mDNS and DNS-SD. Table 6.1 shows a comparison of these three mechanisms.

Homework Problems and Questions

6.1 Describe a scenario where the mDNS responder is a sensor that only advertises unique RRs.

Table 6.1 Resource identification and management protocols

Protocol	Transport	Overhead	Generic?
mDNS/DNS-SD	UDP	Low	Yes
CoAP SD	UDP	Medium	No, specific to CoAP
UPnP	TCP	High	Yes

6.2 Describe a scenario where the mDNS responder is a sensor that only advertises shared RRs.

6.3 Under the SD-DNS the following transaction is carried out. What mDNS query type is likely to be sent by the querier?

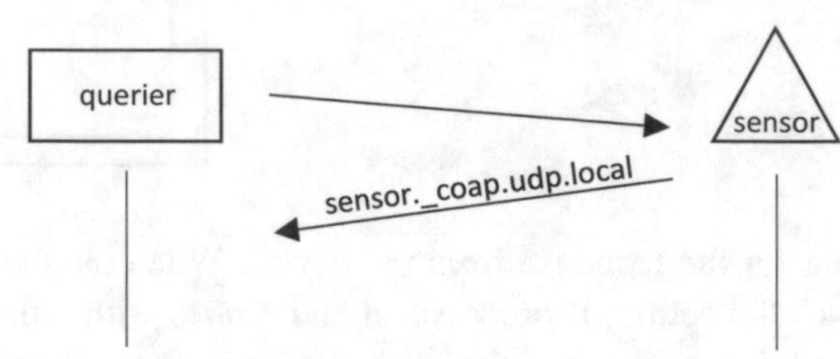

6.4 Under the SD-DNS the following transaction is carried out. What mDNS query type is likely to be sent by the querier?

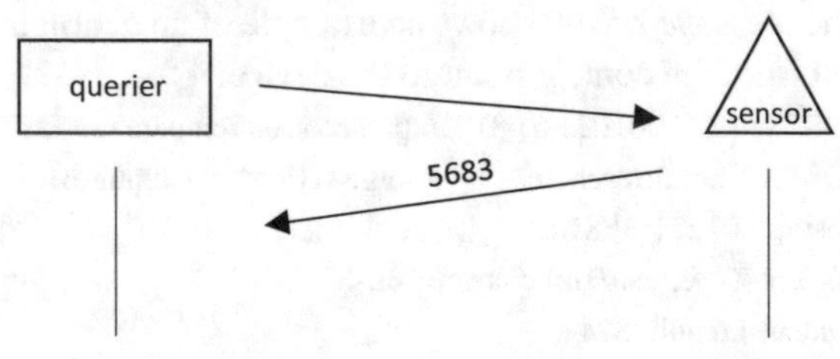

6.5 Under the SD-DNS the following transaction is carried out. What mDNS query type is likely to be sent by the querier?

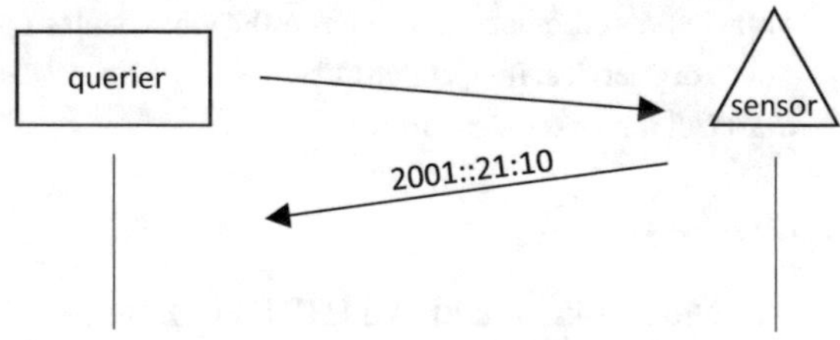

6.6 Consider an SD-DNS client (running on an application) that queries four sensors to discover service instances. If the only delay is the transmission delay and the transmission rate on both directions is 135 Kbps, how long does it take for the client to discover all instances? Assume a uniform packet size of 120 bytes.

6.7 Consider an SD-DNS client that queries the addresses of four sensors. If TTL fields of RRs are multiples of 100, please fill the empty spaces in the figure below:

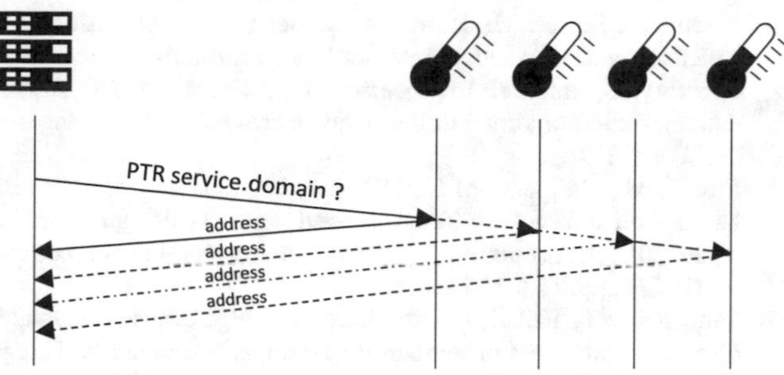

6.8 Show an mDNS message flow example where a sensor responds by transmitting at least one NSEC RR.

6.9 If the RTT is 200 ms, how long does it take for a client to fully discover and transmit a NON CoAP request to a sensor?

6.10 How long does a transaction to delete a sensor entry in the CoAP resource directory take? Assume a bidirectional 130 Kbps transmission rate as well as minimum size IEEE 802.15.4 and 6LoWPAN headers. Use Fig. 6.14 as reference.

6.11 Consider the UPnP scenario shown in Fig. 6.17. Assuming a 120 ms RTT and nonpersistent HTTP connections, how long does it take for all transactions to be completed?

References

1. Al-Fuqaha, A., Guizani, M., Mohammadi, M., Aledhari, M., Ayyash, M.: Internet of things: A survey on enabling technologies, protocols, and applications. IEEE Commun. Surv. Tuts. **17**(4), 2347–2376 (2015)
2. Andrew Banks Ed Briggs, K.B., Gupta, R.: Mqtt version 3.1.1 oasis committee specification (2014). http://docs.oasis-open.org/mqtt/mqtt/v3.1.1/mqtt-v3.1.1.html
3. Banks, A., Gupta, R.: Mqtt version 5.0 oasis committee specification (2019). https://docs.oasis-open.org/mqtt/mqtt/v5.0/mqtt-v5.0.html
4. Bodlaender, M.P.: Upnp/spl trade/ 1.1 - designing for performance compatibility. IEEE Trans. Consumer Electron. **51**(1), 69–75 (2005)
5. Carballido Villaverde, B., Alberola, R.D.P., Jara, A.J., Fedor, S., Das, S.K., Pesch, D.: Service discovery protocols for constrained machine-to-machine communications. IEEE Commun. Surv. Tuts. **16**(1), 41–60 (2014)
6. Cheshire, S., Krochmal, M.: DNS-based service discovery. RFC 6763 (2013). https://doi.org/10.17487/RFC6763. https://rfc-editor.org/rfc/rfc6763.txt
7. Cheshire, S., Krochmal, M.: Multicast DNS. RFC 6762 (2013). https://doi.org/10.17487/RFC6762. https://rfc-editor.org/rfc/rfc6762.txt
8. Datta, S.K., Costa, R., Bonnet, C.: Resource discovery in internet of things: current trends and future standardization aspects. In: IEEE World Forum on Internet of Things (WF-IoT), pp. 542–547 (2015). https://doi.org/10.1109/WF-IoT.2015.7389112
9. Domain names - concepts and facilities. RFC 1034 (1987). https://doi.org/10.17487/RFC1034. https://rfc-editor.org/rfc/rfc1034.txt

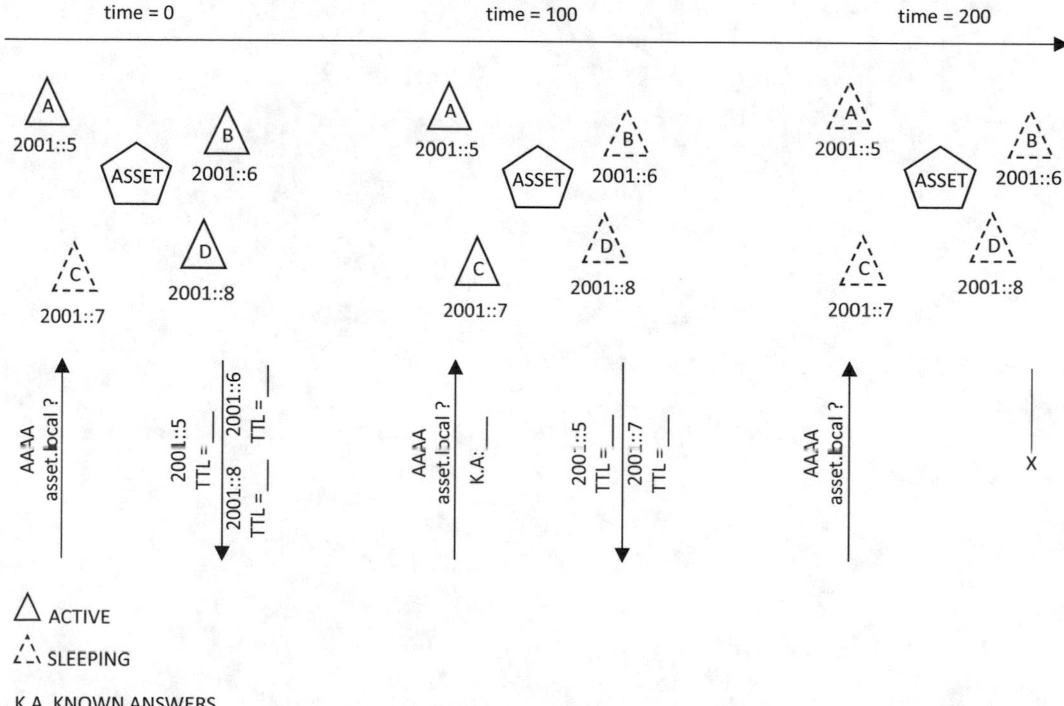

10. Fielding, R., Gettys, J., Mogul, J., Frystyk, H., Masinter, L., Leach, P., Berners-Lee, T.: Rfc 2616, hypertext transfer protocol – http/1.1 (1999). http://www.rfc.net/rfc2616.html
11. Florea, I., Rughinis, R., Ruse, L., Dragomir, D.: Survey of standardized protocols for the internet of things. In: 2017 21st International Conference on Control Systems and Computer Science (CSCS), pp. 190–196 (2017)
12. Foundation, O.C.: Upnp device architecture 2.0 (2020). https://openconnectivity.org/upnp-specs/UPnP-arch-DeviceArchitecture-v2.0-20200417.pdf
13. Jeronimo, M., Weast, J.: UPnP Design by Example: A Software Developer's Guide to Universal Plug and Play. Intel Press (2003)
14. Shelby, Z., Hartke, K., Bormann, C.: The Constrained Application Protocol (CoAP). RFC 7252 (2014). https://doi.org/10.17487/RFC7252. https://rfc-editor.org/rfc/rfc7252.txt
15. Shelby, Z., Koster, M., Bormann, C., der Stok, P.V., Amsuss, C.: CoRE resource directory. Internet-Draft draft-ietf-core-resource-directory-25, Internet Engineering Task Force (2020). https://datatracker.ietf.org/doc/html/draft-ietf-core-resource-directory-25. Work in Progress
16. Stolikj, M., Cuijpers, P.J.L., Lukkien, J.J., Buchina, N.: Context based service discovery in unmanaged networks using mdns/dns-sd. In: 2016 IEEE International Conference on Consumer Electronics (ICCE), pp. 163–165 (2016)
17. Tanganelli, G., Vallati, C., Mingozzi, E.: Edge-centric distributed discovery and access in the internet of things. IEEE Internet Things J. **5**(1), 425–438 (2018)

7 Routing on Constrained Devices

7.1 Routing Concepts

Routing in the context of IoT has its origins in the routing of WSNs due to the similarities between both technologies. As indicated in Sect. 2.5, routing protocols that have been proposed for MANETs are not typically suitable for either WSNs or IoT networks in general [5]. The same applies to traditional routing mechanisms widely used in the Internet as a whole. Moreover, most of these traditional mechanisms rely on routing that is based on network layer addresses. This means that routers will forward datagrams ultimately based on routing tables that look at destination addresses to determine outgoing links. Routing tables, in turn, are directly or indirectly populated by routing algorithms. In WSNs, however, addresses are not always the best candidates to base routing decisions. Alternative approaches are needed.

WSN routing protocols can be flat or hierarchical [9]. This is shown in Fig. 7.1. Flat routing relies on devices interacting with each other without a single device acting as parent that concentrates and aggregates traffic of children devices. Hierarchical routing, on the other hand, groups devices in clusters with clusterheads that act as parent devices to concentrate and aggregate traffic of children devices. Children devices do not typically interact with each other and all traffic is routed through the clusterheads that talk to other clusterheads in order to provide end-to-end connectivity. The advantage of flat routing is lower latency at a cost of lower energy efficiency. Similarly, the advantage of hierarchical routing is higher energy efficiency at a cost of higher latency. Energy efficiency in hierarchical routing is accomplished by means of multihop or mesh routing under the umbrella of capillary networks. A capillary network, shown in Fig. 2.3, consists of devices relying on intermediate devices to route their traffic to the clusterhead. On average, the individual devices consume less power by aggregating traffic on intermediate nodes than when directly taking to their destinations. But traffic aggregation means extra buffering and retransmissions that increase latency. This leads to multipoint clustering where multiple clusterheads aggregate the traffic from multiple devices before it gets forwarded to other clusterheads for wide distribution.

The main alternative to address based routing is data-centric routing. Data-centric routing involves clients sending queries to specific network regions in order to retrieve specific readouts and data associated with specific capabilities. This is illustrated in Fig. 7.2; multiple devices acting as sources collect readouts and forward them to sinks in response to an initial query. In this scenario, attribute based tags replace addresses for the purpose of routing. Temperature readouts are forwarded by network routers until they reach a temperature processing application. Similarly, humidity and pressure readouts are forwarded throughout the network based on tags until they reach applications that can process them. Conceptually, this is similar to the Publish/Subscribe application layer messaging described in Sect. 5.3. In fact, data-centric routing is also performed at the application layer with the queries acting as subscriptions and the responses serving as notifications. However, one important difference between data-centric routing and application layer protocols is that the latter rely on additional brokers to complete the communication path. In fact, under data-centric routing each device acts as a broker keeping track of attribute based names of other devices and forwarding traffic accordingly. Essentially, requests and responses are diffused over the network with mandatory support of aggregation at the routers. Data-centric routing typically goes hand in hand with hierarchical routing as clusterheads serve as ideal routers to aggregate and propagate traffic. Another advantage of the data-centric routing is that attribute naming can be used to reach multiple devices with a single message if the attribute represents a common feature. For example, a client may send a datagram intended to all devices supporting temperature readouts in each region. The datagram will be forwarded by all intermediate devices supporting data-centric routing. Note this is analogous to multicast propagation of address based

R. Herrero, *Fundamentals of IoT Communication Technologies*, Textbooks in Telecommunication Engineering, https://doi.org/10.1007/978-3-030-70080-5_7

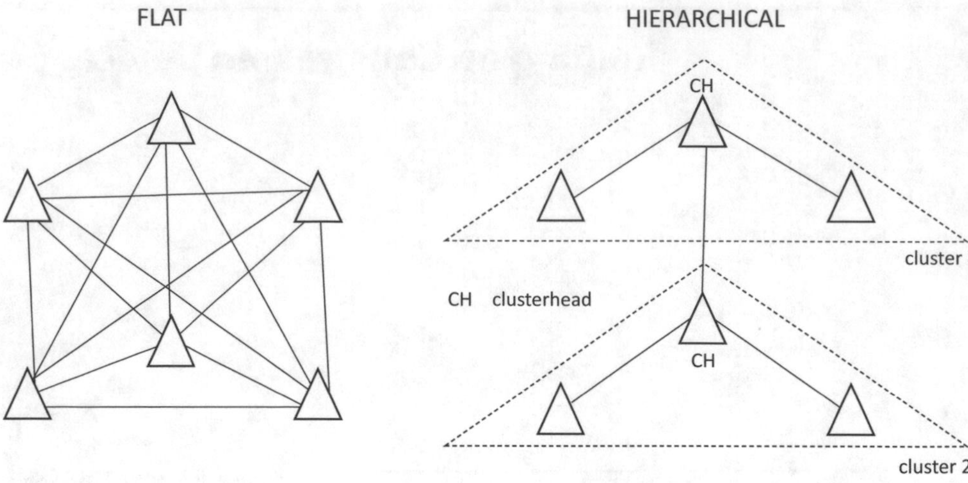

Fig. 7.1 Flat vs hierarchical routing

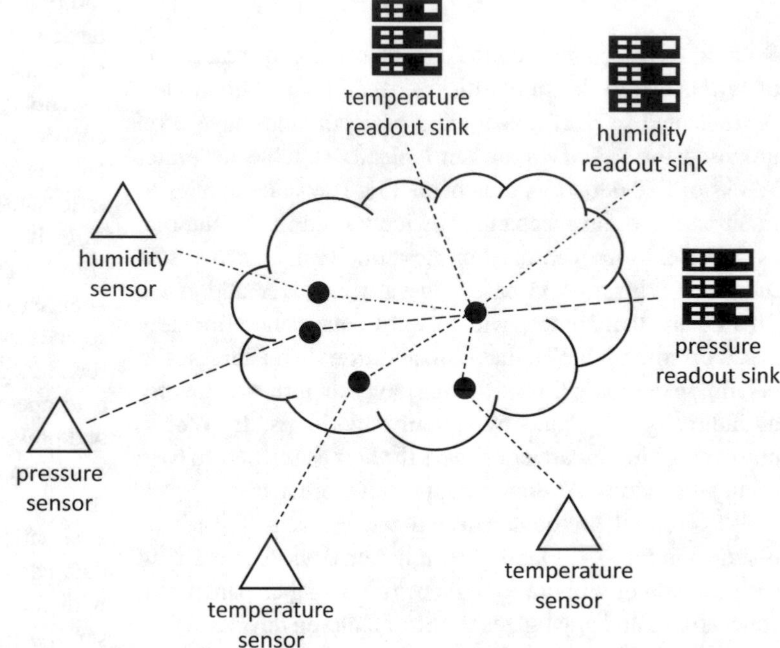

Fig. 7.2 Data-centric routing

routing. Data naming relies on the use of relevant attributes like data types and timestamps. In the context of a fully connected network, Fig. 7.3 shows a scenario with a remote user residing at the edge of the IP cloud that transmits a query to reach a WSN sensor. The query is propagated beyond the gateway all the way to the sensor by means of intermediate nodes forwarding this request. The response, on the other hand, makes it back all the way to the remote user.

Location based routing is an alternative to both address and data-centric routing. It consists of a client or device transmitting datagrams to another client or device by forwarding the traffic based on the geographical or physical location of the destination. To this end, the client and each router along the way estimate the direction of propagation of datagrams to efficiently reach the destination with minimal energy consumption. Figure 7.4 illustrates an example of location based routing; a client attempts to send a datagram to a destination that is known by its geographical coordinates. First the client, that knows its own coordinates, forwards the datagram to router 1 that is the closest one in the direction of the destination. Similarly, router 1, that also knows its own coordinates, forwards the datagram to router 2 that, again, is the closest one in the direction of the destination. Finally, router 2, that has a direct link to the destination, just forwards the datagram. Location based routing can be combined with hierarchical routing to exploit traffic aggregation.

One drawback of traffic aggregation, however, is that it results in increased throughput that limits channel capacity and leads, in turn, to contention as well as additional latency. So, one desired property when performing aggregation is making sure that readouts are processed and compressed at the device whenever possible. In other words, local pro-

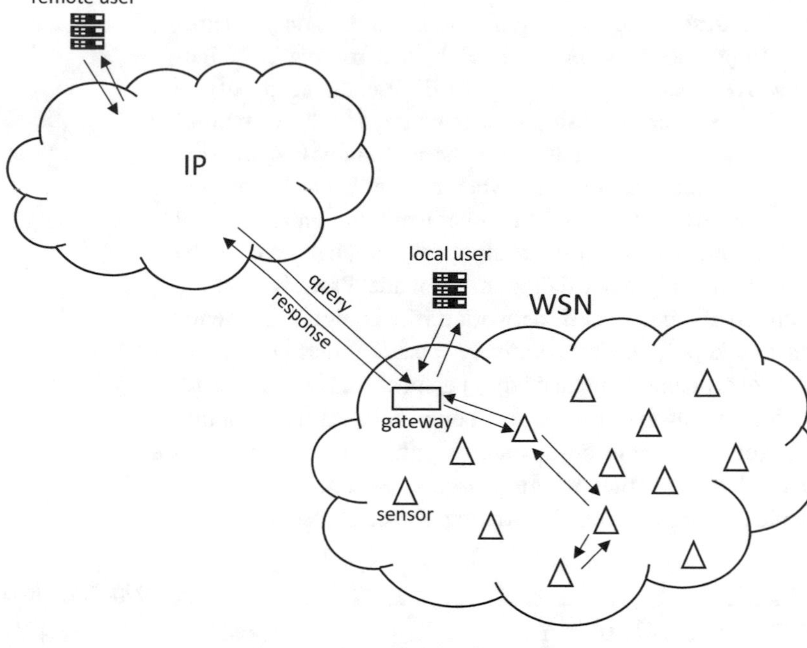

Fig. 7.3 WSN data forwarding

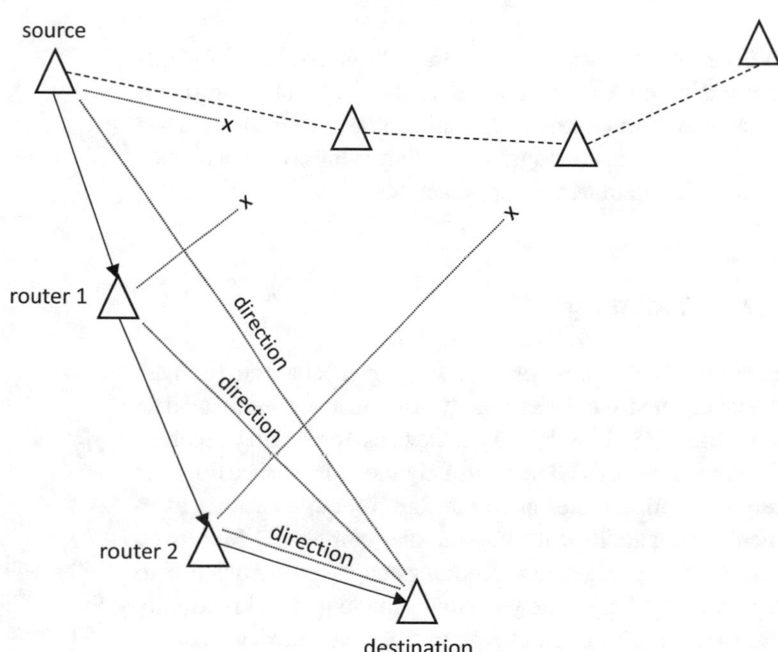

Fig. 7.4 Location based routing

cessing enables the conversion from data to information presented in Sect. 1.3. The price to pay for the conversion is the need for higher computational power that together with the transmission of aggregated data increase battery drainage.

The overall idea of routing mechanisms in the context of both WSNs and IoT is to maximize the network lifetime attempting to distribute routing traffic over as many devices as possible. Aggregation and routing must be highly efficient due to the well known energy limitations and constraints of embedded devices like sensors and actuators. In general, the lower the power consumption, the less responsive the routing mechanism is, as fewer routing control packets are typically transmitted thus reducing the overall network bandwidth utilization. There is, therefore, a balance between responsiveness and power consumption. Moreover, reducing power consumption is typically accomplished by means of power cycles that require full network coordination to minimize latency and other negative effects on the routing algorithms.

Routing is classified in regard to three possible strategies that relate to the balance between responsiveness and power consumption: (1) proactive routing consists of building routing tables that ultimately define forwarding behavior by means of the periodic transmission of routing information across all nodes in the network, (2) reactive routing relies on dynamic on-demand building of routes from sources to specific destinations by relying on route discovery queries that flood the network, and (3) hybrid routing combines proactive and reactive routing on different sections of the network. Proactive routing is characterized by heavy control traffic overhead that results from the continuous transmission of routing information even when device datagrams are not sent. Reactive routing, on the other hand, minimizes control traffic transmission until a routing decision, triggered by the transmission of device datagrams, is made. Proactive routing is effective when overall network traffic is high, and latency must be kept low. Comparatively speaking, reactive routing is more efficient from an energy perspective, but it can result in device traffic transmission being delayed until routing decisions are made on-demand. Hybrid routing is used in hierarchical schemes where proactive routing is performed inside clusters and reactive routing is used across clusters.

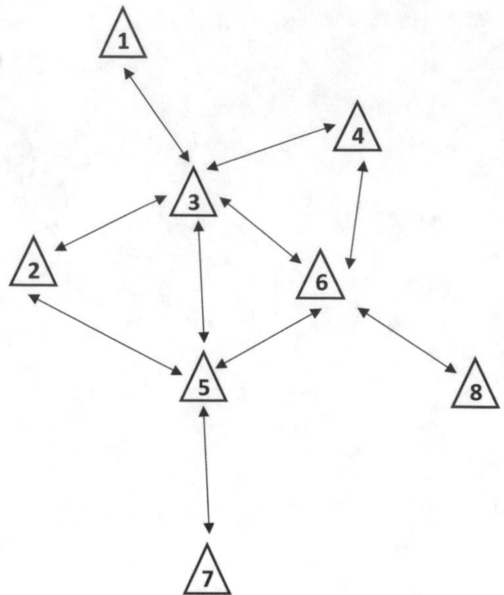

Fig. 7.5 Flooding

7.2 WSN Routing

WSN routing schemes incorporate different algorithms that use and integrate aspects of flat and hierarchical routing combined with data-centric and location based mechanisms. They range from simple datagram flooding to highly sophisticated hierarchical data-centric approaches.

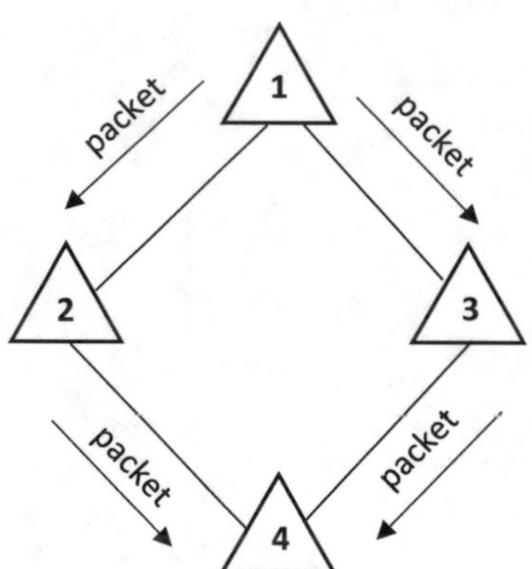

Fig. 7.6 Traffic implosion

7.2.1 Flooding

Flooding is the simplest flat routing mechanism for path discovery that has been widely used in the context of data networks [4]. It is based on devices forwarding received datagrams through all possible neighbors. As such, it does not rely on routing tables and therefore it requires neither extra memory nor additional computational complexity. Moreover, due to its dynamic nature, flooding adapts quite efficiently to network topology changes. Now, simpleness and flexibility are expensive from an energy perspective, making flooding largely impractical in most deployments.

Figure 7.5 presents a flooding example, eight devices generating and forwarding datagrams to provide full end-to-end connectivity. Because flooding generates many copies of the same datagram, both channel utilization and contention are high enough that they lead to some loss and latency that is typically compensated by the intensity of the traffic. Additionally, since a device can forward the same datagram an infinite number of times, flooding usually relies on the use of hop counters to prevent this type of loops. As a datagram circulates throughout the network, its hop counter is decremented by one each time it is forwarded. When the counter becomes zero, the datagram is dropped by the corresponding device. A slightly more resource intensive but more efficient approach consists of assigning, on generation, a random identifier to the datagram. Devices can then build tables to keep track of the different datagrams in order to automatically drop them whenever they are about to be retransmitted. Obviously, this implies extra memory and additional computational complexity.

The energy inefficiencies come manifested as (1) traffic implosion and (2) overlapping. Traffic implosion is shown in Fig. 7.6. The same datagram generated by device 1 arrives as two copies to device 4. This is not only a waste of energy but a waste of network bandwidth and channel capacity. Overlapping, shown in Fig. 7.7, consists of redundant sensor data being transmitted over multiple datagrams that simulta-

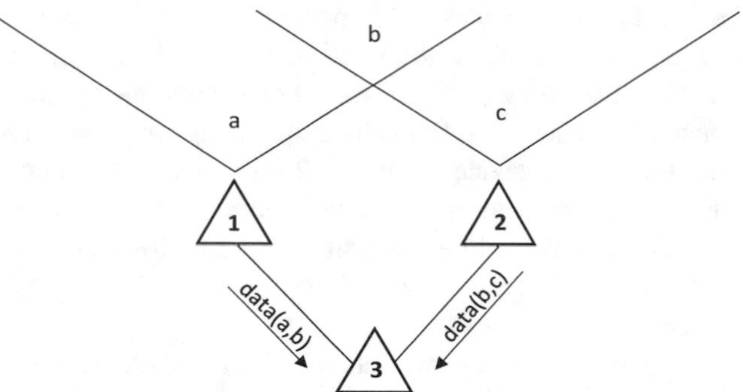

Fig. 7.7 Flooding traffic overlapping

neously arrive at destination. Device 1 transmits a datagram that contains data associated with assets a and b while device 2 transmits a datagram that contains data associated with assets b and c. Then both datagrams arrive at device 3, two copies of the data related to asset b are received. Again, this is not only a waste of energy but a waste of network bandwidth and channel capacity. Both traffic implosion and overlapping eventually lead to resource blindness that result in the device energy being depleted. Since the flooding algorithm is not aware of device power consumption, resource blindness is, sooner than later, unavoidable.

7.2.2 Gossiping

An alternative to flooding that relies on intermediate devices forwarding a received datagram to a single randomly selected neighbor is called gossiping. Gossiping, by being a variation of flooding, is also a flat routing algorithm [11]. The idea is to minimize the power consumption by bounding the number of transmissions associated with each hop. Moreover, as in the flooding case, datagrams also include hop counters that limit how many times they can be forwarded.

Figure 7.8 shows an example of gossiping; device 1 sends a datagram to device 7. The datagram traversal follows seven hops a, b, c, d, e, f, and g. At each hop, the datagram is forwarded to only one neighbor. While the most efficient path consists of three hops through devices 3 and 5, the random nature of gossiping can result, as illustrated in the example, in a lot more inefficient path. Longer paths imply, when compared to flooding, higher latency and imply, when compared to the randomly selected shortest path, higher energy consumption.

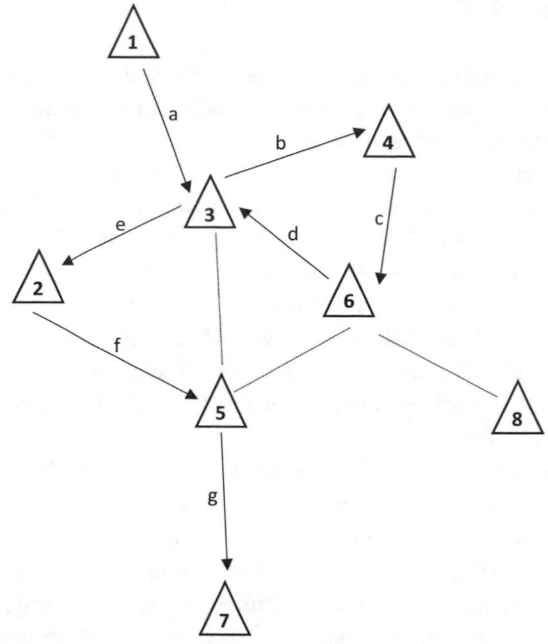

Fig. 7.8 Gossiping

Example 7.1 Consider a WSN based ring topology where devices send packets to each other:

(continued)

(continued)

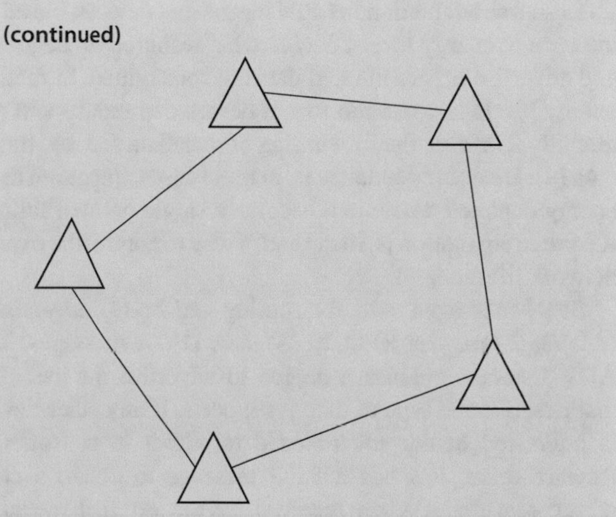

Assume that packets are uniformly transmitted in order to provide full connectivity among all devices. If the transmission rate is 250 Kbps, the packet size is 200 bytes and ignoring collisions as well as delays (other than the transmission delay), what is the expected transmission delay in a flooding scenario?

Solution: The transmission delay of a packet over a single hop is $D_h = \frac{8\times 200 \text{ bytes}}{250000 \text{ bps}} = 6.4$ ms. Each device transmits to $N = 4$ other devices, therefore, the delay with respect to two of those devices is D_h while the delay with respect to the remaining two devices is $2 \times D_h$. The expected delay due to flooding is, therefore, $D_{\text{flooding}} = \frac{2\times D_h + 2\times 2\times D_h}{N} = 9.6$ ms.

7.2.3 SPIN

Sensor Protocols for Information via Negotiation (SPIN) is a flat data-centric routing mechanism that through data negotiation and resource adaptation enables devices to forward datagrams from a source to a sink in a much more controlled and more energy efficient way than flooding and gossiping [14]. Energy efficiency is accomplished by making devices learn about the metadata associated with other devices before any datagrams can be forwarded. Metadata are data that describe other data and, as such, are typically much smaller than the described data themselves. By being smaller, lower transmission rates and less energy are needed for transmission.

SPIN relies on end and intermediate devices advertising metadata that they generate or forward so that other devices interested in the associated data can request them [9]. Data are sent to the devices that are really interested in forwarding or using them in order to minimize the waste of energy that results from sending datagrams to nodes that do not need to be involved in the process. Moreover, SPIN, when compared to flooding, eliminates the transmission of redundant data that causes traffic implosion and overlapping.

Resource adaptation in SPIN means that devices, based on the current energy levels, decide what resources to negotiate and advertise before the real data are transmitted. In fact, as energy levels become too low, a device can reduce and potentially eliminate the forwarding of metadata and real traffic data [8]. Device redundancy is, in these cases, important as an energy depleted device can become a single point of failure. Resource adaptation is, in the end, a way to control the overall network lifetime.

SPIN messages used for routing can be (1) *advertising* (ADV), (2) *request* REQ, or (3) *data* DATA messages. The ADV message enables a device to advertise the metadata that describe the type of data it supports. If any other device is interested in the advertised data, either to consume or forward them, it sends a REQ message to obtain a copy of the real data. When received by the original metadata transmitter, the request triggers the transmission of a DATA message containing the actual data. These transactions are illustrated in Fig. 7.9. Again, since the ADV message is much smaller than the corresponding DATA message, the mechanism guarantees that energy is only spent on the transmission of datagrams that carry relevant information.

Figure 7.10 shows a simple example of SPIN data negotiation and propagation; the idea is for device 1 to transmit its data to other devices in the network even if they are not directly connected. Essentially, device 1 starts by advertising metadata that its neighbor device 2 expresses interest in by transmitting back a request. This request triggers device 1 to send the data to device 2. Device 2 forwards the associated advertising metadata to all its neighbors. From those, only devices 3 and 6 are interested in requesting the data. Upon receiving the requests, device 2 forwards the data to those devices.

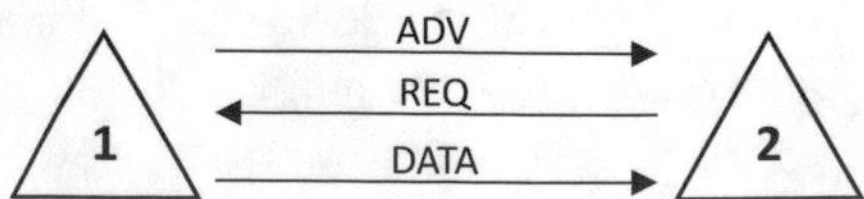

Fig. 7.9 SPIN transactions

SPIN is further improved as *SPIN Energy Conservation* (SPIN-EC) by considering energy consumption. SPIN-EC enables devices to minimize their participation in operations if their energy levels fall below a certain threshold. For example, a device will only send an ADV message if it considers that it has enough energy to process REQ and generate the corresponding DATA messages. Similarly, if a device receives an ADV message, it will only transmit a REQ message if it considers it has enough energy to process the incoming data.

Although SPIN is a P2P mechanism it also introduces a broadcast version, called *SPIN Broadcast* (SPIN-BC) that relies on devices sharing a single channel. If a device sends an ADV, REQ, or DATA message, it is received by all other devices that are located within the transmission coverage area of the sender. The idea is to take advantage of the redundancy generated by the transmission of broadcast (or multicast) messages. Moreover, timing becomes critical under SPIN-BC since a device will delay the transmission of a REQ message to see if another co-located device is also attempting to request the same data. The length of the delay is estimated based on the available device energy; the more energy, the shorter a device will wait to transmit a REQ. The idea is to extend overall network lifetime by giving those devices with the least amount of energy the greatest opportunity to preserve it. Once the advertising device receives the REQ message, it broadcasts the DATA message to all interested devices including those that did not implicitly send the original REQ message.

Figure 7.11 shows SPIN-BC operations; device 1 advertises its metadata to all listening devices 2, 3, 4, 5, and 6. If devices 2, 5, and 6 are interested in the data, device 2 broadcasts a request first, as it comparatively has more energy. Both devices 2 and 5 do not do anything and just wait for the DATA message that is later broadcasted by device 1.

One last SPIN mechanism, known as SPIN-RL, provides reliability when the underlying network is prone to packet loss and other impairments. Essentially, SPIN-RL introduces two ways to accomplish additional reliability: (1) it relies on retransmissions that are triggered whenever a DATA message fails to arrive after the transmission of a request and (2) it transmits periodic ADV messages that are tracked by the receiving devices to make sure that they are not redundant.

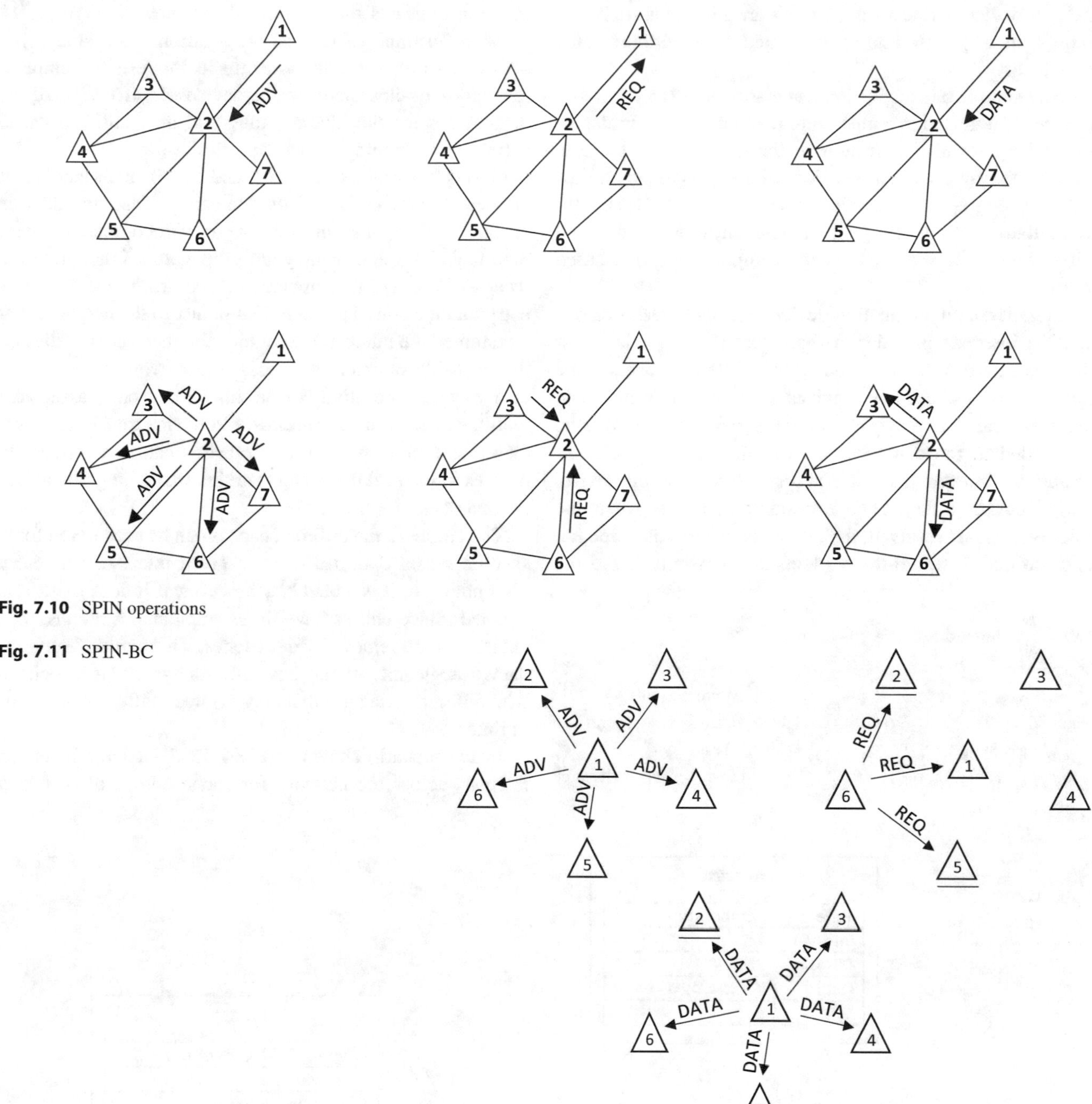

Fig. 7.10 SPIN operations

Fig. 7.11 SPIN-BC

One critical issue with any of the versions of SPIN is that if a single device, for whatever reason, fails to propagate ADV messages, the end-to-end network connectivity collapses. Fortunately, other mechanisms overcome this problem.

7.2.4 Directed Diffusion

Directed Diffusion is another flat data-centric routing,10.5555/1203508 mechanism that relies on response aggregation to accomplish power consumption efficiency [17]. It provides multipath datagram propagation or diffusion throughout a network by primarily focusing on the message exchange of close-by devices. The message exchange is based on four main components: (1) interests that define the nature of data to be requested or subscribed, (2) gradients that specify the spatial direction of specific data types, (3) data messages that provide the sensor readouts, and (4) reinforcements that enable a device to give higher priority to certain paths in a multipath scenario. Directed diffusion provides routing through a publish/subscribe scheme like that of the application layer described in Sect. 5.3. In this case, a device acting as a querier transmits a question that represents a specific interest presented as an attribute-value pair [18].

An example of a device interest is shown in Table 7.1. The device is interested in temperature readouts that are transmitted every second for an hour in the geographical location determined by the square region with opposite corners located at geographical coordinates 41.40,2.17 and 41.41,2.19. Note that by including geographical information, directed diffusion can be seen also as an example of location based routing.

In general, an application device will periodically broadcast its interests in order to figure out if there exist other devices that can service them. As an interest traverses the network, it is cached by devices such that whenever data are propagated down from servicing devices, they can be forwarded to those with an active interest. A cache entry includes a timestamp, a gradient for each neighboring device, and a duration field. The timestamp indicates when the interest was last received, the gradient specifies the direction and the rate at which the readouts are transmitted, and the duration indicates the lifetime of the interest. Figure 7.12 shows an example of interest propagation and caching. The device 1, acting as a sink, transmits its interests for temperature data to its close neighboring devices 2 and 3. When these devices receive the interest, they add an additional cache entry to tie the temperature type to the device 1 sink. The interest is forwarded by device 2 and 3 to their neighbors that now create new cache entries indicating these two devices are sinks of the data. Interests are forwarded unaltered with modified sink fields as they are propagated. When a device receives an interest, it updates or adds to its cache a new entry that includes a gradient that points to the neighbor that transmitted the interest. Essentially, the mechanism relies on the diffusion of interests across the network such that all devices with connectivity to the sink have an entry associated with the interest in their caches. If a device transmits sensor data that comply with an interest, this causes intermediate devices to forward the data to the sink based on gradient and desired rates.

Table 7.1 Interest example

Attribute/value pair	Description
Type = temperature	Temperature readouts
Interval = 1 s	Report events every second
Duration = 3600 s	Report for an hour
Field = [(41.40, 2.17) , (41.41, 2.19)]	Report from sensors in this area

The entries in the different caches can be used to reinforce specific sensor data paths. For a given interest, if an entry does not exist, it is added but, however, if it does exist, it is updated. Since entries have timestamps and durations, they expire and are removed from caches. Through different updates a single interest may have different gradients associated with different data rates that may expire at different instances of time.

In the scenario shown in Fig. 7.13 device 1 diffuses its interests across the network for cache entries to be added

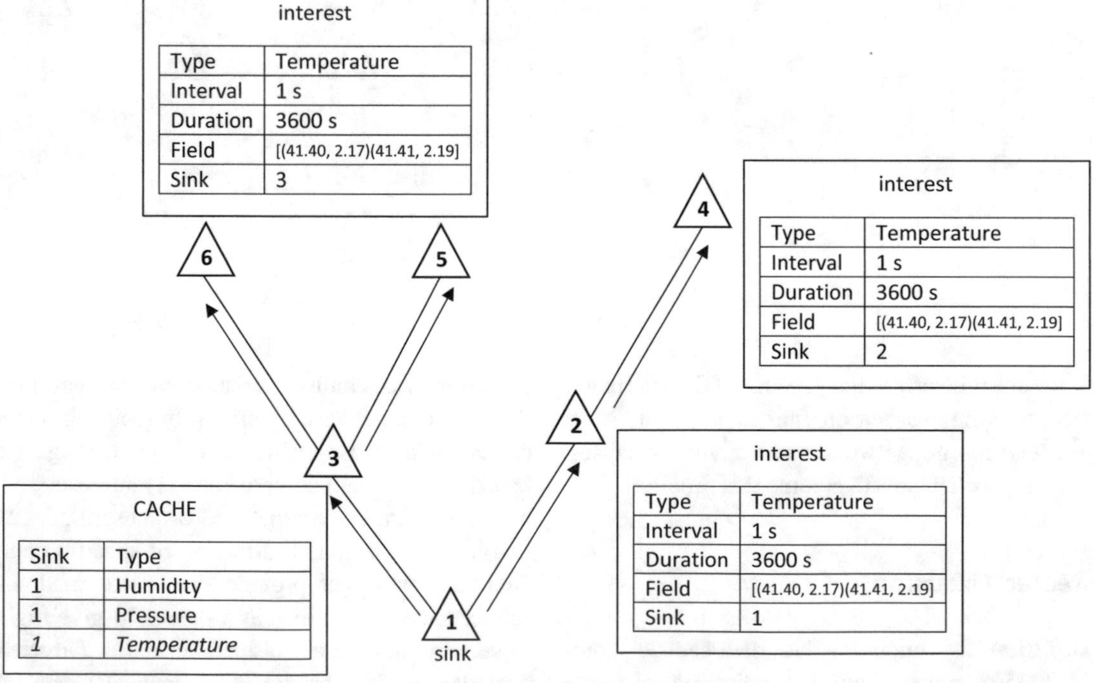

Fig. 7.12 Interest propagation and caching

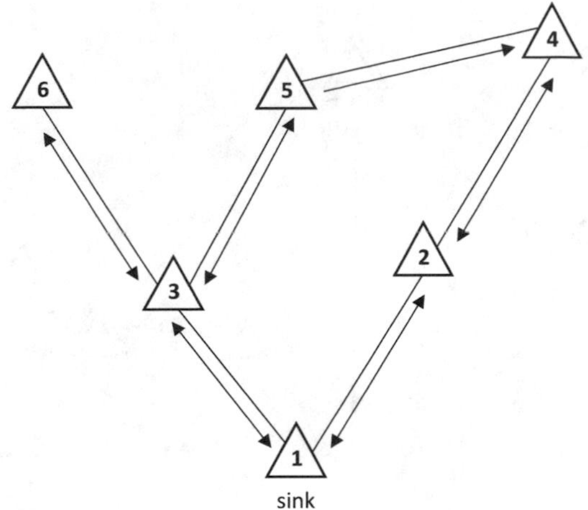

Fig. 7.13 Initial setup

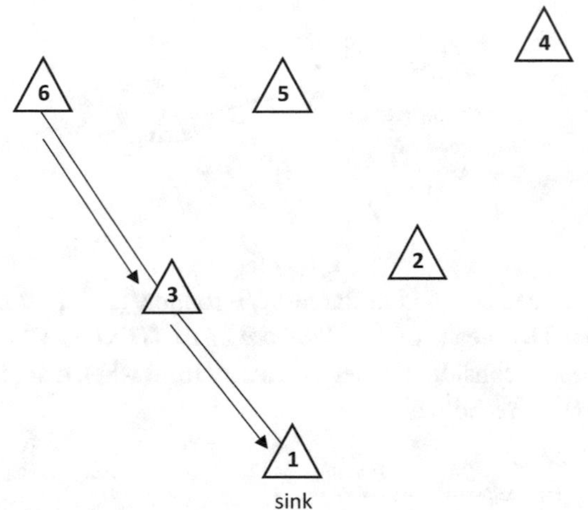

Fig. 7.14 Reinforced path

in all intermediate and end devices. As time progresses, device 1 may reinforce the path to device 3 by sending the original interest at a higher rate such that the source device 5 sends data more frequently. This is illustrated in Fig. 7.14. Reinforcement is typically applied to those paths that serve as alternatives to other paths affected by excessive latency and network loss.

7.2.5 LEACH

Low Energy Adaptive Clustering Hierarchy (LEACH) is a hierarchical routing algorithm that by means of aggregation it transmits device data to an application that acts as a sink [7, 14]. By relying on both hierarchical routing and aggregation, LEACH accomplishes levels of power consumption efficiency that extend the overall network lifetime. As any hierarchical approach, the algorithm segments the network into multiple clusters with clusterheads that aggregate all data before they are forwarded to the sink. In order to minimize throughput and preserve device energy, clusterheads usually eliminate any redundancy that may exist in the aggregated data. This redundancy removal plays the role of the IoT data-information conversion first introduced in Sect. 1.3.

Figure 7.15 shows a LEACH network. Essentially, devices talk to clusterheads, that in turn, talk to the sink. The transmission between the clusterheads and the sink as well as between the devices and the clusterheads is single hop. The media access is TDMA based such that devices are scheduled by clusterheads to transmit the sensor data in specific time slots. In addition, and to reduce the interference, communication between devices inside and outside a cluster is CDMA based.

Since clusterheads typically consume more energy than regular devices, LEACH implements a scheme of rounds where clusterheads are dynamically assigned on a round-by-round basis. This clusterhead rotation distributes the overall energy consumption between devices thus maximizing network lifetime. This is illustrated in Fig. 7.16. Within a single round there are two stages: (1) cluster setup and (2) steady state. In the cluster setup stage, the clusterhead is selected and the cluster itself is formed. Once devices have their roles assigned, in the steady state stage, regular devices collect data that are aggregated at the clusterhead and delivered to the sink. The steady state stage consists of multiple frames with timeslots that give the opportunity to all regular devices to propagate their data to the clusterhead. The setup stage is typically short in order to lower both the scheme latency and the overhead of the mechanism. The steady state stage is longer, but it cannot be too long because it could potentially drain all the energy of the clusterhead.

One of the advantages of LEACH is the highly distributed nature of the setup stage. Specifically, clusterhead selection and cluster formation require little communication between devices. Given several N devices, the clusterhead selection consists of device n with $n \in (1, N)$ selecting a threshold $T(n)$ given by

$$T(n) = \begin{cases} \frac{P}{1-P(r \bmod (\frac{1}{P}))} & \text{if } n \in G \\ 0 & \text{otherwise} \end{cases} \tag{7.1}$$

where $r \geq 0$ the round number, P is the probability that a device will be clusterhead in any given round and G contains the set of devices that has not been selected as clusterhead in the last $1/P$ rounds. Device n then generates a uniformly distributed random number $v \in (0, 1)$ that is compared to $T(n)$. If $v < T(n)$, then device n becomes clusterhead otherwise it just behaves like a regular device. This mechanism guarantees that a device can only be selected as clusterhead once in a period of $1/P$ rounds. Once the clusterheads are selected, the remaining devices select what cluster to join.

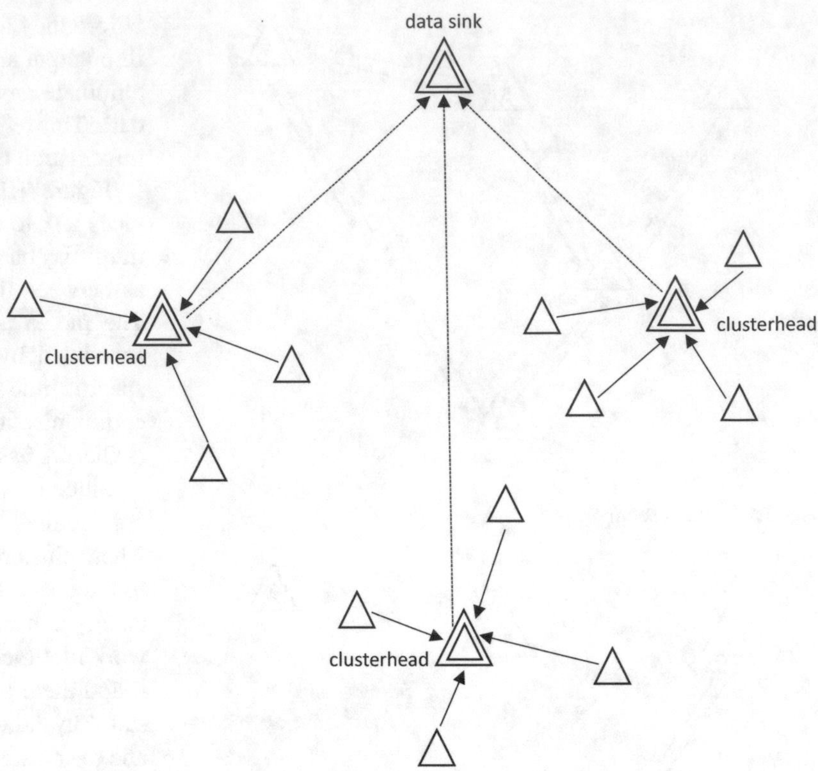

Fig. 7.15 LEACH network

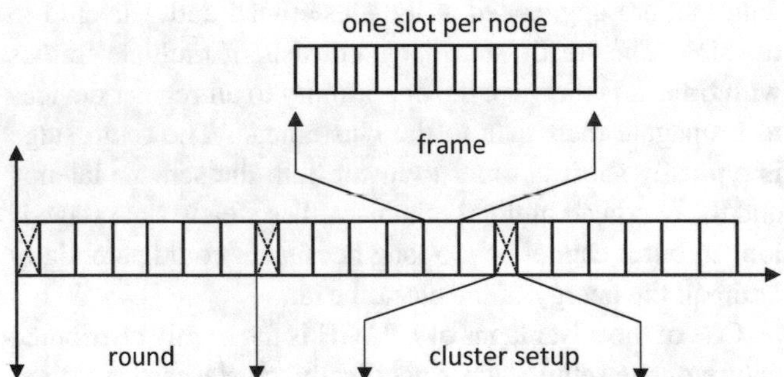

Fig. 7.16 LEACH rounds

Specifically, devices estimate the distance to the clusterheads based on the received signal power and select the closest one. Until this point, all clusterhead selection and cluster formation are done offline. The devices then communicate with their selected clusterheads to initiate the association. The clusterheads, in turn, set up and broadcast the TDMA schedules to their associated devices. After the setup stage is finished, the steady state stage starts. Devices take turns transmitting sensor data to the clusterhead in accordance to the broadcasted schedule. As indicated in Fig. 7.16, each device transmits only once per frame with multiple frames being part of the stage.

One issue with the clusterhead selection process is that it does not have into consideration energy levels of the devices under consideration. Even if a device does not have enough energy, it may still result in being selected as clusterhead if it has not been selected in the last $1/P$ rounds [13]. LEACH is extended by means of the *eXtended LEACH* (XLEACH) [10] protocol to consider the device energy level when calculating the $T(n)$ threshold as

$$T(n) = \frac{P}{1-P\left(r \bmod \left(\frac{1}{P}\right)\right)} \left[\frac{E_{n,\text{current}}}{E_{n,\max}} + \left(r_n \operatorname{div} \frac{1}{P}\right)\left(1 - \frac{E_{n,\text{current}}}{E_{n,\max}}\right)\right] \quad (7.2)$$

where r_{n} is the number of consecutive rounds in which the device has not been a clusterhead and $E_{\text{n,current}}$ as well as $E_{\text{n,max}}$ are respectively the current and initial energy levels of the device. When the value of r_{n} becomes equal to $1/P$, the threshold becomes

$$T(n) = \frac{P}{1-P\left(r \bmod \left(\frac{1}{P}\right)\right)} \quad (7.3)$$

which is the original threshold that does not take into account energy in the computation of its value. When a device becomes clusterhead the value $r_{\text{n,s}}$ is reset.

7.2.6 PEGASIS

Power-Efficient Gathering in Sensor Information Systems (PEGASIS) is another hierarchical routing mechanism that relies on data aggregation [14, 15]. In a similar way to

LEACH and XLEACH, PEGASIS attempts to lower power consumption by distributing the role of clusterhead among all the devices in a section of the network. Moreover, PEGASIS also lowers network latency by performing simultaneous data aggregation on multiple devices. The overall result of these mechanisms is to extend the overall network lifetime.

Under PEGASIS, a device is aware of the geographical location of its neighbors and can control its transmission power in order to just reach the closest one. This latter property is usually enforced using CDMA modulation schemes. PEGASIS relies on a chain structure that results from direct communication of devices with their immediate neighbors. Figure 7.17 shows an example of such a scheme, device *A* first selects its closest neighbor, device *B*, as the first link of the chain. Device *B*, in turn, selects its closest neighbor device *C* as the second link. The process continues with all other devices selecting their closest neighbors in order to set up the different links of the chain. Note that the chain formation is initiated by the farthest device and continuous by sequentially adding devices to the chain. The selection is greedy based on distance that can be inferred from the received signal power.

As in the LEACH case, transmission is based on rounds where devices take turns in becoming clusterheads. The idea is that the energy consumption burden gets distributed among the different devices in the cluster. For any given round, a device in the chain is designated clusterhead. The clusterhead selection is sequential so that if the chain is 8-device long, a device will become, on average, clusterhead every eight rounds. The clusterhead aggregates traffic from all other devices and forwards it to the sink application. The start of each round can be triggered by timing or by the sink application itself that transmits a special beacon to indicate this. Note that depending on the distance to the sink, certain devices may end up, on average, consuming more power per round than others.

But aggregation is not just performed by the clusterhead, under PEGASIS half of the devices in the round perform some type of data aggregation. Figure 7.18 shows an example of a PEGASIS round with 8 devices, *A* through *H*, in positions 0 through 7 within the chain respectively. In this case device *D*, initially in position 3, is the clusterhead. The first stage of the round have all the devices in even positions (i.e. 0, 2, 4, 6) transmit their data to their *next door* neighbors in the odd positions (i.e. 1, 3, 5, 7). At this point, device *B* has device *A* data, device *D* has device *C* data, device *F* has device *E* data, and device *H* has device *G* data. Now devices *B*, *D*, *F*, and *H* are at positions 0, 1, 2, and 3. The second stage of the round consists, again, of all devices left in even positions (i.e. 0 and 2) transmit their data to their neighbors in the odd positions (i.e. 1 and 3). After this stage, device *D* has device *B* data and device *H* has device *F* data. For the last stage, only the clusterhead device *D* and the device *H* are left. Because the clusterhead device always aggregates data regardless of its position, the device *H* forwards all its data to it. Once the clusterhead has all the data from all other devices it transmits them to the sink.

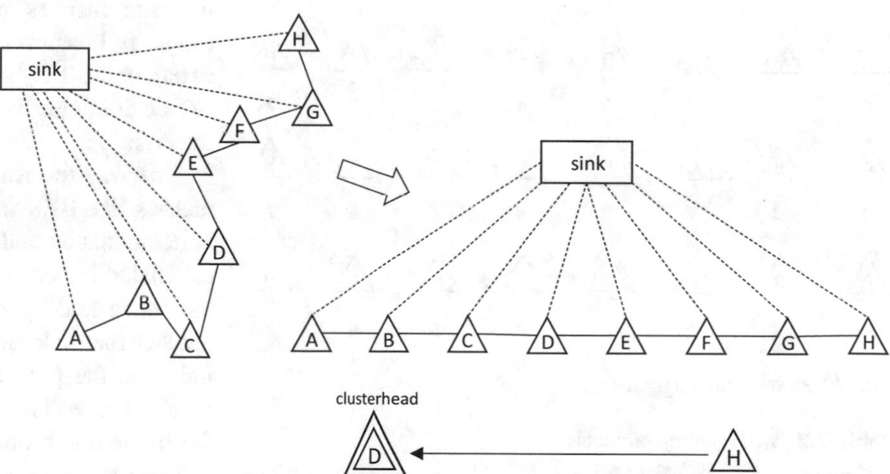

Fig. 7.17 PEGASIS structure

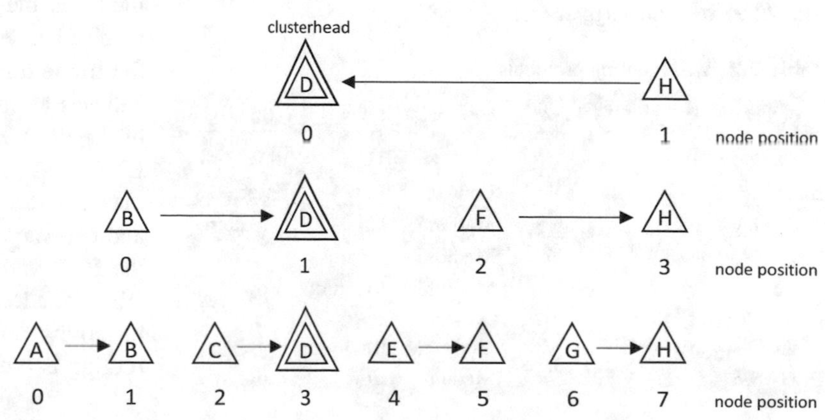

Fig. 7.18 PEGASIS aggregation

The mechanism is developed to make sure that energy consumption is fairly distributed throughout the network by shifting the role of the clusterhead and the devices transmitting data based on their position. For any given round with N devices, the number of stages is around $\lceil \log_2(N) \rceil$. Note that energy is also preserved by limiting the transmission power of a device by making sure it only reaches its uplink neighbor. This results in a higher SNR that enables parallelism that maximizes the number of simultaneous transmissions and minimizes the number of stages. Compare this scenario to a scheme, shown in Fig. 7.19, where interference between devices forces the sequential transmission of data that are aggregated until they reach the clusterhead. In this case, the number of stages is $N - 1$. Table 7.2 shows a comparison of LEACH against other WSN routing mechanisms.

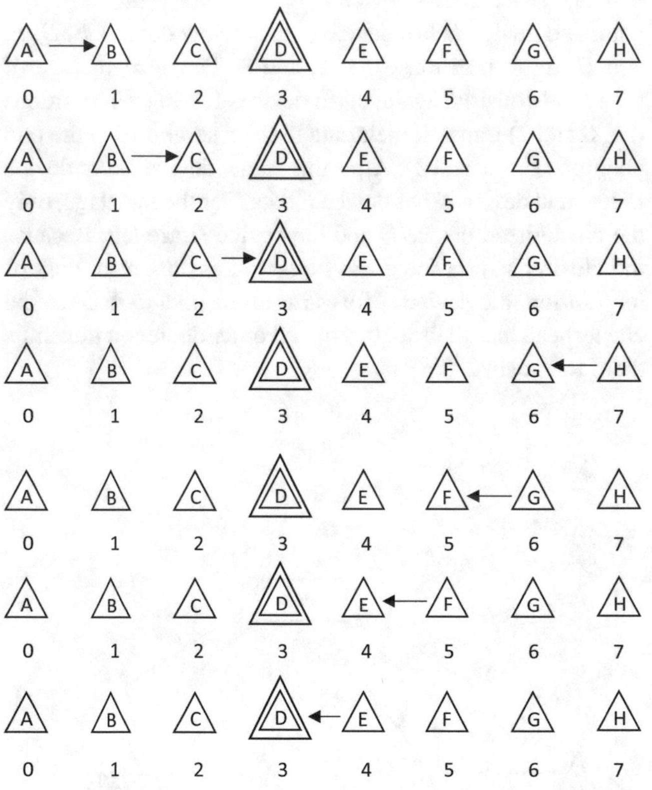

Fig. 7.19 Sequential aggregation

Table 7.2 WSN routing protocols

Mechanism	Hierarchical	Data-centric	Complexity	Latency	Energy consumption
Flooding	No	No	Low	Low	High
Gossiping	No	No	Medium	High	Medium
SPIN	No	Yes	Medium	High	Medium
Directed diffusion	Yes	Yes	Medium	Medium	Low
LEACH	Yes	No	High	Medium	Low
PEGASIS	Yes	No	Medium	Low	Low

7.3 RPL

WSN routing protocols like SPIN, PEGASIS, and LEACH provide elements that are relevant to routing in the context of IoT. Essentially, WSN are, like IoT networks, limited from a power consumption perspective leading to computational, resource, and communication constraints that heavily affect routing. Specifically, these constraints manifest themselves as packet loss that is typically transient and unpredictable. Traditional IP routing falls under two main categories: *Distance Vector* (DV) and *Link State* (LS) [6]. DV consists of devices sending their entire routing tables to their connected neighbors. LS, on the other hand, consists of devices sending the state of their links to all the devices in the network. With LS all devices get the same picture of the network and therefore they can recreate the same global routing table in a distributed fashion. LS routing protocols are, therefore, more reliable than more localized DV routing protocols. Both, DV and LS, react to network lossiness by attempting to quickly reconverge and find alternative routing paths. The assumption is that under conventional high throughput traffic any long term interruption leads to large amounts of data not being received at the far end. Unfortunately, due to the transient nature of the IoT network datagram loss, these traditional routing approaches lead to instabilities and unacceptable control data overhead. IoT traffic is associated with low throughput so slow reconvergence does not necessarily imply a large loss of data. In this context, a more appropriate routing solution for IoT consists of under reacting to the transient changes of connectivity in the network and only trigger full reconvergence when really needed. Additionally, as part of the IPv6 bootstrapping initiated by SAA, routing on IoT devices must be self-manageable and require no human intervention.

One way to measure network reliability is by means of metrics like BER introduced in Sect. 3. Whenever a frame is demodulated and a bit ends up decoded incorrectly, the calculated FCS does not comply with the received FCS and the frame is dropped. This loss is measured by means of another metric known as *Packet Delivery Ratio* (PDR) that indicates the probability that a transmitted frame will be received at destination. Essentially $PDR = 1 - p$ where p is the frame loss probability. A link that is fully disconnected behaves as if it were affected by a 0% PDR. The inverse of the PDR is the *Expected Transmission* (ETX) defined as $ETX = \frac{1}{PDR}$. Therefore as $PDR \rightarrow 0$ then $PDR \rightarrow \infty$. In the context of IoT LLNs, PDR values are inherently unstable and can vary a lot as a function of time. Figure 7.20 shows the PDR variation as a function of time for three different IEEE 802.15.4 links. These variations would cause network instability under traditional IP routing because traditional routing is prepared to react to dynamic changes of network

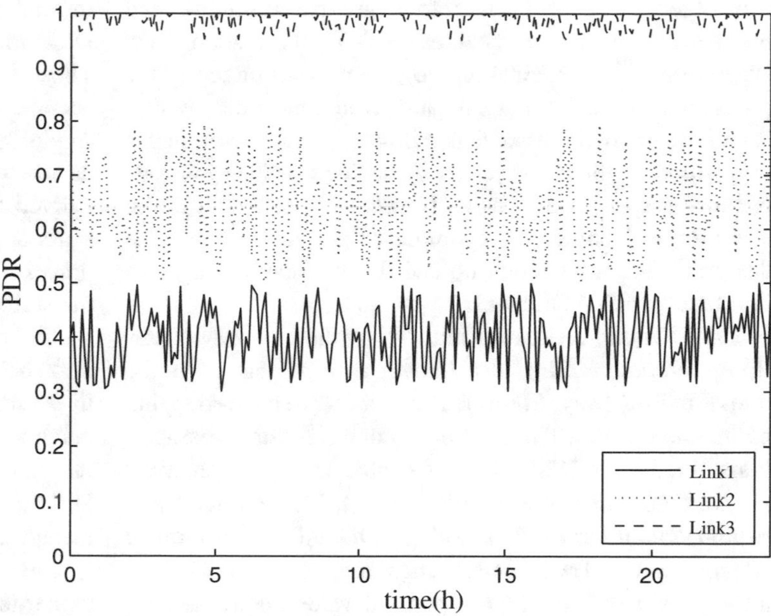

Fig. 7.20 PDR variation

metrics like PDR very fast. Under IoT LLN routing, however, the assumption is that these metrics naturally change over time and it requires the routing algorithm to underreact in order to prevent additional connectivity issues. Only if routing problems are persistent it is expected that the routing mechanism will redirect traffic through alternative routes without triggering global reconvergence.

The IETF *Routing over Low Power and Lossy Networks* (ROLL) work group standardized a routing protocol that follows these principles [16]. The routing protocol, known as *Routing for Low Power* (RPL) or *Ripple*, is a hierarchical IPv6 address centric [1] protocol. RPL was designed to comply with three goals: scalability, reliability, and energy efficiency. RPL builds IPv6 networks with thousands of constrained devices over multiple low power wireless hops that combine reliable transmissions with several years of battery life. The idea behind RPL is that any device in a WPAN can be router so the network can be easily deployed. This, indirectly, limits the characteristics of the routing mechanism to guarantee the support of low complexity devices.

By being based on WSN routing, RPL makes two important assumptions: upward focused routing and reactive route update. Upward focused routing, by means of M2P traffic flow, implies that traffic moves mostly upward due to sensors transmitting readouts. Although RPL supports traffic in both directions, the focus is reliable upward routing. The idea is for each device to select a parent as the next hop for its upward route based on DV from the gateway and set the downward route by reversing the upward route. Reactive route update implies that some routing decisions are made by detecting physical connectivity changes that are reactive to data transmissions. This minimizes the use of proactive control traffic usually triggered by trickle timers described in Sect. 7.3.3.

7.3.1 DODAG Creation

RPL is a routing protocol that operates at the IPv6 network layer and, as such, contrasts with other mechanisms that work at lower layers like 6LoWPAN link layer mesh forwarding described in Sect. 4.3.3. Additionally, RPL is based on the adaptation of a hierarchical DV scheme applied to LLNs. The procedure starts with the creation of a *Destination Oriented Directed Acyclic Graph* (DODAG) that enables devices to keep track of the routing topology. A DODAG is built based on an objective function that operates on a combination of metrics and constraints in order to find the best path. In fact, for the same network topology many different objective functions associated with the different metrics and constraints may lead to routing scenarios that would carry traffic based on QoS requirements. Metrics and constraints can be either device or link based. For example, a device metric can be the device energy level while a link metric can be the latency associated with the link. Similarly, a device constraint can be that the device must support encryption while a link constraint can be that the link must be wireless. Metrics can be applied individually or aggregated along a path. For example, latency of a link can be the latency associated with a single hop or the latency aggregated over multiple hops.

Objective functions may result in DODAGs that attempt to find the best path that (1) minimizes ETXs (metric) over encrypted links (constraint) or (2) lowers latency (metric) over unencrypted links (constraint) or (3) minimizes power consumption (metric) or (4) uses wireless links (constraint).

The objective function, therefore, groups constraints and metrics as a way to specify rules for DODAG creation. Each graph represents a logical topology that exists on top of the physical network topology. At any given time, multiple logical topologies overlap and depending on quality requirements only some of them are active. A device can then join one or more graphs, known as RPL instances, and route traffic accordingly in order to accomplish the requirements. The datagrams are transmitted up and down along the edges of the corresponding DODAG.

The DODAG creation is initiated by the *Low Power PAN Border Router* (LBR), also known as *root*, that plays the role of IoT gateway. Multiple roots can serve an overlapping number of devices. There are three control routing messages, transmitted over ICMPv6, that enable the graph creation: (1) *DODAG Information Object* (DIO), (2) *DODAG Information Solicitation* (DIS), and (3) *DODAG Advertisement Object* (DAO). The ICMPv6 encoding, shown in Fig. 4.6, introduces a new ICMPv6 type field value known as RPL and each message is indicated by different values of the ICMPv6 code field. DIS, DIO, and DAO messages carry respectively ICMPv6 type field values 0, 1, and 2. The LBR starts by periodically transmitting DIO messages that advertise information about the DODAG to its neighbors. Those devices that receive the DIO message and decide to join the graph become children of the LBR. For any child device, the LBR becomes its parent and assigns itself a rank that specifies its position in the graph hierarchy relative to the root and plays the role of DV path cost. In general, the smaller the rank the closer the device is to the root. Conceptually the idea of the rank is to enable the routing algorithm to determine the presence of loops. If the device is configured with routing capabilities it can transmit, in turn, its own DIO to its neighbors in order to advertise the route to the LBR. The process is then repeated, and those neighbors interested in joining the graph calculate their own rank and transmit DIOs in order to hierarchically extend the DODAG range. Those devices joining a graph but without routing capabilities become leaves as they cannot transmit DIOs. Note that a single device may receive DIOs from multiple nodes and decide whether to join or not a DODAG based on the rules derived from its objective function. The route propagation starts at the root and ripples down to the leaves where the process finishes. The route enables all devices in the graph to transmit datagrams to the LBR by means of the M2P upward routing mechanism shown in Fig. 7.21. Note that the Figure also shows DIS messages can be used by devices to proactively request upward routing information to its reachable neighbors. Essentially, this provides an on-demand routing mechanism. Specifically, when a non-leaf device receives a DIS message it must respond by transmitting DIO messages that advertise the supported graph it is associated with.

Upward routing is representative of traffic going from a higher to a lower ranked node. Similarly, downward routing results from datagrams propagating from lower to higher ranked devices. Specifically, for a parent to be able to send traffic to a child, it must know first how to reach it. RPL, as illustrated in Fig. 7.22, introduces DAO messages to accomplish this task. They are transmitted by a device to its parent as soon as the former joins a DODAG. Many times, DAO transmission can be induced by means of an indication in the incoming DIO message. In general, DAO messages carry and aggregate prefix information from a device and its children. This information is used by lower rank nodes to add routing table entries that enable downward routing. Essentially, prefix data eventually reach the LBR and a complete downward path becomes available.

Multi-Topology Routing (MTR) is supported by RPL to overlap multiple routing instances over a single physical topology. Figure 7.23 shows an example of two instances

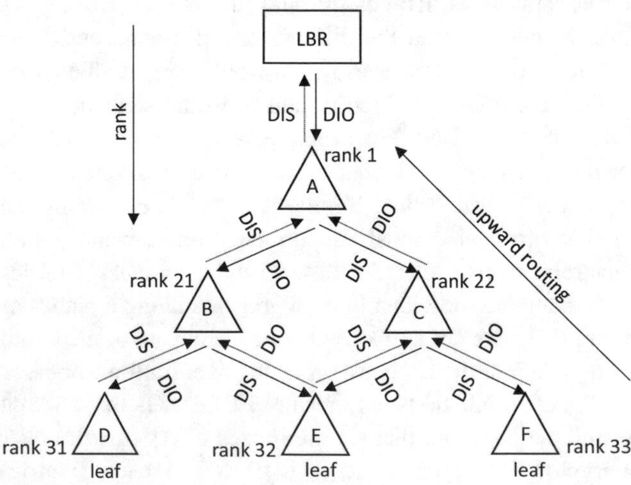

Fig. 7.21 Upward routing

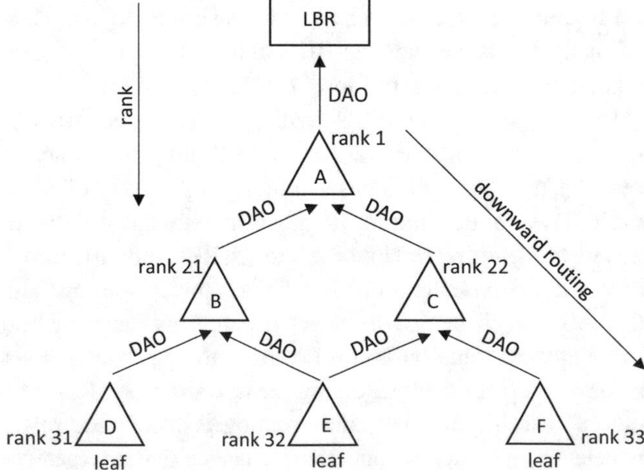

Fig. 7.22 Downward routing

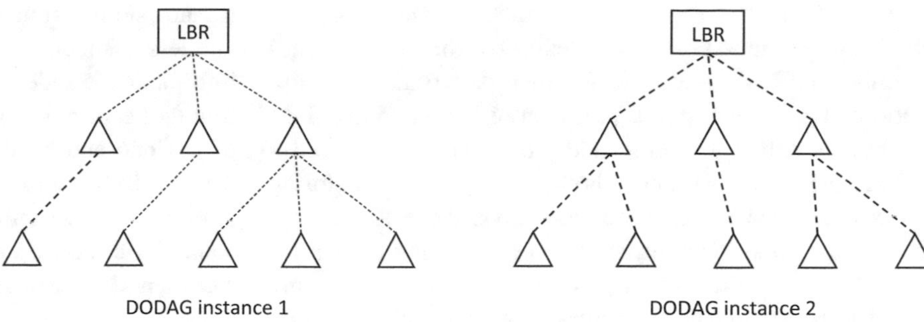

Fig. 7.23 MTR

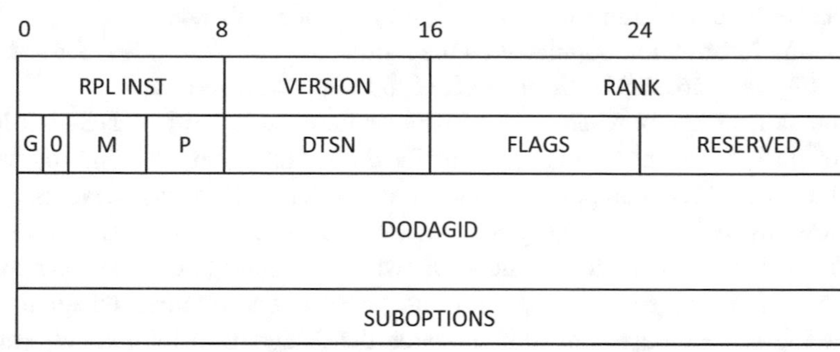

Fig. 7.24 DIO message format

identified by two different instance identifier values. Each instance is associated with an objective function that defines routing rules and each device is associated, in turn, to multiple instances. The device must then select the right instance based on the type of traffic being transmitted. For example, real time data may need to be transmitted through an instance that supports low latency while data to be processed offline may need to be transmitted through an instance that relies on encrypted links. Because of the dynamic nature of the network, the graph that complies with the objective function of an instance also mutates. Therefore, a device that is transmitting datagrams within the context of an instance may trigger the reassociation with a new parent to support graph and rank changes.

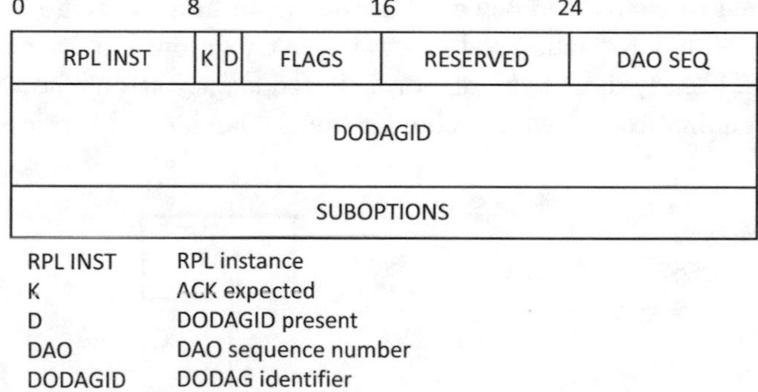

Fig. 7.25 DAO message format

Figures 7.24 and 7.25 respectively show the DIO and DAO message structures. The DIO message includes an 8-bit *RPL instance* field that identifies the routing instance, a 16-bit *rank* field that specifies the rank of the device sending the message, a 1-bit *ground* flag that indicates whether the DODAG root is attached to the public Internet, a 3-bit *mode of operation* field that specifies whether storing or non-storing nodes are in use (described in next Section) and a 2-bit *preference* field that specifies the root preference. The 8-bit *Destination Advertisement Trigger Sequence Number* (DTSN) is set by the message sender to maintain downward routes. The 128-bit *DODAG identifier* uniquely identifies the DODAG and it typically carries the IPv6 of the root. A DAO message contains some of those fields but it also includes a 1-bit *acknowledgment* flag that indicates whether the DAO sender expects a DAO-ACK message back, a 1-bit *DODAG present* flag that specifies whether the DODAGID is carried in the DAO message and an 8-bit *DAO Sequence Number* field used to keep track of acknowledgments.

7.3.2 Storing and Non-storing Nodes

Besides keeping track of its own parent, a device that has received DAO messages from higher rank devices must have enough memory to store prefix information corresponding to these devices' subgraphs [3]. Since memory requirements and computational complexity can be critical issues in embedded IoT devices, RPL supports two modes of operation: if a device stores prefix information from all its children is called a storing node, otherwise it is called a non-storing node. Although both, storing and non-storing devices, can route datagrams uplink, only the former can route datagrams downlink as they keep routing entries for each of their children subgraphs. In non-storing mode, only the LBR can route datagrams downstream and other devices can forward datagrams downward if they include embedded routing information populated by the root.

Figure 7.26 shows the difference between storing and non-storing nodes when device *D* transmits a datagram to device *E*. In storing mode, shown by the dashed lines, device *D* forwards the datagram to its parent device *B* that, in turns, by being a storing node routes the datagram down to device *E*. In non-storing mode, shown by the solid lines, device *D* forwards the datagram to its parent *B*, that by not having routing capabilities, it forwards the datagram to its own parent device *A*. Device *A*, another non-storing node, finally forwards the datagram to the LBR. The LBR makes routing decisions based on its routing table and embeds source routing information in the datagram to indicate the path the datagram must follow downward. The datagram is sent to device *A* that processes the routing information and forwards it to device *B*. Device *B*, in turn, looks at the routing information and forwards it to the destination device *E*. Clearly there is a trade-off between latency and memory requirements as well as computational complexity; the price for not storing routing information is both higher latency and less channel utilization efficiency. Since source routing information is added to the datagrams by means of a specific routing header known as RH4, the ratio between the sizes of payload and headers decreases leading to less efficient use of the channel. Note that all devices must be configured as either storing or non-storing nodes. Mixed scenarios with some devices configured as non-storing nodes and others as storing nodes are not allowed.

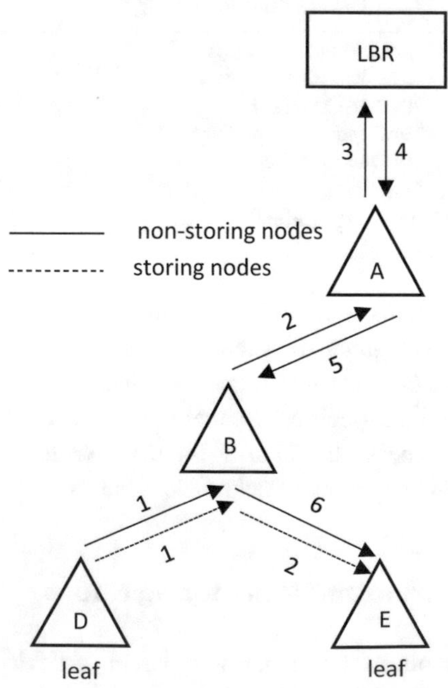

Fig. 7.26 Storing vs non-storing nodes

7.3.3 Loop Detection and Avoidance

Network loops typically result from both topology changes and routing information synchronization problems between devices. As such they are a temporary phenomenon that persists until the routing mechanism solves the inconsistencies associated with the loop. Their detection is critical because datagrams in a loop cause not only additional congestion but also network loss when the datagrams are dropped due to them reaching the maximum hop count. In traditional networking, high transmission rates magnify the effect of loops in the overall effect of the network performance. In this scenario, routing algorithms tend to overreact in order to detect and remove loops, as fast as possible, in order to minimize the effects of congestion and loss. In IoT networks, lower transmission rates, however, limit the impact of loops to network routing where an under reactive approach is more appropriate as many of the loops are caused by transient link reliability issues. In this situation, overacting may lead to routing oscillations that end up generating unpredictable latency and energy consumption.

RPL attempts to avoid loops by enforcing two basic rules: a device (1) cannot select as parent another device that is deeper in the graph and it (2) is not allowed to move deeper in a graph in order to increase the number of parents. Besides avoiding loops, RPL enables devices to detect them through data path validation. Specifically, the mechanism relies on information carried by a series of bits in the RH4 header that are used to determine the whereabouts of a given datagrams. These bits are set and read as datagrams move up and down along the edges of the graph. Whenever a node sends a datagram to its children, it sets the *down* bit in the header. If a datagram is received from a child and its RH4 header has the *down* bit set, it indicates the presence of a loop. This causes, in turn, the device to drop the datagram.

Routing protocols rely on repairs to fix routing topologies when failures are detected. RPL supports two complementary mechanisms: (1) local repairs and (2) global repairs. If due to a node or link failure a device cannot forward datagrams in the upward direction, it must initiate a local repair to find an alternative parent or link. The effect of local repairs is cumulative, and they can cause a graph to diverge from its optimal shape. In this case, a global repair serves to rebuild a DODAG. Global repairs are started by the root with the transmission of DIO messages. From that point on, the graph is rebuilt with devices selecting parents based on objective functions. The overall cost of a global repair is additional latency and control traffic that affect channel efficiency.

Example 7.2 Consider the following RPL scheme where sensor A transmits a readout to sensor B:

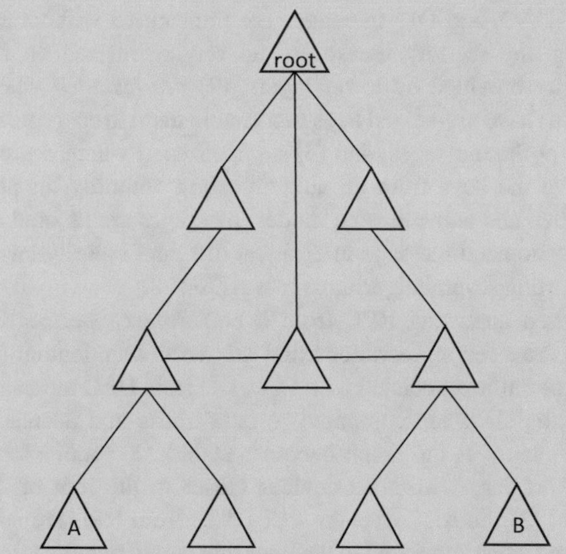

If the transmission rate is 250 Kbps, the packet size is 100 bytes and ignoring collisions as well as delays (other than transmission delay), what is the expected transmission delay when propagation is over storing

(continued)

(continued)

and non-storing nodes?

Solution: The transmission delay of a packet over a single hop is $D_h = \frac{8 \times 100 \text{ bytes}}{250000 \text{ bps}} = 3.2$ ms. There are two possible paths:

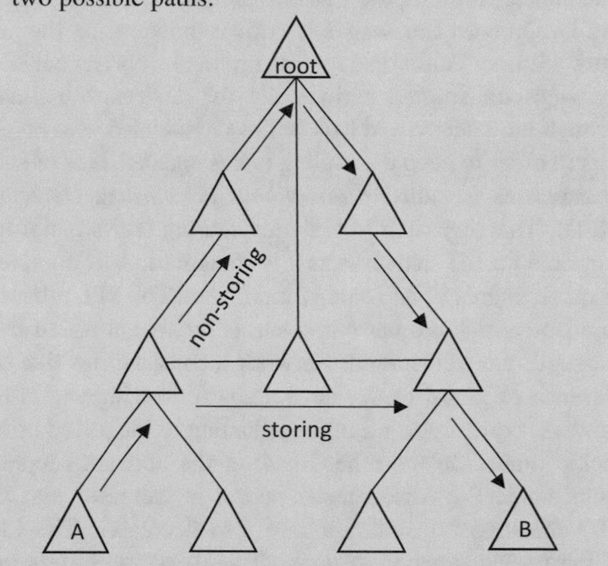

For non-storing nodes, the delay results from six hops $D_{\text{non-storing}} = 6 \times D_h = 19.2$ ms. Similarly, for storing nodes, the delay results from four hops $D_{\text{storing}} = 4 \times D_h = 12.8$ ms.

Figure 7.27 shows a scenario of a topology with seven devices with neighbor connectivity indicated by the dashed lines. For a given objective function an initial DODAG is generated where device 2 selects device 1 as parent. As time

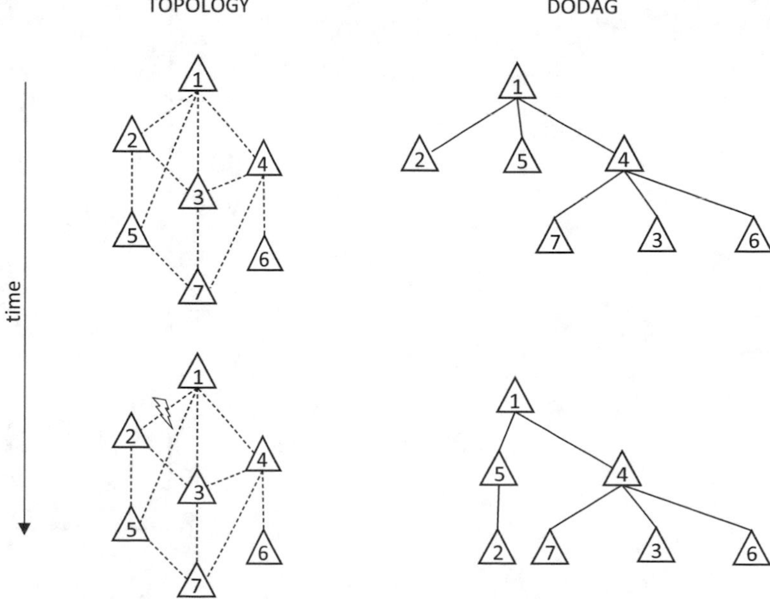

Fig. 7.27 Local repair

progresses, the link between these two devices is perturbed and connectivity is lost. This triggers a local repair that results on device 2 now selecting device as parent. Note that many times there is no strong correlation between physical topology links and routing connectivity.

Because embedded devices in IoT networks are energy constrained, limiting the transmission of routing DIO, DIS, and DAO control messages is critical to extending the network lifetime. Traditional routing protocols rely on keepalive messages transmitted periodically for devices to estimate throughput, latency, and loss of links associated with neighbors in order to keep the routing tables updated. One of such mechanisms is called *Bidirectional Forwarding Detection* (BFD). This approach to updating routing tables is not appropriate for IoT networks as it is computationally complex, resource intensive, and energy inefficient. The RPL relies on an adaptive tickle timer that controls the rate at which DIO messages are transmitted. Network inconsistencies like the presence of loops, or devices joining or moving within the network, typically result in the reduction of the period of the trickle timer. On the other hand, as the network becomes more stable the trickle timer period is increased and the DIO message transmission rate also decreases. The idea of the trickle timer is to transmit network parameters and attributes while minimizing at the same time the load of the network.

Figure 7.28 shows both data and control RPL traffic. While the data traffic rate stays pretty constant, the control traffic rate has peaks whenever the trickle timer triggers the graph to be rebuilt. As the network becomes more stable, the peaks are less common. The implementation of a trickle timer as a mechanism for routing control traffic propagation is simple enough that can be easily implemented on constrained devices without affecting computational complexity and other resources.

Security in embedded IoT devices is computationally expensive and resource demanding. Most RPL friendly link layer technologies like IEEE 802.15.4 and BLE include their own security mechanisms so RPL security is enabled by means of optional extensions. Essentially, RPL supports three modes of operations: (1) *unsecured* where routing control DIO, DAO, and DIS messages are transmitted without enabling any security measures and relying instead on the security provided by lower layers, (2) *pre-installed* where devices have pre-shared keys that enable them to encrypt and decrypt the messages, and (3) *authenticated* where devices receive the keys from an authentication authority. In pre-installed and authenticated modes, messages are secured by encryption and message integrity at different security levels supporting several algorithms like AES-128.

When analyzing RPL from a performance perspective there are a few observations that are worth mentioning: (1) control traffic associated with DIO, DIS, and DAO messages is negligible when compared to data traffic and decreases significantly as the graph becomes stable, (2) routing table sizes are larger in those devices closer to the root of the graph, (3) the routing paths that result from RPL are sub-optimal when compared to the ones that would obtained from an ideal routing mechanism, and (4) by increasing the interval between global repairs, local repairs become more prevalent and control traffic throughout the network is reduced at a cost of longer lasting connectivity outages.

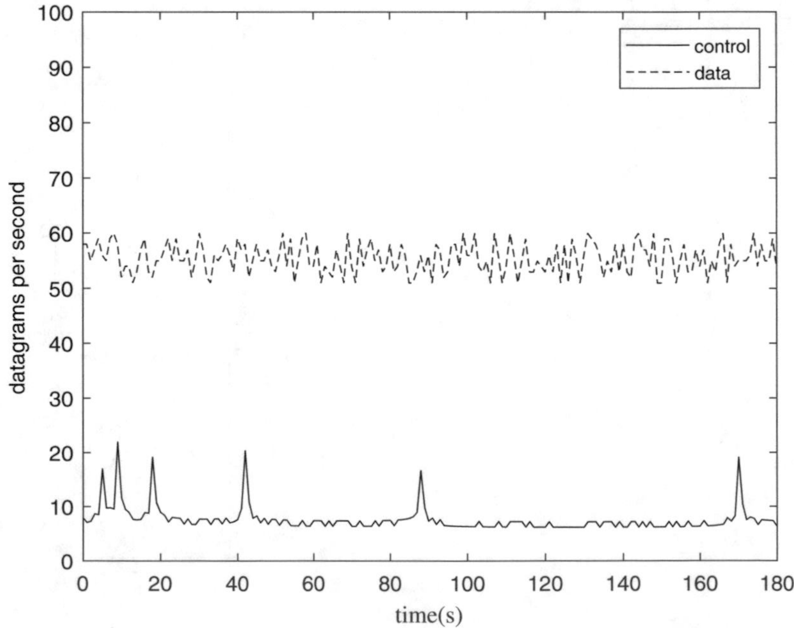

Fig. 7.28 RPL traffic

▶ **RPL Implementations** There are several implementations of the RPL routing suite. Most of them are intended for RTOSs. The list below shows some of the most popular RPL implementations.

Implementation	Description
Tripple	Apache licensed open source C implementation
	Available for general purpose OSs like Linux.
	Adapted to support RTOS through lwIP.
Contiki RPL	BSD licensed open source C implementation.
	Implemented as part of the Contiki RTOS.
	Very low memory requirements (order of kilobytes).
TinyOS RPL	BSD licensed open source C implementation.
	Implemented as part of WSN support of the TinyOS RTOS.
RIOT RPL	LGPLv2 licensed open source C implementation.
	Implemented as part of the RIOT RTOS.
Unsprung	BSD licensed open source C implementation.
	Implemented as lightweight implementation of RPL for Linux.

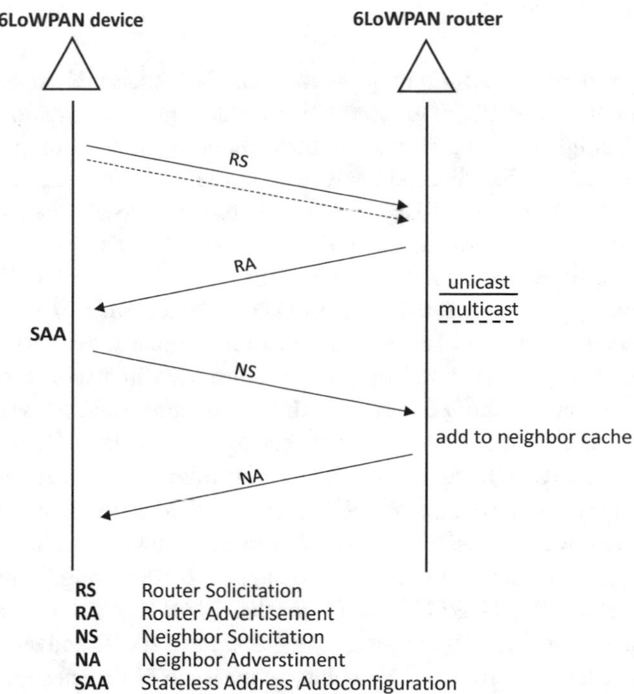

Fig. 7.29 ND bootstrapping

7.3.4 RPL, 6LoWPAN, and ND

RPL is typically used in the context of 6LoWPAN route-over routing. In general, any link layer technology that supports 6LoWPAN mechanisms can rely on RPL for routing [12]. Moreover, multiple different link layer technologies can be routed under a single RPL domain. Edge routers that connect to traditional IP cores and access 6LoWPAN networks operate as RPL roots. Devices rely on RPL to build multiple routing topologies to enable connectivity to their destinations. Host devices indicate their presence to neighboring devices in two different ways: (1) by transmitting DAO messages upwards and processing, but not forwarding, incoming DIO messages or (2) by relying on ND to discover neighboring routers and notify them about the device existence. In this later case, hosts devices attach to 6LoWPAN routers by means of traditional ND bootstrapping without participating in routing.

Figure 7.29 shows the exchanged messages between a device and router as part of ND bootstrapping. The 6LoWPAN device starts by transmitting an RS message as both multicast and unicast transmissions. The router responds back with a unicast RA message. The prefix information carried by the RA is used by the device to initiate SAA, described in Sect. 4.2, and derive an IPv6 address. The device then registers itself at the router by transmitting a unicast NS message that includes the address registration option. The router responds back with a NA message. Note that in addition to the prefix information, an RA message may also include 6LoWPAN compression context data associated with stateful compression. Since the RA message is received by all devices on a link, all devices are synchronized with respect to the compression context in use. This is important, since device synchronization is key to the proper behavior of 6LoWPAN stateful compression. Interaction between devices and the router is progressive as they exchange ND messages. In turn, routers supporting RPL, generate DAO messages that propagate the information obtained from the registration by means of ND bootstrapping.

7.4 LOADng

The *Lightweight On-demand Ad hoc Distance Vector* (LOAD) and its successor the *LOAD Next Generation* (LOADng) protocols are technologies that provide reactive flat routing in LLNs [2]. Although neither LOAD nor LOADng have been standardized, LOADng serves as valuable alternatives to proactive RPL routing. LOADng is designed for simplicity supporting a minimal core that can be extended for additional functionality. This is possible due to a modular design that enables the addition of new messages and data types encoded by means of TLVs. LOADng is quite flexible, supporting addresses of variable length between 8 and 128 bits long as well as different metrics besides the standard hop count.

7.4.1 Minimal Core

As a reactive routing protocol LOADng transmits *Route Requests* (RREQs) to discovery paths to a given destination. A single RREQ is forwarded throughout the network until it reaches the destination device which replies by transmitting a *Route Reply* (RREP). If a path is broken a local repair is typically performed, otherwise, a *Route Error* (RERR) is sent back to the sender of the RREQ. In general, an RREQ message is generated by a LOADng device when it does not have a valid route to destination. Similarly, an RREP message is generated by a LOADng device in response to a received RREQ message with a destination address local to that of the device. RREP messages are acknowledged by the RREQ original sender by transmitting back a *Route Reply Acknowledgment* (RREP-ACK) to indicate that the route was successfully received. The encoding of these messages, transmitted over UDP to support broadcasting, is by means of IETF RFC 5444 *Generalized Mobile Ad Hoc Network (MANET) Packet/Message Format* as a fixed number of header fields followed by a block of message TLVs. These are then followed by a block of addresses with associated address TLVs.

Route discovery, carried out in a similar way to WSN flooding presented in Sect. 7.2.1, consists of devices broadcasting RREQ messages along the way to reach the destination. This is shown in Fig. 7.30 where device 1 is the originator and device 4 is the destination. When the request arrives at the destination, a unicast RREP, that contains the path information, is transmitted back to the originator as illustrated in Fig. 7.31. From this point on, the originator will send all datagrams to reach the destination through this path until a failure is detected. This process, called path maintenance, makes sure that if a datagram cannot be forwarded for any reason, a RERR message is sent back to the originator. At this point the RERR message retriggers a new route discovery.

Upon receiving RREQ and RREP messages, devices update their routing tables to adjust the cost of the path to the message sender. The simplest cost metric is hops but more complex metrics like latency, throughput, and frame loss are also possible. A routing scenario may support multiple metrics that are specified in the RREQ and RREP messages. In all cases, all devices must support the basic hop metric that must be used if the other metrics are not available.

7.4.2 Smart Routing

One issue with flooding is that it leads to issues like overlapping and traffic implosion, that in turn, cause a waste of energy, network bandwidth, and channel capacity. LOADng introduces a mechanism called smart routing that takes advantage of known routes to the destination in intermediate devices to minimize the number of broadcast transmissions that are generated during route discovery.

The procedure starts by broadcasting a special RREQ message that includes a flag that converts it into a RREQ_SMART message. When the message arrives to an intermediate device if this device has a valid path to destination and the corresponding next hop if different from the previous hop associated with the message, then RREQ_SMART message is transmitted unicast to the next hop otherwise the message is broadcasted. This procedure is shown in Fig. 7.32 where device 1 broadcasts the RREQ_SMART message that is received by device 2 that has a direct route to destination device 4. Device 2 then forwards the RREQ_SMART message through a unicast transmission to device 3. Device 3 that is also a direct route to device 4 also forwards the message to the destination device 4 through a unicast transmission. Although reactive flat LOADng smart routing improves performance, still it is not as efficient as proactive hierarchical mechanisms like RPL.

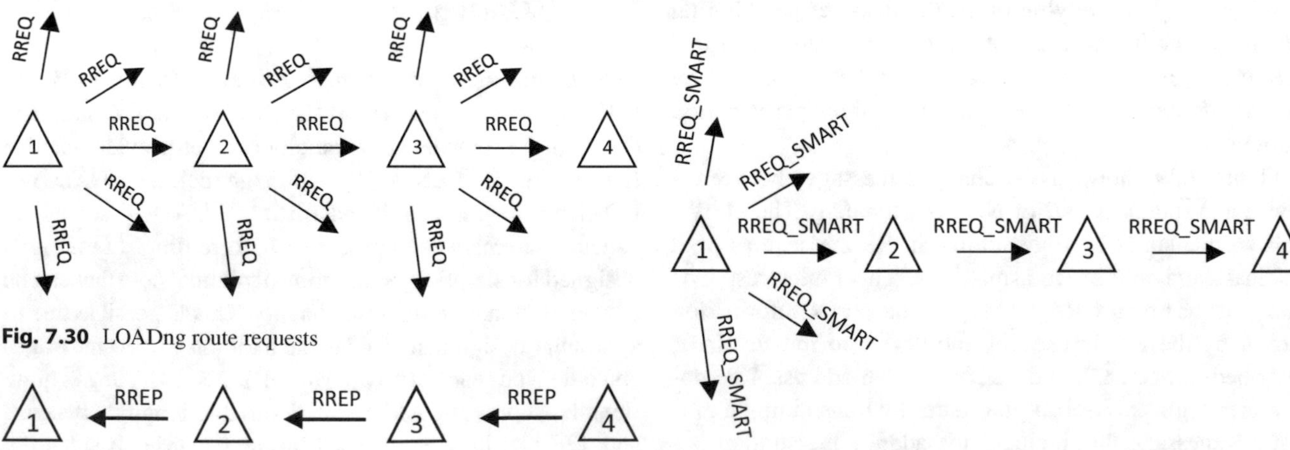

Fig. 7.30 LOADng route requests

Fig. 7.31 LOADng route replies

Fig. 7.32 LOADng smart route requests

Table 7.3 RPL vs LOADng

Protocol	Hierarchical	Proactive	Transport	Latency	Energy consumption
RPL	Yes	No	ICMPv6	Low	Low
LOADng	No	Yes	UDP	Medium	Low

Summary

Routing is key to enable IoT devices to communicate with applications. This chapter started by presenting the most relevant concepts of IoT routing including their types and classifications. As an introductory approach to IoT routing, well known WSN routing mechanisms were presented first. This included flat flooding, gossiping, SPIN, and directed diffusion as well as hierarchical LEACH and PEGASIS. The chapter then presented RPL and details of its hierarchical approach including DODAG creation, maintenance, and support of storing and non-storing nodes. LOADng was then presented as a valid reactive alternative to proactive RPL. Table 7.3 compares some of the most important features of RPL and LOADNg.

Homework Problems and Questions

7.1 Given the following topology and if transmitting/propagating a single datagram over each individual hop takes 1 s:

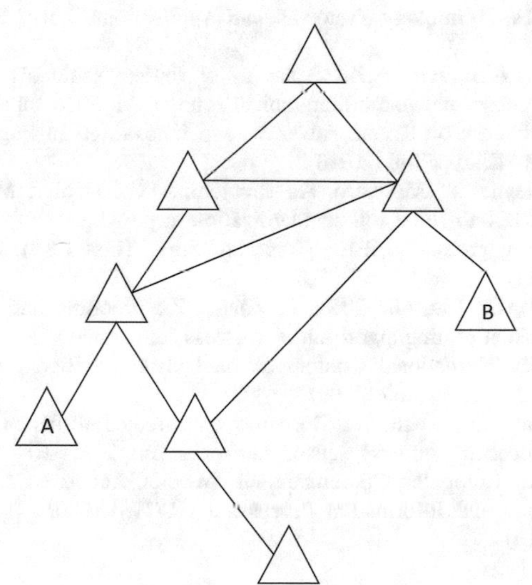

How long does it take for a datagram to go from device A to device B if routing is WSN flooding?

7.2 What WSN routing protocols rely on endpoint negotiation to overcome implosion?

7.3 A WSN has five sensors (A, B, C, D, E) with batteries that are charged at $P_A = 94\%$, $P_B = 97\%$, $P_C = 99\%$, $P_D = 91\%$, and $P_E = 4\%$, respectively. In the last 4 rounds sensors A, B, C, D has been selected as clusterheads. For the following round, and assuming LEACH routing, which of the sensors is most likely selected as clusterhead? Clusterhead probability is $P = \frac{1}{5}$?

7.4 Given the following RPL topology:

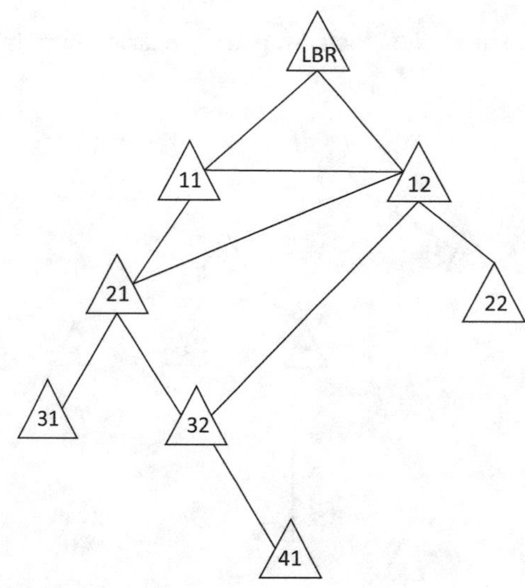

(a) What is the best route from node 41 to node 22 when non-storing nodes are in place?
(b) What is the best route from node 41 to node 22 when storing nodes are in place?

7.5 How does an actuator request routing information from an LBR?

7.6 Given a PEGASIS chain with 8 sensors, how many stages are needed for the data to arrive at the leader?

7.7 Why are extra computational capabilities typically associated with storing nodes under RPL?

7.8 Why is source routing typically associated with non-storing nodes under RPL?

7.9 What type of nodes are more likely to cause application packet loss? Non-storing or storing nodes?

7.10 Given the following RPL topology:

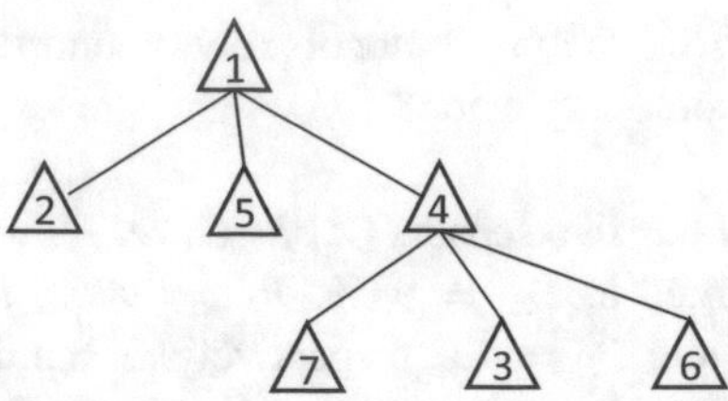

if the transmission rate is 120 Kbps and assuming no impairments and an average packet size of 56 bytes, what is the end-to-end delay when five packets are transmitted from node 6 to node 7 for both storing and non-storing node support? Consider no delay other than the transmission delay.

7.11 Consider the local repair scenario introduced in Fig. 7.27:

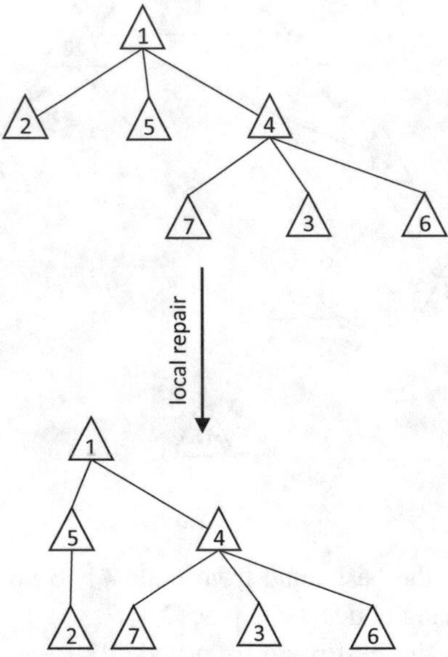

If the transmission rate is 120 Kbps and assuming no impairments and an average packet size of 56 bytes, what is the average latency degradation of node 2 transmissions after the local repair? Consider no delay other than the transmission delay.

References

1. Alexander, R., Brandt, A., Vasseur, J., Hui, J., Pister, K., Thubert, P., Levis, P., Struik, R., Kelsey, R., Winter, T.: RPL: IPv6 routing protocol for low-power and lossy networks. RFC 6550 (2012). https://doi.org/10.17487/RFC6550. https://rfc-editor.org/rfc/rfc6550.txt
2. Clausen, T.H., Yi, J., Herberg, U.: Lightweight on-demand ad hoc distance-vector routing - next generation (LOADng): protocol, extension, and applicability. Comput. Netw. **126**, 125–140 (2017). https://doi.org/10.1016/j.comnet.2017.06.025
3. Gan, W., Shi, Z., Zhang, C., Sun, L., Ionescu, D.: MERPL: a more memory-efficient storing mode in RPL. In: 2013 19th IEEE International Conference on Networks (ICON), pp. 1–5 (2013)
4. Glisic, S.: Advanced Wireless Networks: Technology and Business Models, 3rd edn. Wiley (2016)
5. Holler, J., Tsiatsis, V., Mulligan, C., Avesand, S., Karnouskos, S., Boyle, D.: From Machine-to-Machine to the Internet of Things: Introduction to a New Age of Intelligence, 1st edn. Academic Press, Cambridge (2014)
6. Khan, I., Qureshi, I., Shah, S.B., Akhlaq, M., Safi, E., Khan, M., Nazir, A.: Comparative analysis and route optimization of state of the art routing protocols. Mobile Netw. Appl. **3**, 6 (2018). https://doi.org/10.4108/eai.22-3-2018.154385
7. Kodali, R.K., Sarma, N.: Energy efficient routing protocols for WSN's. In: 2013 International Conference on Computer Communication and Informatics, pp. 1–4 (2013)
8. Jing, L., Liu, F., Li, Y.: Energy saving routing algorithm based on spin protocol in WSN. In: 2011 International Conference on Image Analysis and Signal Processing, pp. 416–419 (2011)
9. Misra, S., Goswami, S.: Network routing: Fundamentals, applications, and emerging technologies. Wiley (2017)
10. Saadat, M., Saadat, R., Mirjalily, G.: Improving threshold assignment for cluster head selection in hierarchical wireless sensor networks. In: 2010 5th International Symposium on Telecommunications, pp. 409–414 (2010)
11. Semchedine, F., Atmani, M., Ouaret, S.: Gossiping with probabilistic selection for routing in wireless sensor networks. In: 2013 World Congress on Computer and Information Technology (WCCIT), pp. 1–5 (2013)
12. Shelby, Z., Bormann, C.: 6LoWPAN: The Wireless Embedded Internet. Wiley, Hoboken (2010)
13. Singh, K.: Wsn leach based protocols: a structural analysis. In: 2015 International Conference and Workshop on Computing and Communication (IEMCON), pp. 1–7 (2015)
14. Sohraby, K., Minoli, D., Znati, T.: Wireless Sensor Networks: Technology, Protocols, and Applications. Wiley, Hoboken (2007)
15. Toor, A.S., Jain, A.K.: A survey of routing protocols in wireless sensor networks: hierarchical routing. In: 2016 International Conference on Recent Advances and Innovations in Engineering (ICRAIE), pp. 1–6 (2016)
16. Watteyne, T., Molinaro, A., Richichi, M.G., Dohler, M.: From MANET to IETF roll standardization: a paradigm shift in WSN routing protocols. IEEE Commun. Surv. Tuts. **13**(4), 688–707 (2011)
17. Zhao, L., Liu, G., Chen, J., Zhang, Z.: Flooding and directed diffusion routing algorithm in wireless sensor networks. In: 2009 Ninth International Conference on Hybrid Intelligent Systems, vol. 2, pp. 235–239 (2009)
18. Zheng, M., Zhao, X.: Research on directed diffusion routing protocol in wireless sensor networks. In: 2013 10th International Computer Conference on Wavelet Active Media Technology and Information Processing (ICCWAMTIP), pp. 53–57 (2013)

8 LPWAN Technologies

8.1 LPWAN in IoT

IoT technologies seen so far fall under the realm of PAN scenarios, that because they are usually wireless, they are representative of WPANs. Although these are low-power technologies, they are still powerful enough to provide transmission rates that natively enable IPv6 support [5]. However, the direct hop-to-hop transmission range of WPANs is typically quite short, rarely extending more than a few hundred meters. LPWAN technologies, briefly introduced in Sect. 1.3, attempt to increase the device coverage, but unfortunately, they are affected by a lower SNR that further reduces transmission rates and MTU sizes [13]. These limitations prevent these devices from fully supporting IPv6 and derived protocols like CoAP and RPL. In other words, almost all LPWAN physical and link layers do not support the upper layer protocols introduced in Chaps. 4 through 7. Most LPWAN mechanisms are therefore hybrid technologies with proprietary access networking stacks that rely on IoT gateways to enable application IP support. Moreover, most LPWAN stacks only carry a subset of available layers of the layered architecture.

Figure 8.1 shows the relationship between throughput and coverage, of physical layers of different IoT technologies when compared to LPWAN schemes. Best-case scenario is for traditional mobile cellular technologies like 5G that provide high throughput and coverage, but they are highly inefficient from a power perspective as they do not scale well for support of IoT applications. With a similar level of throughput, IEEE 802.11 derived technologies drain, in their majority, too much power to be effective in IoT environments. IoT WPAN schemes like IEEE 802.15.4 and BLE are power-efficient but they exhibit both low-throughput and very short coverage. LPWAN technologies complement WPAN by improving the distance range while keeping lower consumption to a minimal level in order to offer a multi-year battery lifetime. LPWAN devices typically transmit very small packets only a few times per hour over long distances. Different LPWAN solutions are characterized by different factors, namely, (1) network topology, (2) device coverage, (3) battery lifetime, (4) resilience to interference, (5) node density, (6) security, (7) unidirectional vs bidirectional communication, and (7) nature of the applications. The following section introduces several state-of-the-art LPWAN technologies that are relevant to IoT.

8.2 LoRa

Long range (LoRa) is the generic name for a full protocol stack that provides LPWAN capabilities that enables devices to run on a single battery for more than 10 years [18,33,28]. The LoRa stack is shown in Fig. 8.2. It is a small subset of the layered architecture that includes physical, link, and application layers. This is fine since LoRa devices only interact with other LoRa devices. There is no need for a network or transport layer since LoRa does not natively support IP to begin with. The figure also shows that each layer includes multiple sublayers that provide different functionality [7].

Note that LoRa provides security at the link and application layers. Link layer security provides authentication and AES-128-based device encryption in the network by relying on a 64-bit IEEE EUI-64 unique network key. On the other hand, application layer security provides AES-128 based encryption of device data.

8.2.1 Physical Layer

The LoRa physical layer relies on transmission over different ISM bands, specifically, the 433 MHz and the 915/868 MHz ISM bands. LoRa modulation is based on CSS. In the context of the physical layer of LoRa, CSS, as an SS mechanism, is not only power-efficient, but it also enables very long-range communications with a coverage that exceeds 10 km at

R. Herrero, *Fundamentals of IoT Communication Technologies*, Textbooks in Telecommunication Engineering, https://doi.org/10.1007/978-3-030-70080-5_8

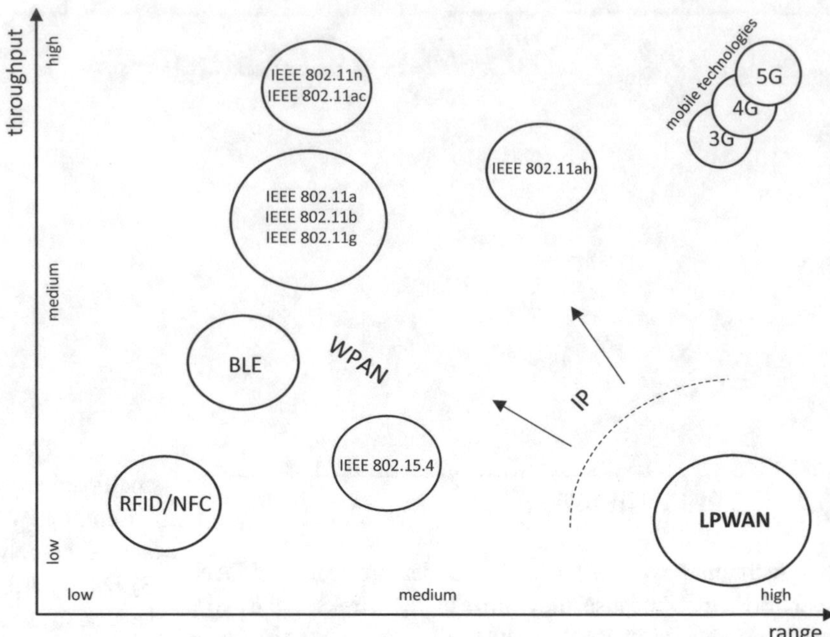

Fig. 8.1 IoT throughput and range

very low transmission rates. Of course, estimations of LoRa device coverage greatly depend on the physical obstructions. For example, a single LoRa base station or gateway can cover an area of hundreds of square kilometers. Moreover, studies show that based on the number of devices associated with a single gateway, it is possible to estimate the maximum transmission rate and coverage. For example, for an 8-km range at a maximum nominal rate of 250 bps, it is possible to support up to 50 devices. Similarly, for a 2.5-km range at a maximum nominal rate of 5 Kbps, it is possible to support up to 700 devices. In general, the data rate depends also on the ISM band under consideration with rates between 250 bps and 50 Kbps for the 868 MHz band in Europe and between 980 bps and 21.9 Kbps for the 915 MHz band in the United States. This has to do with the fact that the different ISM bands support a different number of subchannels; the 868 MHz band supports 10 subchannels while the 915 MHz band supports more than 64 subchannels. Moreover, the bandwidth of each subchannel depends on the spectral band; for both bands the uplink bandwidth is between 125 and 500 KHz, while for the 868 and the 915 MHz bands, the downlink bandwidths are 125 and 500 KHz, respectively.

Figure 8.3 shows the layout of gateways of the *The Things Network* (TTN), one of the few free and open LoRa providers, in New York City. In general, with minimal infrastructure, LoRa networks can completely cover cities and countries. Although LoRa hardware is proprietary, it has been licensed to multiple manufacturers. Moreover, there are several open source LoRa protocol stacks that support different embedded device platforms.

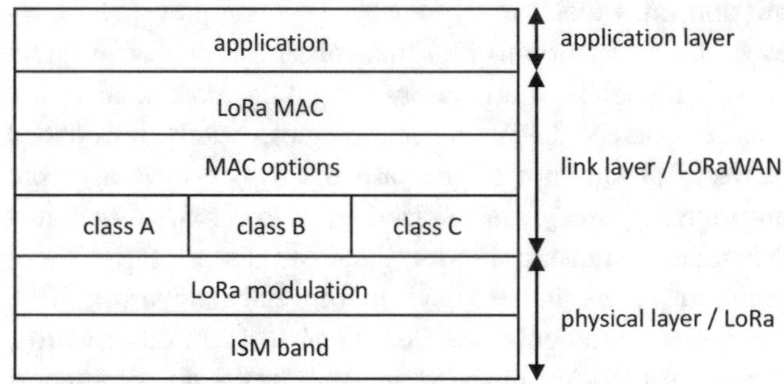

Fig. 8.2 LoRa stack

8.2.2 Link Layer

The link layer that enables devices to access the channel is the building block of the LoRa networking mechanism known as LoRaWAN [6]. Specifically, LoRaWAN is responsible for defining not only the media access algorithm but also the network architecture. Most WPAN technologies accomplish long-range communications by relying on mesh networking provided by capillary connectivity. Mesh networking, however, requires coordination to support data aggregation through intermediate nodes. Moreover, aggregation reduces both battery life and transmission capacity of these intermediate nodes even when carried data samples are not relevant to them. This is particularly important for low-power embedded devices like those involved in most LoRa scenarios. LoRaWAN introduces a much simpler approach that relies on exchanging transmission rate for coverage. Specifically,

Fig. 8.3 TTN coverage in NYC

LoRaWAN relies on devices talking to gateways directly without relying on intermediate nodes. To provide redundancy a single device is associated with multiple gateways. Therefore, for the same power restrictions and compared to WPAN technologies like IEEE 802.15.4, LoRaWAN transmits data at a much slower rate in order to reach farther distances while attempting to defeat channel interference by means of highly available gateways. LoRa gateways have two interfaces, (1) a LoRa interface and (2) an IP interface; packets sent by LoRa devices arrive at the LoRa interface and they are then forwarded via IP to a server. If one of the copies of a packet arrives at the server, the packet is not lost. Therefore, a server not only verifies that a packet complies with security settings but also makes sure to drop redundant copies. Since gateways are embedded devices themselves, LoRa pushes complexity and decision-making to the server. In mobile environments, the simple LoRa approach removes the need for complex handover mechanisms that occur when a device switches from one gateway to another.

LoRaWAN MAC is based on an ancestor of CSMA/CA known as ALOHA that enables devices to send datagrams without having to wait for the channel to be free. If during transmission the device detects a collision, it postpones the transmission by waiting a random amount of time before retrying. ALOHA is so simple that it is ideal for LoRaWAN MAC as it removes the need for network synchronization therefore minimizing power requirements. Moreover, because LoRa modulation is CSS based, device collisions are not common. ALOHA is also compatible with LoRaWAN topologies where multiple devices transmit sensor readouts to a gateway. A gateway, in turn, must be able to have enough capacity to support the throughput from many devices. This capacity is accomplished by a combination of techniques including multichannel modulation that supports multiple simultaneous transmissions, payload lengths, and adaptive data rates. This later mechanism consists of devices, with access links that have comparatively higher SNR that can transmit faster in order to lower channel access intervals and allow more devices to send traffic. The transmission rate is also affected by other issues like the spreading factor of the CSS modulation. In general, devices that are close to gateways, and therefore have good links, transmit faster than those devices that are farther away. Moreover, by lowering transmission rates, the battery lifetime of a device can be improved. One characteristic of LoRaWAN is that it is highly scalable, a network can be deployed with a single gateway and, as more capacity is needed, more gateways can be added.

Example 8.1 Consider a LoRa scenario with a device and a gateway transmitting 13-byte frames:

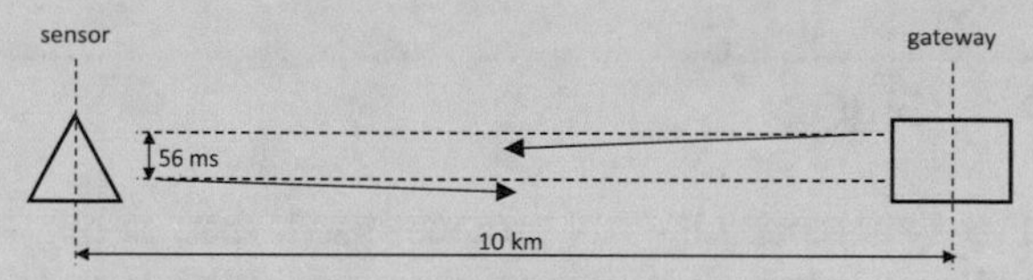

First the gateway attempts to send a frame, and then, 56 ms later, the device sends another frame. Assume a 1 Kbps transmission rate, a separation between sensor and gateway of 10 km and no delays other than propagation and transmission delays. How long does it take for both frames to be sent and received under LoRa classes A and C?

Solution Under class C, the gateway and the device transmit frames as soon as possible. The overall delay

(continued)

Table 8.1 LoRa classes

	Power consumption	Downlink
Class A	Low	After transmission
Class B	Medium	At scheduled times
Class C	High	Always

(continued)

is $D_C = 2 \times (D_t + D_p)$ where $D_t = \frac{8\times L}{R} = 104$ ms with $L = 13$ bytes and $R = 1000$ bps and $D_p = \frac{10{,}000\ \text{m}}{c} = 33.34\,\mu\text{s}$ with $c = 300\ 10^6$ m/s (speed of light). Because $D_p \ll D_t$, $D_C \approx 2 \times D_t = 208$ ms. Under class A, the gateway only transmits a frame in response to a frame sent by the device. In this scenario, this means that the delay D_A is given by $D_A = D_C + 56$ ms $= 264$ ms. Class A, when compared to class C, provides better energy consumption but increases latency.

Based on power consumption and level of interaction with gateways, device operation falls within the umbrella of three different classes A, B, and C shown in Table 8.1. Specifically, each class affects the trade-off between battery life, latency, and throughput. Class A is supported by all devices and enables gateways to send packets 6LoWPAN as soon as device traffic is received. As such class A devices consume the least amount of energy and extend battery life. Class B enables gateways to send packets downlink on a scheduled basis and therefore exhibit an intermediate level of power consumption and battery lifetime. Finally, class C enables gateways to send packet downlink always unless traffic is being received. Class C is the least power-efficient mechanism and exhibits the shortest battery lifetime.

8.3 SigFox

SigFox is another LPWAN protocol stack that, as opposed to LoRa, it relies on a unique commercial network called SigFox [14, 22, 25]. SigFox hardware, however, is open with several manufacturers making SigFox chips. Although SigFox was originally designed as a unidirectional traffic technology where sensors propagate readouts uplink, the stack was later updated to support downlink data transmission. SigFox devices are characterized by very low power consumption that supports extended battery life and low transmission rates.

Figure 8.4 shows the SigFox protocol stack; it includes a (1) physical layer, a (2) link layer that provide MAC, and a (3) network layer known as the frame layer that wraps the application data and adds a sequence number to the frames.

application	application layer
frame	network layer
MAC	link layer
ISM band	physical layer

Fig. 8.4 SigFox stack

The SigFox protocol stack does not include any encryption mechanism and relies on the applications to encrypt device data. SigFox messages, however, can be signed with a unique device key.

8.3.1 Physical Layer

SigFox, as LoRa, relies on transmission over the 915/868 MHz ISM bands. For uplink, each band is in turn divided into 333 *Ultra Narrow Band* (UNB) 100 Hz subchannels where DBPSK modulation is performed. For the downlink traffic, the subchannels are 600 Hz wide and GFSK modulation is used instead. Because both modulations are binary, the nominal throughput is 100 and 600 bps for uplink and downlink, respectively. Additionally, SigFox uses FHSS across 3 out of the 333 uplink subchannels. This is done to increase channel diversity and improve fading protection. Frequency hopping, based on a pseudo-random sequence, is combined with a random time delay, somewhere between 500 and 525 ms, on what it is called *random frequency and time division multiple access* (RFTDMA). The transmission power is specified to be 14 and 22 dBm for the 868 MHz and the 915 MHz bands respectively. Similarly, the receiver sensitivity is −120 and −142 dBm for the 868 MHz and the 915 MHz bands, respectively. SigFox estimates the allowable transmission rates based on the number of devices connected to the network and duration of the contract.

SigFox is designed for the transmission of infrequent bursts of small sensor readouts. As such the payload size for uplink and downlink is 12 bytes and 8 bytes per message, respectively. Note that these payload sizes are good enough to represent sensor data with a pretty good resolution. Since a single precision floating point number is typically stored in 4 bytes, a SigFox message can carry uplink up to three floating point numbers. Higher-precision floating point numbers, stored in 8 and 12 bytes, reduce the message capacity to transmit sensor readouts. In order to preserve battery life and limit power consumption, the maximum number of messages that a device can transmit and receive daily is 140 and 4, respectively. No device can transmit more than six messages per hour. In many cases, devices can minimize the amount of data sent by using events that are carried as messages with no payload. Of course, this does not circumvent the daily transmission limit, but it reduces the overall power consumption.

Example 8.2 What is the best-case scenario for uplink and downlink payload transmission rates under SigFox?

Solution For uplink and downlink, the payload sizes are $L_{\text{uplink}} = 8 \times 12 = 96$ bits and $L_{\text{uplink}} = 8 \times 8 = 64$ bits, respectively. The uplink rate is given by $R_{\text{uplink}} = \frac{N_{\text{uplink}} \times L_{\text{uplink}}}{T} = 0.15$ bps where $N_{\text{uplink}} = 140$ and $T = 60 \times 60 \times 24 = 86{,}400$ s. The downlink rate is given by $R_{\text{downlink}} = \frac{N_{\text{downlink}} \times L_{\text{downlink}}}{T} = 0.003$ bps where $N_{\text{downlink}} = 4$.

8.3.2 Link Layer

As LoRa, SigFox MAC is based on ALOHA. Moreover, devices are not synchronized with the network. SigFox supports a transmission mechanism similar to that of LoRa class A; when a message from a device arrives at the base station acting as an IoT gateway, the base station has a 20-s window of opportunity to transmit a message down to the device. In turn, the device waits for up to 25 s for an incoming 4-byte message from the base station to arrive.

Figure 8.5 illustrates RFTDMA, where three copies of the payload are transmitted over three random carriers with different random delays. The window for downlink transmission only opens after the last frame is transmitted. Because messages are transmitted in a *fire-and-forget* fashion, where no acknowledgment from the receiver is sent, three copies attempt to provide enough redundancy to guarantee successful delivery. This redundancy acts as a FEC mechanism.

Devices carry a device identifier that is used for routing, message signing, and authentication. Figures 8.6 and 8.7 show the SigFox frame format for both downlink and uplink communication between a device and a base station. Frames start with a preamble used for synchronization that is followed by a synchronization field that indicates the type of frame being transmitted. Uplink frames include the device identifier, the variable length payload, the authentication field, and a FCS field used for error detection. Similarly, downlink frames include flags, the FCS, the authentication field, the error code field, and the variable length payload. Note that since communication is from multiple devices to a base station, no addressing mechanism is needed other than the inclusion of the device identifier in uplink frames. For the worst-case scenario of a 200-bit frame, it takes around $\approx \frac{200\,\text{bit uplink packet}}{100\,\text{bps}} = 2$ s to transmit it uplink. Since three copies of each message are transmitted, the transmission time per individual message is typically 6 s. Because SigFox devices have a 1% duty cycle, they are active only 36 s every hour, and therefore they can only send up to six messages per hour.

Figure 8.8 shows the topology of a SigFox network. It consists of base stations that play the role of gateways, providing an interface between devices in the access side and backend server in the core side. The communication with the devices is by means of the SigFox physical and link layer mechanisms while the communication with the application is by means of traditional IP networking. The overall scheme follows a star topology where up to one million devices can interact with a single base station. In general, the number of supported devices is associated with the number of messages transmitted throughout the network. As for the case with LoRa, SigFox devices transmit their messages simultaneously to multiple base stations, and it is up to the backend server to remove duplicates before storage. Applications rely on SigFox cloud service APIs to retrieve device data from backend servers that

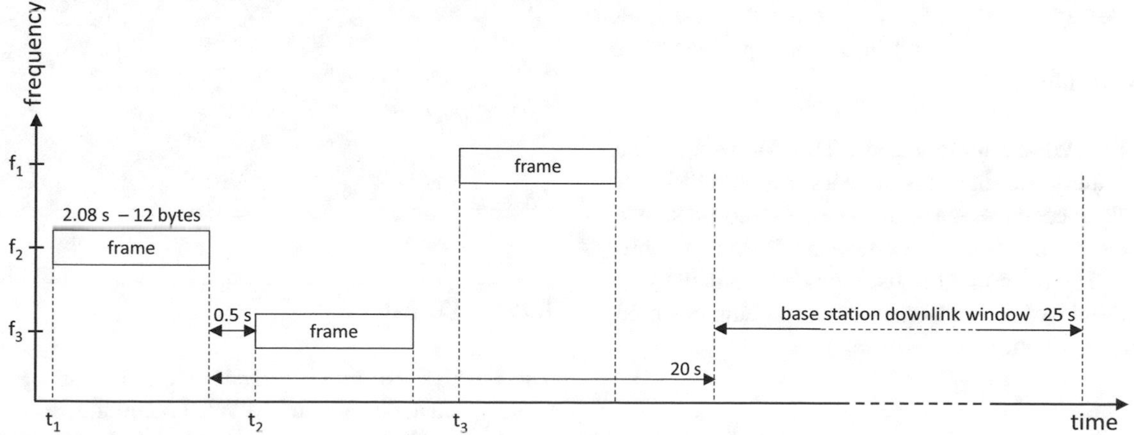

Fig. 8.5 SigFox transmission

Fig. 8.6 Upstream frame

32	16	32	0–96	8	16
PREAMBLE	SYNC	ID	PAYLOAD	AUTH	FCS

SYNC frame synchronization
ID device identifier
AUTH authentication
FCS frame checksum

Fig. 8.7 Downstream frame

32	13	2	8	8	8	0–64
PREAMBLE	SYNC	FLAGS	FCS	AUTH	EC	PAYLOAD

SYNC frame synchronization
FCS frame checksum
AUTH authentication
EC error code

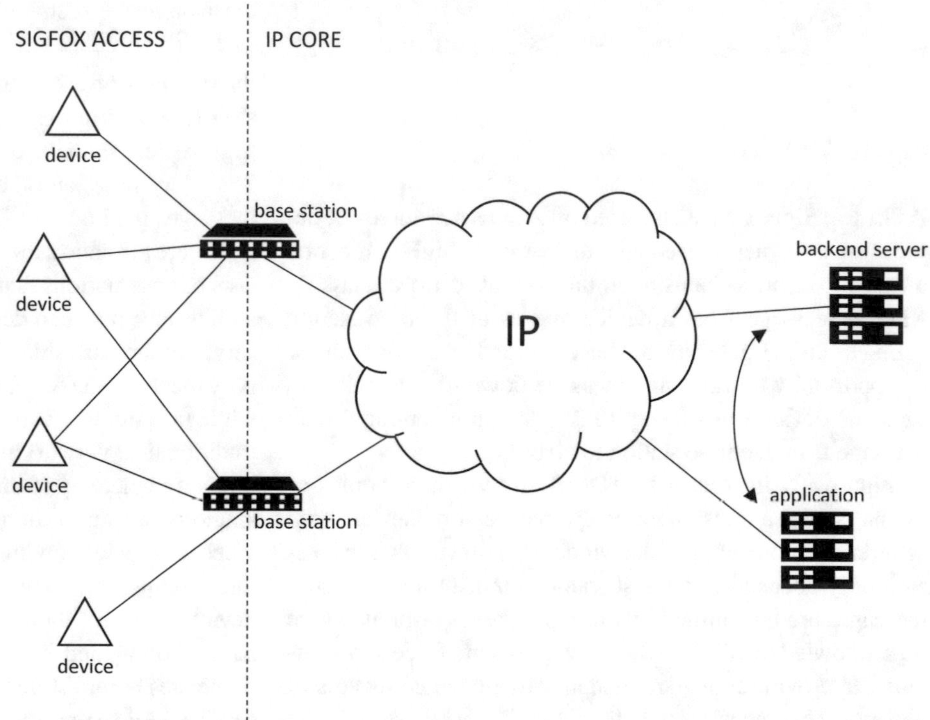

Fig. 8.8 SigFox topology

store the device readouts themselves. In turn, backend servers transmit messages to devices by selecting the base stations with the best connectivity and in the best position to send messages downlink.

▶ **LPWAN vs. WPAN: Battery Life** The following table shows battery life (in terms of years) for both LPWAN (long-distance coverage) and WPAN (short-distance coverage) technologies as a function of different applications [21]. Note that the list includes 1% duty cycle SigFox and LoRa. It assumes devices being powered by two AAA batteries.

Application	WPAN					LPWAN	
	BLE	IEEE 802.15.4	IEEE 802.15.4e	IEEE 802.11b	IEEE 802.11ah	SigFox	LoRa
Media (64 Kbps)	≈ 0.25	≈ 0.05	≈ 0.06	≈ 0.05	≈ 0.015	N/A	N/A
Sensor (400 bps)	10.5	5	5.5	3	1	N/A	N/A
Sensor (4 bps)	23.5	23	20.5	8	9	N/A	1
Sensor (400 bits/day)	24	24	20.5	8	10	21	27

8.4 D7AP

The *DASH7 Alliance Protocol* (D7AP) is an open source protocol stack that enables LPWAN communications [37]. Its physical and link layers are derived from the *ISO/IEC 18000-*

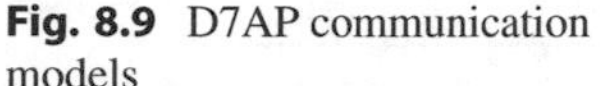

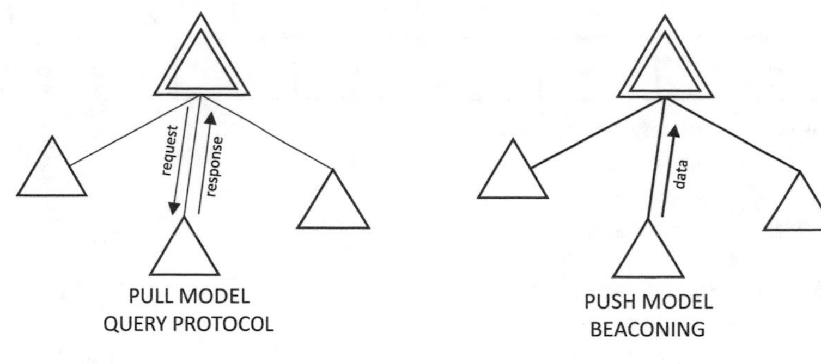

Fig. 8.9 D7AP communication models

Fig. 8.10 Foreground frame

PREAMBLE	SYNC	L	S	C	TADR	PAYLOAD	CRC16
32	16	8	8	8	≤ 64	≤ 251	16 bits

SYNC frame synchronization
L length
S subnet
C control parameter
TADR target address
CRC16 error detection

7 standard for active (as opposed to passive) RFID. The *7* in DASH7 refers to the section -*7* of the original standard. D7AP was originally designed for military purposes with batteries lasting for years and supporting low-latency long-range wireless mobility [8, 9]. It was adopted by the US Department of Defense in 2009 to support asset tracking, but in 2011 it was repurposed for location based services associated with smartcards and device tracking and extended to support a huge number of WSN applications ranging from mobile advertising and billboards to cargo monitoring in transport vehicles. As an RFID mechanism, it compliments traditional short-range NFC technologies by providing mobile asset tracking and enabling devices to communicate with each other.

D7AP defines four device classes: (1) blinker that only transmits traffic, (2) endpoint that transmits and receives data and supports wake-up events, (3) subcontroller that is a full featured device and also supports wake-up events, and (4) gateway that is always active and provides the interface with other networks as any IoT gateway.

D7AP, as shown in Fig. 8.9, supports two modes of operation based on a star topology; the pull model where the interaction between devices and gateways is bidirectional by means of requests as well as responses and the push model where devices periodically transmit data to gateways unidirectionally. When those messages are beacons, the mechanism is called beacon transmit series.

8.4.1 Physical Layer

D7AP specifies a full stack that, in certain scenarios, can rely on other physical and link layer technologies like LoRa or SigFox for transmission. The native physical layer of D7AP relies on transmission over the 433 MHz and 915/868 MHz ISM bands. Each band is subdivided in several channels: 69, 280, and 1040, respectively, for the 433 MHZ, the 868 MHz, and the 915 MHz bands. The channels can be configured under three different classes: (1) Lo-Rate, (2) Normal, and (3) Hi-Rate that support transmission rates at 9.6 Kbps, 55 Kbps, and 166 Kbps, respectively. In all scenarios signal modulation is by means of GFSK combined with FHSS that supports a transmission power of around 10 dBm. Most deployments are characterized by a maximum 2-s latency for an adjustable transmission range between 10 m and 10 km. In asset tracking applications, the precision is within 4 m.

Before modulation, binary symbols are encoded by means of several different coding schemes; the main mechanism relies on PN9 that produces a predictable pseudo-random sequence of values. PN9 is complemented by a FEC convolutional encoder that provides a 1/2 code rate. The physical layer is also capable of performing RSSI measurement with 6 dBm accuracy.

8.4.2 Link Layer

The link layer of D7AP defines two types of frames: (1) foreground frame and (2) background frame. Foreground frames are common messages used to carry data and data requests. On the other hand, the background frames are very small messages used for advertising and for group synchronization.

Figure 8.10 shows a foreground frame; it includes a 32-bit preamble and 16-bit synchronization word followed by an 8-bit length field that specifies the size of the frame. The 8-bit subnet field is used for filtering of incoming frames. The 8-bit control parameter that is used to specify the target id identifier type is followed by a variable length target address. Similarly, the variable length payload is followed by a 16-bit CRC field used for error detection. Figure 8.11 shows a

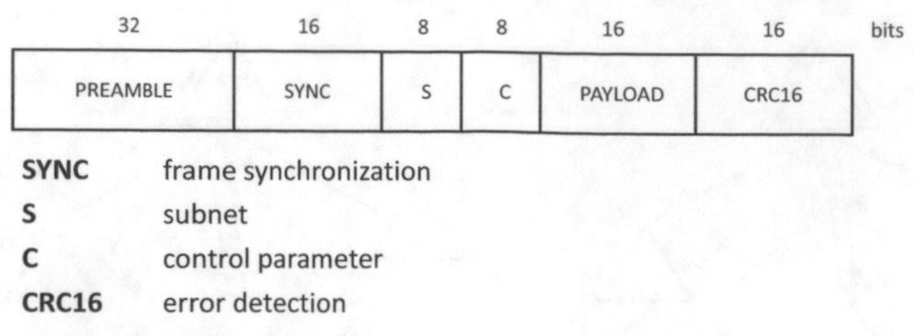

Fig. 8.11 Background frame

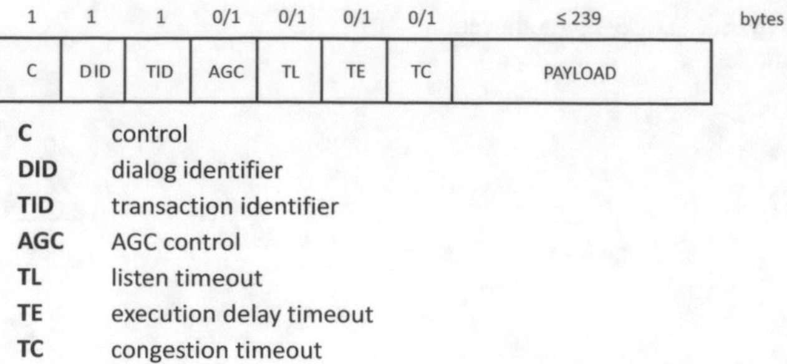

Fig. 8.13 D7ATP request segment

background frame; as for the foreground frame case; it starts with a 32-bit preamble and 16-bit synchronization word that is followed by an 8-bit subnet field is used for filtering of incoming frames and an 8-bit control parameter that is used to specify the target identifier type. The 16-bit payload is followed by a 16-bit CRC field used for error detection.

To match each of the frame types, the link layer supports background and foreground scan and message reception. From a transmission perspective, channel contention is resolved by means of an adaptation of CSMA/CA where consecutive transmission attempts are randomly scheduled based on different modes of operation.

8.4.3 Other Layers

D7AP includes, in addition, network, transport and session layers. The network layer defines two protocols: a background *D7A Advertising Protocol* (D7AAvP) and a foreground *D7A Network Protocol* (D7ANP). D7AAvP has priority over D7ANP. D7AAvP is used with background frames for rapid, ad hoc group synchronization. D7ANP, on the other hand, is used with foreground frames to provide a single datagram and addressable, routable protocol that relies on the structure shown in Fig. 8.12. The datagram contains several fields; a 1-byte control field used to specify the origin identifier type and the network security method, a variable length origin identifier, a 1-byte hopping control field used to specify the destination identifier type and hopping information, a variable length destination identifier, a 1-byte hopping network layer timer value used for multi-hop routing, an optional 5-byte security header use to encode the security state, a variable length payload, and a variable length security footer use to encode the MIC. Network security includes combinations of encryption and authentication that rely on several mechanisms of cipher block chaining and variable length MICs.

The *D7A Transport Protocol* (D7ATP) is based on a scheme where a one-time request is sent by a requester to one or more responders. The interaction between the requester and the responders is performed in the context of a transaction. The transaction, in turn, is made of three distinct periods: (1) the request period which accounts for the transmission of the segment including the contention due to CSMA/CA, (2) the execution delay period that is associated with the propagation delay, and the (3) congestion timeout that tells the responders how long their transmissions should take. Figures 8.13 and 8.14 show the segment structures of requests and responses, respectively. Both requests and responses include a 1-byte control field, a 1-byte dialog identifier that is unique per dialog, a 1-byte transaction identifier that specifies that transaction within the dialog, an optional 1-byte AGC control field, an optional 1-byte listen timeout, and an optional 1-byte execution delay timeout. Requests include, in addition, a 1-byte congestion timeout field and a variable length payload, while responses include a variable length acknowledgment field that specifies the transactions being acknowledged and a variable length payload. D7AP also includes a session layer, known as *D7A Session Protocol* (D7ASP), that defines the events that may result in session initialization, scheduling as well as QoS strategy and power control.

8.5 Weightless

Weightless is another LPWAN protocol stack that relies on transmissions over the white space spectrum [36, 25].

Fig. 8.12 D7ANP datagram

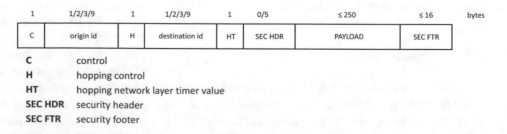

1	1/2/3/9	1	1/2/3/9	1	0/5	≤ 250	≤ 16	bytes
C	origin id	H	destination id	HT	SEC HDR	PAYLOAD	SEC FTR	

C control
H hopping control
HT hopping network layer timer value
SEC HDR security header
SEC FTR security footer

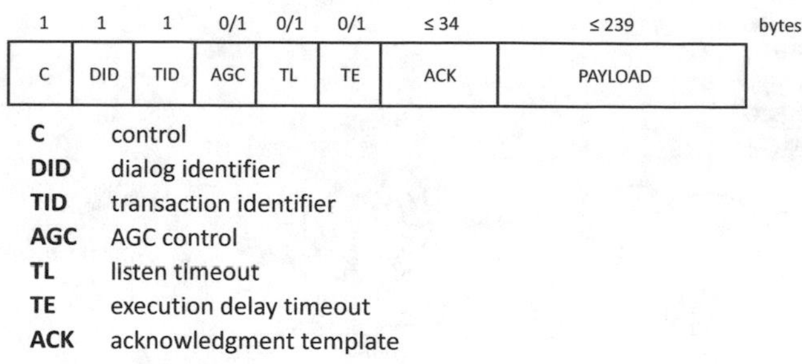

Fig. 8.14 D7ATP response segment

Specifically, the white space spectrum refers to those frequencies that result from the allocation of separation guards between bands and channels in order to minimize interference. Although these frequencies are typically reserved and not intended for transmission, in certain scenarios they can be used if power levels and modulation schemes are carefully selected to prevent any interference. The use of white space frequencies enables Weightless to provide widespread signal coverage [24].

The standard is called Weightless because it relies on a lightweight protocol stack that exhibits very little overhead in order to maximize the performance of small packet transmissions. Weightless applications include home and building automation, smart metering, health care, transportation, point of sale, inventory tracking, and physical security among others.

Weightless, as many LPWAN technologies, is characterized because it provides a low cost solution that is compatible with most large size IoT device deployments. In fact, Weightless is intended to support a very large number of devices with a single cell supporting more than one million devices. A network could easily include around one billion devices. In all cases, low costs imply that both device and service costs are low. Additionally, Weightless is ultralow power and enables devices to run on a single battery for up to 10 years. Moreover, Weightless provides reliable and secure delivery of unicast and broadcast messages. Messages are transmitted in bursts with the Weightless network prepared to support a typical message size of around 50 bytes.

Weightless is an open standard that enables devices to exchange data with a single base station that acts as an IoT gateway. As such, the base station provides connectivity to a network manager and database through traditional IP mechanisms. Weightless relies on the star topology shown in Fig. 8.15. Additionally there are several elements that enable the operation of a Weightless network: (1) the base station controller provides a point of contact for a group of base stations, (2) the authentication database holds the authorization data associated with base stations and devices as well as the encryption keys, (3) the location register keeps track of the location of the devices in order to determine the best base station for message routing, (4) the broadcast register keeps track of the groups the devices belong to in order to support broadcast messages, (5) the Operation and Maintenance Center (OMC) provides the monitoring of the network to detect failures, (6) the billing entity keeps track of device activity to support billing information, and (7) white space database keeps track of the available channels based on the servicing location.

As an open standard, multiple manufacturers produce Weightless hardware. The Weightless standard is the result of the *Weightless Special Interest Group* (Weightless SIG) that dictates the different layers of the protocol stack. Moreover, as any other standardization body, the Weightless SIG is responsible for specifying interop testing technologies as well as mechanisms of certification for the industry. The Weightless SIG serves as an entity that interacts with government regulators to free up the use of the white space spectrum.

Figure 8.16 shows the Weightless stack where layers talk to each other by means of channels. Moreover, the stack is partitioned into two different planes: a user plane and a control plane. The physical layer has three physical channels: the *downlink channel* that provides communication from the base station to the device, conversely, the *uplink channel* that provides communication from the device to the base station, and the uplink contended access channel that enables multiple devices to simultaneously transmit traffic to the base station. Within the physical layer, there are transport channels that provide the interface between the physical and link layers. The transport channels provide the framing of the downlink and uplink channels as well as the contended access associated with the uplink contended access channels. In the link layer, the logical channels provide retransmission control as well as fragmentation and reassembly. Specifically, the logical channels enable the data and control traffic to be transmitted reliably or unreliably based on whether the individual frames are acknowledged or not. The connectivity between base stations and devices is controlled by the radio resource manager (RRM) that relies on control channels to send messages downlink. Alternatively, user channels provide the connectivity between base stations and devices to enable the transmission of *unicast*, *multicast*, *interrupted*, and *acknowledgment* frames. While unicast traffic is bidirectional, multicast traffic is unidirectional between device and base station.

8.5.1 Physical and Link Layers

As indicated above, Weightless relies on the transmission over the white space spectrum. Specifically, the UHF band, located between the 300 MHz and 3 GHz range and shown in Fig. 3.1, is an ideal candidate for white space frequency

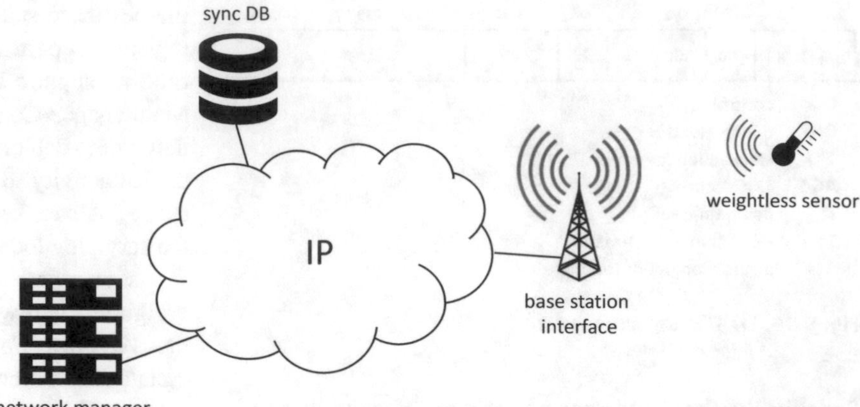

Fig. 8.15 Weightless topology

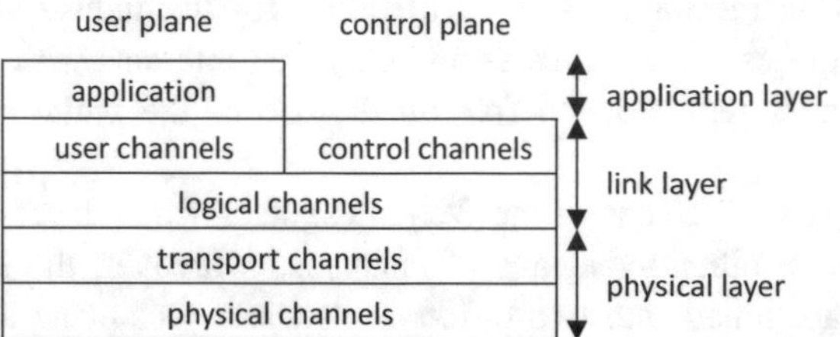

Fig. 8.16 Weightless stack

Table 8.2 Weightless modulation summary

Modulation summary			
Scheme	Code rate	Spreading factor	Rate (Mbps)
16-QAM	1	1	16.0
16-QAM	¾	1	12.0
16-QAM	½	1	8.0
QPSK	¾	1	6.0
QPSK	½	1	4.0
BPSK	½	1	2.0
BPSK	½	4	0.5
BPSK	½	16	0.125
BPSK	½	63	0.040
BPSK	½	225	0.010
DBPSK	½	1023	0.0025

modulation since it is characterized by very good propagation. Moreover, Weightless modulation is typically performed over underutilized white space spectrum channels that exhibit very little interference and require small size antennas. These channels are set up over unoccupied frequency guardbands between TV channels. Again, since UHF TV signals are generated with high-power transmitters and high gain antennas located at high altitude, they are unlikely to interfere with Weightless low-power transmitters and low gain antennas at low altitude.

Weightless relies on FHSS with the carrier being modulated by means of DBPSK, BPSK, QPSK, or 16-QAM. Table 8.2 shows the different modulation schemes and their corresponding nominal transmission rates associated with Weightless. The modulation is adaptive, and the propagation range is related to the negotiated transmission rate. Very high transmission rates are only possible at very short distances. Note that each scheme is linked to a specific code rate and spreading factor. The code rate and spreading factor are, respectively, associated with FEC and resilience against channel noise. Since the same frequency is used for both uplink and downlink communications, TDD is used by device and base station signals to share the channel.

Figure 8.17 shows the processes involved in the Weightless signal generation. The binary data from the link layer is first subjected to FEC that inserts control redundancy and results in a specific code rate. The output data is then processed by a whitening stage that by means of a scrambler XORs the data against a pseudo-random sequence in order to randomize the signal and introduce transitions that are essential to signal synchronization. The output stream is then modulated by the schemes and spread by means of FHSS. Finally *cyclic prefix insertion* (CPI) is used to minimize, among other things, *intersymbol interference* (ISI) and multipath fading, by inserting at the start of a frame a portion of the end of the frame. A preamble is then inserted to provide synchronization between devices and base stations. The resulting signal is passed through a root-raised cosine (RRC) filter that provides pulse shaping to further minimize ISI due the sharp wave transitions it introduces.

Weightless include three main variations known as *Weightless-W*, *Weightless-P*, and *Weightless-N*. Weightless-W is based on the specifications presented so far in this section. Weightless-P is based on Weightless-W, but it is cheaper since it removes the need for a temperature-compensated crystal oscillator (TCXO) by changing some of the modulation scheme parameters. Weightless-N is also based on Weightless-W, but it relies on transmission over the 868/915 MHz ISM bands providing only uplink communication. It exhibits lower power consumption that extends battery life for up to 10 years but limits both the propagation range and the transmission rate.

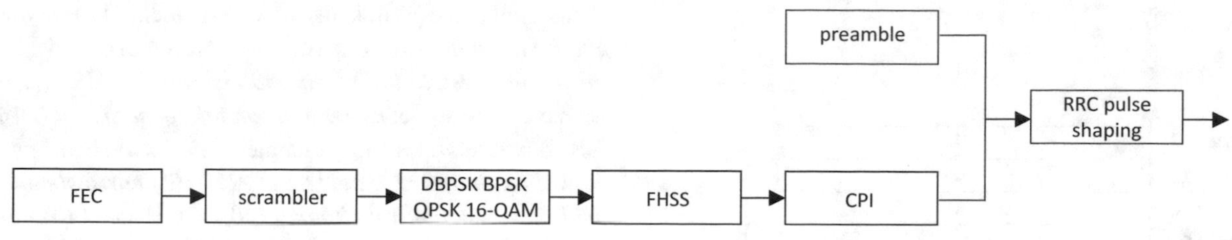

Fig. 8.17 Signal generation

8.6 NB-IoT

Narrowband IoT (NB-IoT) is an LPWAN technology based on cellular communications and first introduced in the *3rd Generation Partnership Project* (3GPP) Release 13 and enhanced through successive releases with Release 17 scheduled to be available in 2021 [22, 12, 3]. Specifically, 3GPP adds several IoT features to the existing *Global System for Mobile Communications* (GSM) and *4G Long-Term Evolution* (LTE) network architectures. NB-IoT is, therefore, also known as *LTE Cat M2*. As with most LPWAN mechanisms, the main goal is to provide wide area coverage at a very low cost per device. Moreover, additional goals like reduction of device complexity and backwards compatibility are also a focus of NB-IoT. To accomplish this latter objective, NB-IoT relies on a very small portion of the existing spectrum used by the cellular technologies. NB-IoT backward compatibility implies that if cellular networks are upgraded in accordance with NB-IoT requirements, they must still support existing 3GPP *user equipment* (UE). This does not mean, however, that NB-IoT devices can communicate in existing legacy networks 8.6.

NB-IoT, when compared to other LPWAN technologies, attempts to improve indoor coverage while supporting a massive number of low-throughput and low-latency devices. NB-IoT supports several IoT use cases including home as well as home and building automation, asset tracking, and industrial control. The main requirements of NB-IoT are very low cost devices with a unit price below $5, a battery life of over 10 years with little human intervention, and the support of up to 50,000 devices per cell.

NB-IoT adopts the protocol stack of legacy LTE with modifications to comply with the requirements including the support of a huge number of devices connected over a very long range.

8.6.1 Physical Layer

NB-IoT relies on half duplex *frequency division duplex* (FDD) communication to support nominal uplink and downlink transmission rates of 60 Kbps and 30 Kbps, respectively. As with TDD, FDD involves, in order to minimize interference, the receiver and transmitter sending traffic on two different nonoverlapping channels. Under the worst network conditions, NB-IoT typically achieves a transmission rate of around 200 bps. Moreover, NB-IoT, as a narrowband IoT technology, requires 180 KHz channels for both directions. This bandwidth allocation is associated with three different deployment scenarios over licensed bands: (1) standalone allocation where a GSM network operator can replace a GSM 850–900 MHz carrier with NB-IoT, (2) inband allocation where an LTE network operator can allocate a *physical resource block* (PRB) to deploy inband inside NB-IoT, and (3) guardband allocation where an LTE network operator can also deploy an NB-IoT carrier in a guardband between two LTE channels.

NB-IoT reuses a lot of the mechanisms introduced by LTE. This has enabled the development of specifications in a fairly short amount of time. Among the many mechanisms that NB-IoT borrows from LTE, the modulation schemes are the most important. Specifically, NB-IoT downlink traffic relies on OFDMA, while uplink traffic relies on *single-carrier FDMA* (SC-FDMA). The reason for two different schemes is that although SC-FDMA is more complex than OFDMA, it is also a lot more power-efficient and therefore more appropriate for the transmission of constrained devices. The main difference between OFDMA and SC-FDMA is how they arrange user data streams in frequency and time. Figure 8.18 shows how, under OFDMA, traffic associated with 12 users *a, b, …, l* is simultaneously transmitted over 12 different subchannels. Similarly, Fig. 8.19 shows how, under SC-FDMA, the same traffic is sequentially transmitted on individual timeslots. As it can be seen, the overall throughput is the same for both scenarios. In order to support a massive number of devices, NB-IoT assigns *resource units* (RUs) to multiple UEs. Under NB-IoT, the carrier separation of both schemes is 15 KHz, accounting for a subcarrier count of 12 that covers the 180 KHz bandwidth. The uplink configuration also supports a 3.75 KHz carrier spacing. With 15 KHz spacing, NB-IoT allocates a single 8-ms tone, 3 4-ms tones, 6 2-ms tones, or 12 1-ms tones. Similarly, with 3.75 KHz spacing, NB-IoT allocates 48 32-ms tones. In each channel the selected modulation scheme is either BPSK or QPSK for both uplink and downlink. The maximum transmission power for uplink and downlink transmissions is 23 dBm and

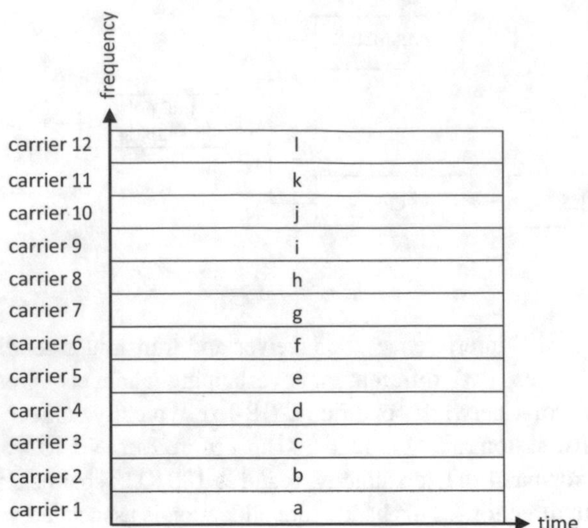

Fig. 8.18 OFDMA

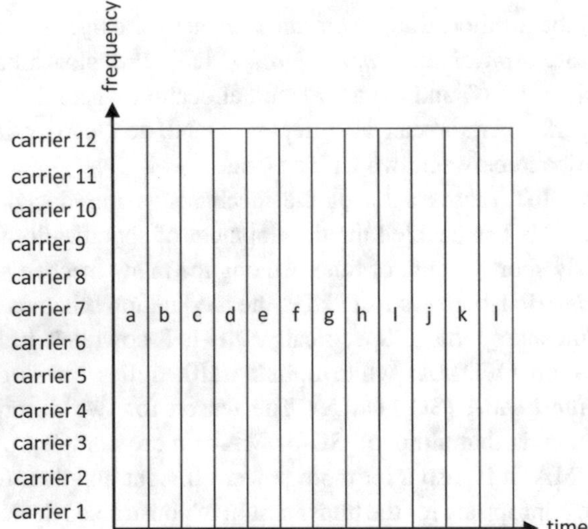

Fig. 8.19 SC-FDMA

46 dBm, respectively. From a battery life preservation perspective, NB-IoT relies on two different mechanisms: *power-saving Mode* (PSM) and *Extended Discontinuous Reception* (eDRX). Under PSM, although a device is registered with the network, it typically sleeps for up to 413 days and cannot be reached by the base station. Similarly, under eDRX a device is typically inactive for up to a few hours only.

NB-IoT follows the frame structure of LTE with 1024 *hyperframes* where each hyperframe, in turn, contains 1024 frames. The duration of each *hyperframe cycle* is 104,85.76 s. For a 15 KHz carrier spacing, associated with both uplink and downlink shown in Fig. 8.20, each frame has ten subframes where each subframe contains two 500-μs slots. Similarly, for a 3.75 KHz carrier spacing associated with only a single uplink shown in Fig. 8.21, each frame has five 2-ms slots. The uplink uses two channels (1) *narrowband physical random access channel* (NPRACH) and (2) *narrowband physical uplink shared channel* (NPUSCH), and it also uses one signal *demodulation reference signal* (DMRS). The downlink uses three channels (1) *narrowband physical downlink shared channel* (NPDSCH), (2) *narrowband physical downlink control channel* (NPDCCH), and (3) *narrowband physical broadcast channel* (NPBCH), and it also uses three signals (1) *narrowband reference signal* (NRS), (2) *narrowband primary synchronization signal* (NPSS), and (3) *narrowband secondary synchronization signal* (NSSS).

Example 8.3 Consider an IPv6 adaptation mechanism encapsulated as part of NB-IoT that results in the transmission of 20-byte packets. What is the worst-case scenario MAC latency that can be expected from such mechanism?

Solution Because under NB-IoT the transmission rate can be as low as $R = 200$ bps, for an $L = 20$ bytes packet, the transmission delay D is given by $D = \frac{8\times L}{R} = 8$ s. Note that the 8-s transmission delay is also a MAC access delay that prevents the support of most relevant IoT real-time applications.

For uplink, the NPRACH channel enables a device to initially access the network and every time after failure as well as to request transmission resources. Similarly, the NPUSCH channel is used to carry uplink data packets. The DMRS signal provides uplink channel estimation accuracy. For downlink, the NPDSCH channel is used to carry downlink data packets. Control traffic is carried down to the device by means of the NPBCH channel and the NPDCCH. The *master information block* (MIB) is transmitted over the NPBCH channel, while the *system information block* (SIB) is transmitted over the NPDCCH channel. The NPSS and NSSS signals are used by the device for timing and frequency synchronization with the base station. Similarly, the NSR signal enables cell search and initial system acquisition.

Figure 8.22 depicts the downlink frame composed of ten subframes; subframe 0 carries the NPBCH, and subframes 1 through 4 and 6 through 8 carry the NPDCCH channel for control traffic and the NPDSCH channel for data traffic, while subframes 5 and 9 carry the NPSS and the NSSS signals. The *downlink control information* (DCI) that is carried by the NPDCCH channel provides the device with information related to the cell identity and available resources to map NPDSCH channel symbols.

NB-IoT devices follow the same procedure as LTE UEs to synchronize and allocate the timing of slots and frames during cell acquisition. After decoding MIB and SIB data,

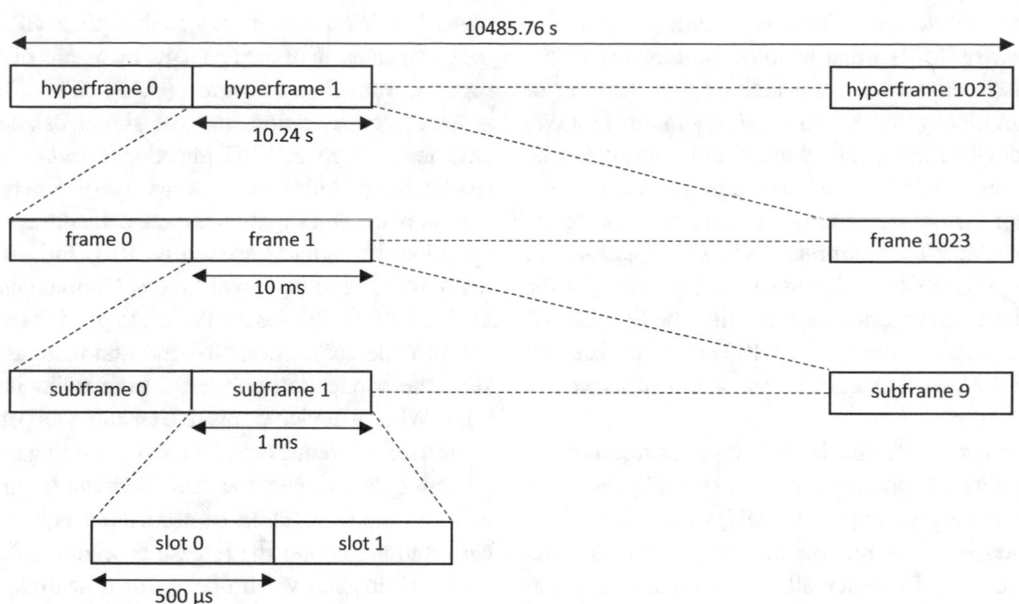

Fig. 8.20 Uplink/downlink hyperframe structure for 15 KHz spacing

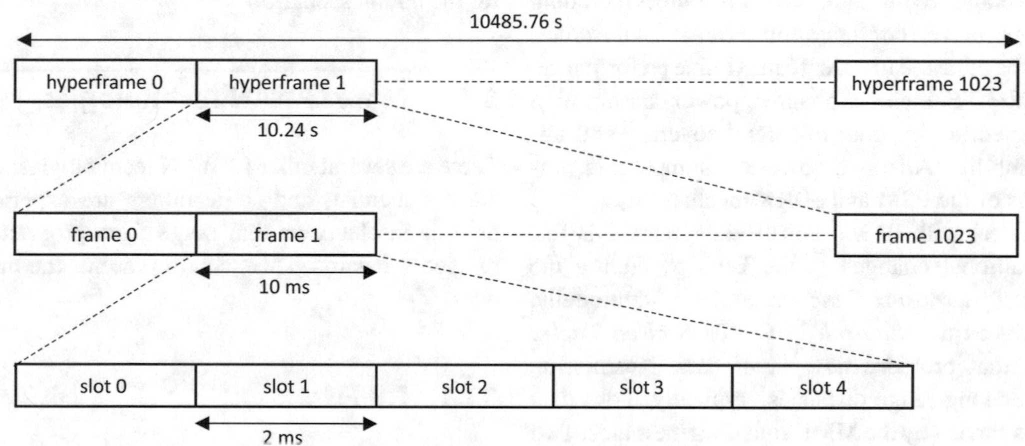

Fig. 8.21 Uplink hyperframe structure for 3.75 KHz spacing

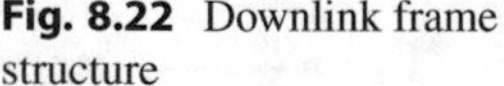

Fig. 8.22 Downlink frame structure

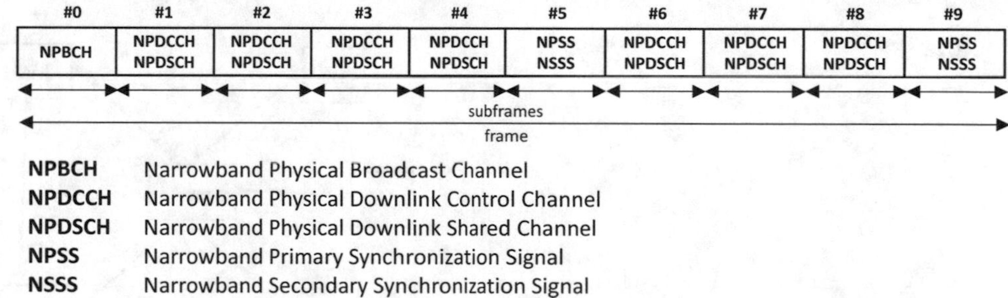

the cell identifier, subframe number, scheduling information, and system bandwidth become available to the device. Additionally, each device acquires a UE identifier as part of the *random access procedure* (RAP) that is performed during initial uplink synchronization and any time synchronization needs to be regained due to a long period of inactivity.

8.6.2 Link and Upper Layers

The NB-IoT network topology is based on legacy LTE infrastructure. It includes (1) a network core called *evolved packet core* (EPC) shown in Fig. 8.23, (2) an *evolved UMTS terrestrial radio access Network* (E-UTRAN), and (3) UE that consists of IoT devices. The EPC includes a *mobility management entity* (MME) that handles the mobility of the devices but also performs several actions like tracking of devices, session management, and *serving gateway* (S-GW) selection for devices initial attachment and authentication. The S-GW routes the data packets through the network and serves as anchor for device handover when transitioning between different *eNode base stations* (eNBs). The *packet data node gateway* (P-GW) provides the interface between the 3GPP network and an external network like the IP network. Finally, the *home subscriber server* (HSS) is a database that is used for mobility, session and user management, as well as access authorization.

The MAC sublayer within the link layer is responsible, among other things, of handling retransmissions by means of a *hybrid automatic repeat request* (HARQ) mechanism, timing advance, multiplexing, random access, priority management, and scheduling. Resource allocation must ensure that the maximum number of devices is served in any given cell while accomplishing a specific throughput level, spectral efficiency, and coverage. To this end, several resources including tone allocations, power configurations, frames, subframes, and slots must be adjusted in order to maximize performance. NB-IoT, as with LTE, includes adaptive power consumption and symbol repetition in order to extend coverage and improve link reliability. Adaptive power consumption is provided by means of the PSM and eDRX mechanisms.

In order to deal with IP and non-IP datagrams, NB-IoT introduces additional changes to the EPC by adding the *Service Capability Exposure Function* (SCEF). Additionally, NB-IoT introduces a *cellular IoT* (CIoT) *Evolved Packet System* (EPS) that provides new small data transmission procedures over long-range distances. Typically, more than one data path is carried by the MME signaling messages. Two main procedures are optimized to support small data transfer: (1) *mandatory control plane CIoT EPS* and (2) *optional user plane CIoT EPS*. Mandatory control plane CIoT EPS enables the encapsulation of data packets by means of signaling messages in *Non-Access Stratum* (NAS). When device mobility is under consideration, loss of signal can lead to a great QoS degradation. NB-IoT introduces *radio resource control* (RRC) that provides the transmission of data packets over the control plane and supports the reestablishment of connection upon loss of signal. Essentially, RRC hides the loss of the radio interface to the upper layers. Optional user plane CIoT EPS relies on RRC for the device to get *Access Stratum* (AS) and provide connection suspend and resume procedures to keep the connection context during radio interface transitions. When a device wants to transmit data packets uplink, it starts a service request procedure by sending a random access preamble that triggers the establishment of an RRC connection and the reservation of associated radio resources. The base station initiates the release procedure after an extended period of inactivity. Similarly, for downlink transmissions, if the device is in eDRX mode, when it receives a paging message, it starts a service request procedure as described for the uplink scenario.

8.7 More LPWAN Technologies

There are several other LPWAN technologies; many of them are commercial, and some others are experimental. They provide mechanisms that range from proprietary protocols to open standards. This section presents the most important ones.

8.7.1 NB-Fi

Narrowband fidelity (NB-Fi) is an open full stack protocol that is the base of a commercial LPWAN turn-key solution

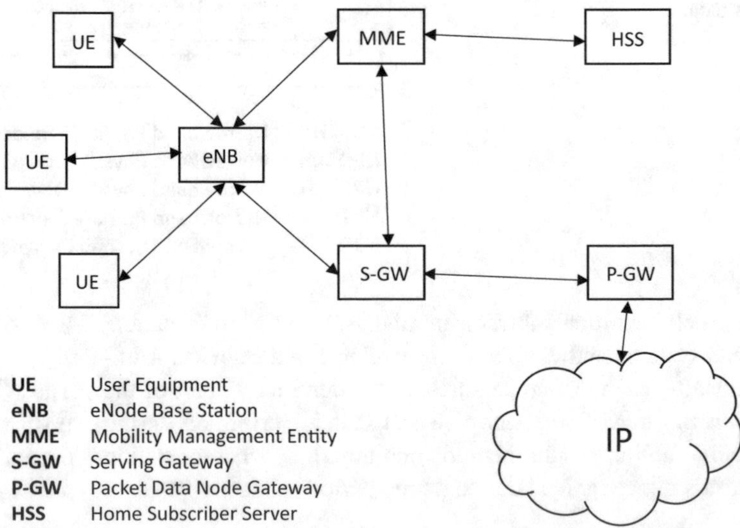

Fig. 8.23 EPC

that includes devices and networks developed by *WAVIoT* [11, 16]. Its goal is to provide a robust and reliable interaction between devices and base stations acting as traditional IoT gateways. As most other LPWAN technologies, NB-Fi provides long battery life combined with low hardware cost. Moreover, NB-FI attempts to lower both deployment costs and deployment times by simplifying the network topology. The typical transmission ranges are around 10 and 30 km for urban and rural environments, respectively. NB-Fi applications range from smart metering and smart cities to home and building automation.

NB-Fi's physical layer relies on transmission over the 433 and 915/868 MHz ISM bands to accomplish transmission rates between 50 bps and 25.6 Kbps. Modulation is by means of DBPSK but optimized to improve spectral efficiency and performance through *software-defined Radio* (SDR). TDD is used by device and base station signals to share the channel. The NB-Fi link layer relies on both TDMA and FDMA for media access. This enables network scenarios that include both star topology and mesh schemes where a single base station can support up to two million devices.

8.7.2 IQRF

IQRF is an open LPWAN framework that includes devices, gateways, and applications addressing scenarios ranging from telemetry and industrial control to home and building automation [10]. OQRF provides a full protocol stack with a physical layer that relies on transmission over the 433 and 915/868 MHz ISM bands. As with many other LPWAN technologies, modulation supported by GFSK enables transmission rates up to 20 Kbps with frames that are 64 bytes long. The nominal signal coverage of IQRF is 500 m. IQRF has a network layer that supports mesh networking through a protocol called *IQMESH* that supports up to 240 hops. IQRF includes an optional transport layer called *direct peripheral access* (DPA) that provides dataflow-oriented communications. The maximum number of devices per gateway is 65,000 but it reduces to 240 if DPA is in place.

8.7.3 RPMA

Random Phase Multiple Access (RPMA) is a media access scheme, developed and patented by Ingenu, that serves as the base of a robust LPWAN technology [15, 25]. RPMA operates over the 2.4 GHz ISM band as opposed to most other LPWAN technologies that rely on sub-GHz frequencies like those of the 433 and 915/868 MHz bands. Although sub-GHz bands exhibit better propagation properties, the 2.4 GHz ISM band is less restrictive and does not impose duty cycle limitations that, in turns, limit transmission rates. The available 2.4 GHz spectrum is divided into 1 MHz channels that can be grouped to support different deployments.

RPMA is a type of a DSSS scheme used only for uplink communications and, as such, represents a variation of traditional CDMA. One of the main differences between DSSS and RPMA is that although both mechanisms allow transmitters to share a single timeslot, under RPMA the duration of the timeslot is increased, and the starting transmission delay (or phase) is randomized in order to minimize intertransmitter interference. Moreover, RPMA supports a mechanism for transmission rate adaptation regulated by the spreading factor estimated from the received signal strength. The base stations rely on multiple receivers tuned to demodulate the signals arriving at different starting times and exhibiting different spreading factors. Downlink communication relies on CDMA. In all cases, the signal modulation is by means of DBPSK. RPMA devices also adapt their transmission power by communicating with the closest base station in order to minimize interference. Overall, all these features together with a higher bandwidth that most other LPWAN technologies give RPMA higher transmission rates. Specifically, nominal rates of 624 and 156 Kbps for uplink and downlink, respectively, are common. RPMA encrypts all traffic and introduces FEC through convolutional codes.

8.7.4 Telensa

Telensa is a fully proprietary LPWAN technology that focuses mainly on smart city applications with emphasis on smart lighting and smart parking as it does not support indoor communications [23, 35]. Millions of Telensa devices have been deployed in more than 50 smart cities networks worldwide. Although Telensa supports bidirectional traffic associated with both sensors and actuators, in the context of smart cities, streetlight poles are typically arranged with several devices including pollution, temperature, humidity, radiation level, and noise level sensors. Moreover, Telensa is not just a technology but a framework for building smart city applications by means of a smart city API. As part of this framework, it provides mechanisms for integration with support services like billing systems and metering. Telensa is a member of the Weightless SIG board, and it is involved with the TALQ consortium that oversees defining control and monitor standards for outdoor lighting. Moreover, although Telensa is a proprietary technology, there is an ongoing effort for standardization through ETSI *low-throughput networks* (LTN) specifications.

Telensa's physical layer relies on UNB transmission over the 915/868 MHz ISM band that provides very low transmission rates. Specifically, Telensa supports 62.5 and 500 bps uplink and downlink, respectively. Note that, as opposed to most other LPWAN technologies, the uplink typically used to the transmission of sensor traffic is considerably slower than the downlink used for actuation. Telensa has a *Central Management System* (CSM) called *Telensa PLANet* that coordinates end-to-end operations and minimizes energy consumption as well as enabling automatic fault detection. A Telensa base station, playing the role of an IoT gateway, can communicate with up to 5000 devices covering a physical range of 2 and 4 km for urban and rural locations, respectively.

8.7.5 SNOW

Sensor network over white spaces (SNOW) is an experimental LPWAN technology that, as Weightless, relies on transmission over white space spectrum with modulation over unoccupied frequency guardbands between TV channels typically between 547 and 553 MHz. In general, a base station determines white space frequencies for devices by means of a database on the Internet [31, 29]. The available bandwidth is divided into multiple channels that enable transmission over multiple subcarriers where a variation of OFDM called *distributed OFDM* (DOFDM) is used. DOFDM, as opposed to OFDM, enables multiple transmitters to simultaneously send traffic over multiple subcarriers. The modulation itself is by means of a variation of ASK called *on-off keying* (OOK) that consists of transmitting a signal whenever a binary one is sent and not transmitting any signal otherwise. The frame size is 40 bytes including a 28-byte header that enables a nominal transmission rate of 50 Kbps. SNOW supports a *star topology* with a single base station that acts as an IoT gateway communicating with multiple devices located as far as 1.5 km away. The link layer operates in two phases, a long period of time for uplink transmissions and a lot short period for downlink communication. FEC through redundant transmission of packets increases reliability and removes the need for acknowledgments [30].

8.7.6 Nwave

Nwave is a commercial LPWAN solution intended for mobile devices associated with smart parking [11]. The physical layer provides UNB transmission over the 915/868 MHz ISM band. Nwave supports a single hop *star topology* with a coverage of up to 10 km in urban environments. Long range and low power are responsible for transmission rates of around 100 bps and a maximum device battery life of 9 years. Nwave provides its own real-time sensor data collection and analytics applications that enable city planners to allocate resources.

8.7.7 Qowisio

Qowisio is another UNB turnkey commercial LPWAN solution supporting a wide range of applications ranging from asset tracking and management to lighting and power monitoring [25]. It is compatible with LoRa by providing dual model technology support. It provides coverage of up to 60 km and 3 km in rural and urban environments, respectively.

8.7.8 IEEE 802.15.4k

The *IEEE 802.15.4k Task Group* (TG4k) introduced a new standard for *low energy, critical infrastructure monitoring* (LECIM) applications that, as IEEE 802.15.4, relies on the 2.4 GHz, the 915/868 MHz, and the 433 MHz ISM bands for transmission [27, 25]. The spectrum is divided into discrete channels with bandwidths ranging from 100 KHz to 1 MHz. It is an attempt to provide LPWAN communications by increasing the transmission range of traditional WPAN-based IEEE 802.15.4. At the physical layer, it supports three different modulations that combine DSSS with BPSK, OQPSK, or FSK. Depending on device and communication constraints, one scheme is chosen over the others. Further adjustments are performed by means of modifications to the spreading factor between 16 and 32,768. IEEE 802.15.4k also introduces a FEC scheme based on convolutional codes. The link layer provides MAC through conventional CSMA/CA and ALOHA with *priority channel access* (PCA). PCA is a mechanism by which devices and base stations support QoS levels that can be used to specify the priority of the traffic. IEEE 802.15.4k operates in a star topology with a nominal coverage of 3 km. Transmission rates of up to 50 Kbps are possible [2].

8.7.9 IEEE 802.15.4g

The *IEEE 802.15.4g Task Group* (TG4g) introduced a new standard known as *wireless smart utility networks* (Wi-SUN) targeting smart metering applications like, for example, gas metering [26, 17]. As with IEEE 802.15.4k, IEEE 802.15.4g addresses some of the limitations of traditional WPAN IEEE 802.15.4. Specifically, IEEE 802.15.4, as indicated in Sect. 3.3.3.4, is highly affected by interference and multipath fading that reduce communication reliability. Moreover, IEEE 802.15.4 relies on complex and costly multihop transmissions for long-range communication that are not viable in smart metering scenarios. IEEE 802.15.4g modulation is over the 2.4 GHz and the 915/868 MHz ISM bands and supported by three different physical layers that provide a trade-off between transmission rates and power consumption: (1) myindexFSK combined with FEC, (2)

DSSS with *OQPSK*, and (3) OFDMA in scenarios affected by multipath fading. Although nominal transmission rates, depending on the modulation scheme under consideration, can range from 6.25 to 800 Kbps, the standard defines FSK as the default mandatory mode of transmission supporting a transmission rate of 50 Kbps. As transmission rates are higher than those of IEEE 802.15.4, the corresponding MTU size is also larger without affecting the transmission delay. Specifically, IEEE 802.15.4g supports a maximum frame size of 1500 bytes that enable the transmission of a complete IPv6 datagram over a single frame without needing to apply any fragmentation. The link layer can be configured to provide MAC based on IEEE 802.15.4 or on IEEE 802.15.4e introduced in Sect. 3.3.3 [1].Typical signal coverage of IEEE 802.15.4g is around 10 km.

8.7.10 LTE-M

Standardized together with NB-IoT as part of the 3GPP Release 13, *LTE-M*, also known as *LTE Cat M1* or *enhanced machine-type communication* (eMTC), is a simplified version of 4G LTE that attempts to provide CIoT support by reducing power consumption while extending signal coverage [11, 3]. Specifically, battery life is increased to over 10 years, while modem costs are reduced by 25%. When compared to LTE, LTE-M lowers nominal transmissions rates to around 1 Mbps by reducing the available bandwidth from 20 to 1.4 MHz (as opposed to 200 KHz of NB-IoT). Most LTE functionality is available under LTE-M including *Voice over LTE* (VoLTE). In order to increase battery life, LTE-M supports optional half duplex transmissions. Some features of NB-IoT like eDRX and PSM, introduced in Sect. 8.6, are also available under LTE-M. In most scenarios upgrading an LTE network to support LTE-M is as simple as a software upgrade. LTE-M does not compete with NB-IoT, and the selection of one technology versus the other depends on rate, coverage, and power consumption considerations. Device density is also increased by supporting over 100,000 devices in each cell. For deep coverage deployments associated with CIoT scenarios, latency is within 10 s, while battery life is never longer than 10 years, but for normal coverage, similar to that of a smartphone, latency lowers to 0.1 s while the battery life extends to 35 years [16].

8.7.11 EC-GSM-IoT

As LTE-M, *extended coverage GSM IoT* (EC-GSM-IoT) was standardized together with NB-IoT as part of the 3GPP Release 13. As opposed to LTE-M and NB-IoT that are based on LTE, EC-GSM-IoT is based on *enhanced GPRS*, also known as 2.75G. EC-GSM-IoT does to the GSM spectrum what LTE-M does to the LTE spectrum [11, 16, 3]. Specifically, it increases coverage and lowers power consumption. As a CIoT technology, it operates on the GSM 850–900 and 1800–1900 MHz bands providing a similar coverage and power consumption to that of NB-IoT. The channel bandwidth is 200 MHz, where like in most GSM networks, MAC is performed by means of combining FDMA with TDMA and FDD. Modulation is carried out by *Gaussian Minimum Shift Keying* (GMSK) and 8PSK providing nominal transmission rates between 70 Kbps and 240 Kbps, respectively. Latency is typically less than 2 s being lower than that of both LTE-M and NB-IoT. As LTE-M, EC-GSM-IoT is enabled in existing GSM networks by means of a software upgrade. Because GSM networks are in the process of decommission, EC-GSM-IoT is a lot less popular than LTE-based technologies. EC-GSM-IoT also exhibits comparatively high capacity supporting around 50,000 devices in each cell. Moreover, EC-GSM-IoT is designed to provide coverage in locations where radio conditions are challenging such as indoor basements where many sensors are usually installed. Battery life is around 10 years, and security and privacy, as with LTE-M and NB-IoT, are guaranteed by mutual authentication, confidentiality, and encryption provided by legacy mechanisms [34].

8.7.12 5G and B5G Considerations

4G LTE IoT solutions like NB-IoT and LTE-M intended for fixed/low-rate and mobile/high-rate applications, respectively, have transitioned into 5G scenarios as part of the evolution of the 3GPP releases. This is further simplified by the fact that 5G is typically deployed as *5G non-standalone* (5G NSA) where 5G is supported as a software upgrade that updates protocols on legacy 4G LTE hardware. This latter deployment is usually known as 5G LTE as opposed to *5G standalone* (5G SA) where additional hardware also known as *new radio* (NR) is needed to support frequency bands above 6 GHz [4, 32]. In this context, 3GPP Release 15 addresses the NR modifications needed to support both 5G NB-IoT and 5G LTE-M. As part of this evolution, one of the main requirements for the future IoT applications is massive access. *Beyond 5G* (B5G) including future 6G networks will attempt to address this and other requirements in order to increase density to over a million devices per square kilometer, increase battery life to 20 years, support space-air-ground-sea coverage, reduce latency to less than a second, and increase positioning precision to 1 m [19].

8.7.13 IPv6 Support Considerations

LPWAN technologies use most of the device energy to provide extended coverage in order to minimize infrastructure

Table 8.3 LPWAN technologies

Technology	Bands	Max nominal rate	Max coverage
LoRa	ISM 1 GHz	50 Kbps	20 km
SigFox	ISM 1 GHz	600 bps	40 km
D7AP	ISM 1 GHz	166 Kbps	10 km
Weightless	TV guardbands	16 Mbps	5 km
NB-IoT	LTE/5G	200 Kbps	10 km
NB-Fi	ISM 1 GHz	25.6 Kbps	30 km
IQRF	ISM 1 GHz	20 Kbps	500 m
RPMA	ISM 1 GHz, 2.4 GHz	624 Kbps	25 km
Telensa	ISM 1 GHz	500 bps	16 km
SNOW	TV guardbands	50 Kbps	1.5 km
Nwave	ISM 1 GHz	100 bps	10 km
IEEE 802.15.4k	ISM 1 GHz, 2 GHz	50 Kbps	3 km
IEEE 802.15.4g	ISM 1 GHz, 2 GHz	50 Kbps	10 km

investment. The cost of this extended coverage, however, is very little energy left for computation, resource, and protocol complexity. This leads to several challenges that prevent LPWAN from implementing IPv6 adaptation mechanisms like 6LoWPAN and 6Lo: (1) MTU and transmission rates are several other of magnitude below those of WPAN technologies, (2) lack of link layer fragmentation support, and (3) uplink and downlink asymmetry. In this scenario, any adaptation that attempts to overcome the 6LoWPAN and 6Lo limitations results in energy consumption levels that are unacceptable for LPWAN.

There have been, however, some attempts to provide IPv6 support in LPWAN by adapting some of the lightweight features of 6LoWPAN and 6Lo. Specifically, IETF RFC 8724 *SCHC: Generic Framework for Static Context Header Compression and Fragmentation* introduces *Static Context Header Compression* (SCHC) a mechanism that provides both static header compression and fragmentation in the context of LPWANs [20]. Static header compression relies on a common static context on both device and network infrastructure sides. SCHC is used to compress IPv6 and UDP headers. There are, in addition, several proposals to specialize the use of SCHC under LoRa, NB-IoT, and SigFox.

Summary

LPWAN solutions provide long-range signal coverage between devices and gateways at a cost of comparatively low transmission rates. As such most LPWAN schemes do not fully support end-to-end IPv6 connectivity, and they rely on gateways and border routers to convert protocol stacks. The chapter started by briefly describing IoT support of LPWAN technologies and then switched to presenting LoRa, SigFox, D7AP, Weightless, and NB-IoT. The chapter then introduced other less common technologies like NB-Fi, IQRF, RPMA, Telensa, SNOW, NWave, Qowisio, IEEE 802.15.4k, IEEE 802.15.4g, LTE-M, and EC-GSM-IoT including 5G, B5G, and 6G considerations. The chapter ended by exploring IPv6 support of LPWAN mechanisms. Table 8.3 compares the features of some of the most relevant LPWAN technologies.

Homework Problems and Questions

8.1 If an application in the cloud is sending actuation commands to a LoRa device, what LoRa media access class configuration is likely to minimize the delay of these commands arriving at the device?

8.2 What LPWAN technologies do not rely on ISM bands for transmission?

8.3 What LoRa MAC class enables the longest battery life?

8.4 If it were possible to transmit a CoAP message over LoRa and assuming:

(a) LoRa configured with 50 devices transmitting over 8 km at a nominal maximum rate of 250 bps.
(b) An 8-byte payload over an unencrypted NON CoAP message transmitted over an adaptation mechanism that encodes IPv6 and UDP headers as a single 21-byte header. Assume the LoRaWAN header is included in the adaptation layer header.
(c) No network loss and no delays other than the transmission delay.

How long does it take for a sensor to send a single readout? How does it compare to the best-case scenario of IEEE 802.15.4?

8.5 Many LPWAN technologies like SigFox support different transmission rates for uplink and downlink traffic. (a) Why is this important? (b) How does this relate to IPv6 support?

8.6 What are the best and worst-case latency scenarios for a 45-byte long packet transmitted over NB-IoT?

8.7 What are some differences between Weightless and other LPWAN technologies?

8.8 Is there any LPWAN technology that provides IPv6 support?

8.9 Many LPWAN technologies like Weighless-N only enable the transmission of uplink traffic. (a) How does this affect performance? (b) What are the implications for IoT applications?

8.10 Assuming a symbol rate of 4×10^6 symbols per second, how is the 500 Kbps transmission rate accomplished under Weightless? Assume a 4 MHz channel.

References

1. IEEE standard for local and metropolitan area networks–part 15.4: Low-rate wireless personal area networks (LR-WPANS) amendment 3: Physical layer (PHY) specifications for low-data-rate, wireless, smart metering utility networks. IEEE Std 802.15.4g-2012 (Amendment to IEEE Std 802.15.4-2011) pp. 1–252 (2012)
2. IEEE standard for local and metropolitan area networks—part 15.4: Low-rate wireless personal area networks (LR-WPANS)-amendment 5: Physical layer specifications for low energy, critical infrastructure monitoring networks. IEEE Std 802.15.4k-2013 (Amendment to IEEE Std 802.15.4-2011 as amended by IEEE Std 802.15.4e-2012, IEEE Std 802.15.4f-2012, IEEE Std 802.15.4g-2012, and IEEE Std 802.15.4j-2013) pp. 1–149 (2013)
3. 3GPP: 3GPP release 13 (2015). https://www.3gpp.org/release-13
4. Akpakwu, G.A., Silva, B.J., Hancke, G.P., Abu-Mahfouz, A.M.: A survey on 5G networks for the internet of things: communication technologies and challenges. IEEE Access **6**, 3619–3647 (2018)
5. Al-Sarawi, S., Anbar, M., Alieyan, K., Alzubaidi, M.: Internet of things (IoT) communication protocols: Review. In: 2017 8th International Conference on Information Technology (ICIT), pp. 685–690 (2017)
6. Alliance, L.: Lorawan 1.1 specification (2017). https://lora-alliance.org/sites/default/files/2018-04/lorawantm_specification_-v1.1.pdf
7. Augustin, A., Yi, J., Clausen, T., Townsley, W.M.: A study of lora: Long range & low power networks for the internet of things. Sensors **16**(9), 1466 (2016). https://doi.org/10.3390/s16091466. https://pubmed.ncbi.nlm.nih.gov/27618064. 27618064[pmid]
8. Ayoub, W., Nouvel, F., Samhat, A.E., Prévotet, J.C., Mroue, M.: Overview and measurement of mobility in DASH7. In: 2018 25th International Conference on Telecommunications (ICT), pp. 532–536. IEEE, St. Malo (2018). https://doi.org/10.1109/ICT.2018.8464846. https://hal.archives-ouvertes.fr/hal-01991725
9. Ayoub, W., Samhat, A.E., Nouvel, F., Mroue, M., Prevotet, J.: Internet of mobile things: overview of LoRaWAN, DASH7, and NB-IoT in LPWANs standards and supported mobility. IEEE Commun. Surv. Tutor. **21**(2), 1561–1581 (2019)
10. Calvo, I., Gil-Garcia, J., Recio, I., Lopez, A., Quesada, J.: Building IoT applications with raspberry Pi and low power IQRF communication modules. Electronics **5**, 54 (2016). https://doi.org/10.3390/electronics5030054
11. Chaudhari, B., Zennaro, M., Borkar, S.: LPWAN technologies: emerging application characteristics, requirements, and design considerations. Future Internet **12**, 46 (2020). https://doi.org/10.3390/fi12030046
12. Chen, M., Miao, Y., Hao, Y., Hwang, K.: Narrow band internet of things. IEEE Access **5**, 20557–20577 (2017)
13. Farrell, S.: Low-Power Wide Area Network (LPWAN) Overview. RFC 8376 (2018). https://doi.org/10.17487/RFC8376. https://rfc-editor.org/rfc/rfc8376.txt
14. Ferré, G., Simon, E.P.: An introduction to Sigfox and LoRa PHY and MAC layers (2018). https://hal.archives-ouvertes.fr/hal-01774080. Working paper or preprint
15. Finnegan, J., Brown, S.: A Comparative Survey of LPWA Networking. arXiv (2018), arXiv:1802.04222
16. Foubert, B., Mitton, N.: Long-range wireless radio technologies: a survey. Future Internet **12**, 13 (2020). https://doi.org/10.3390/fi12010013
17. Harada, H., Mizutani, K., Fujiwara, J., Mochizuki, K., Obata, K., Okumura, R.: IEEE 802.15.4g based wi-sun communication systems. IEICE Trans. Commun. **E100.B**, 1032–1043 (2017). https://doi.org/10.1587/transcom.2016SCI0002
18. Lavric, A., Popa, V.: Internet of things and lora low-power wide-area networks: A survey. In: 2017 International Symposium on Signals, Circuits and Systems (ISSCS), pp. 1–5 (2017)
19. Malik, H., Sarmiento, J.L.R., Alam, M.M., Imran, M.A.: Narrowband-internet of things (NB-IoT): Performance evaluation in 5G heterogeneous wireless networks. In: 2019 IEEE 24th International Workshop on Computer Aided Modeling and Design of Communication Links and Networks (CAMAD), pp. 1–6 (2019)
20. Minaburo, A., Toutain, L., Gomez, C., Barthel, D., Zuniga, J.C.: SCHC: Generic Framework for Static Context Header Compression and Fragmentation. RFC 8724 (2020). https://doi.org/10.17487/RFC8724. https://rfc-editor.org/rfc/rfc8724.txt
21. Morin, E., Maman, M., Guizzetti, R., Duda, A.: Comparison of the device lifetime in wireless networks for the internet of things. IEEE Access **5**, 7097–7114 (2017). https://doi.org/10.1109/ACCESS.2017.2688279.https://hal.archives-ouvertes.fr/hal-01649135
22. Mroue, H., Nasser, A., Hamrioui, S.: Mac layer-based evaluation of IoT technologies: Lora, sigfox and NB-IoT. In: IEEE Middle East and North Africa Communications Conference (MENACOMM) (2018). https://doi.org/10.1109/MENACOMM.2018.8371016
23. Naik, N.: LPWAN technologies for IoT systems: Choice between ultra narrow band and spread spectrum. In: 2018 IEEE International Systems Engineering Symposium (ISSE), pp. 1–8 (2018)
24. Oliveira, L., Rodrigues, J., Kozlov, S., Rabelo, R., Albuquerque, V.: Mac layer protocols for internet of things: a survey. Future Internet **11**, 16 (2019). https://doi.org/10.3390/fi11010016
25. Raza, U., Kulkarni, P., Sooriyabandara, M.: Low power wide area networks: An overview. IEEE Commun. Surv. Tutor. **19**, 855–873 (2016)
26. Righetti, F., Vallati, C., Comola, D., Anastasi, G.: Performance measurements of IEEE 802.15.4g wireless networks. In: 2019 IEEE 20th International Symposium on “A World of Wireless, Mobile and Multimedia Networks” (WoWMoM), pp. 1–6 (2019)
27. Roth, Y., Dore, J.B., Ros, L., Berg, V.: A comparison of physical layers for low power wide area networks. In: Cognitive Radio Oriented Wireless Networks. CrownCom 2016, pp. 261–272 (2016). https://doi.org/10.1007/978-3-319-40352-6_21
28. Saari, M., bin Baharudin, A.M., Sillberg, P., Hyrynsalmi, S., Yan, W.: Lora-a survey of recent research trends. In: 2018 41st International Convention on Information and Communication Technology, Electronics and Microelectronics (MIPRO), pp. 0872–0877 (2018)
29. Saifullah, A., Rahman, M., Ismail, D., Lu, C., Chandra, R., Liu, J.: Snow: Sensor network over white spaces. In: SenSys ’16: Proceedings of the 14th ACM Conference on Embedded Network Sensor Systems CD-ROM, pp. 272–285 (2016). https://doi.org/10.1145/2994551.2994552
30. Saifullah, A., Rahman, M., Ismail, D., Lu, C., Liu, J., Chandra, R.: Enabling reliable, asynchronous, and bidirectional communication in sensor networks over white spaces. In: Proceedings of the 15th ACM Conference on Embedded Network Sensor Systems, SenSys ’17. Association for Computing Machinery, New York (2017). https://doi.org/10.1145/3131672.3131676
31. Saifullah, A., Rahman, M., Ismail, D., Lu, C., Liu, J., Chandra, R.: Low-power wide-area network over white spaces. IEEE/ACM Trans. Netw. **26**(4), 1893–1906 (2018). https://doi.org/10.1109/TNET.2018.2856197
32. Salva, P., Alcaraz-Calero, J., Wang, Q., Bernal Bernabe, J., Skarmeta, A.: 5g NB-IoT: Efficient network traffic filtering for

multitenant IoT cellular networks. Secur. Commun. Netw. **2018**, 1–21 (2018). https://doi.org/10.1155/2018/9291506

33. Shanmuga Sundaram, J.P., Du, W., Zhao, Z.: A survey on lora networking: research problems, current solutions, and open issues. IEEE Commun. Surv. Tutor. **22**(1), 371–388 (2020)
34. Silva, P., Kaseva, V., Lohan, E.S.: Wireless positioning in IoT: a look at current and future trends. Sensors **18**, 2470 (2018). https://doi.org/10.3390/s18082470
35. Walden, M.C., Jackson, T., Gibson, W.H.: Development of an empirical path-loss model for street-light telemetry at 868 and 915 MHz. In: 2011 IEEE International Symposium on Antennas and Propagation (APSURSI), pp. 3389–3392 (2011)
36. Webb, W.: Weightless: the technology to finally realise the m2m vision. Int. J. Interdiscip. Telecommun. Netw. **4**, 30–37 (2012). https://doi.org/10.4018/jitn.2012040102
37. Weyn, M., Ergeerts, G., Berkvens, R., Wojciechowski, B., Tabakov, Y.: Dash7 alliance protocol 1.0: Low-power, mid-range sensor and actuator communication. In: 2015 IEEE Conference on Standards for Communications and Networking (CSCN) (2015)

9 Thread Architecture

9.1 Thread and IoT

In the domain of WPAN solutions, the IETF protocol stack plays a very important role in providing a generic framework for IoT wireless applications. One of the most common implementations of this stack is shown in Fig. 9.1. Specifically, the stack is a combination of some of the layers presented in Sect. 4.3. Physical and link layers are carried out by IEEE 802.15.4 [2], while the network layer relies on IPv6 adapted by means of 6LoWPAN [8] with routing provided through RPL [3]. UDP encrypted with DTLS [9] provides transport to the application layer that is, in turn, subdivided into two sublayers. One session sublayer is associated with CoAP messages, while the other carries application sensor and actuator traffic. This quick introduction serves to present Thread because *Thread* is an open IoT architecture that relies on a protocol stack that is quite similar to the one shown in Fig. 9.1. The Thread protocol stack is shown in Fig. 9.2; the main big difference with the IETF stack is that routing instead of being based on IoT specific RPL relies on traditional DV mechanisms. Additionally, Thread uses plain IEEE 802.15.4 with CSMA/CA MAC and therefore does not take advantage of the IEEE 802.15.4e [1] improvements introduced by TSCH MAC. Moreover, Thread does not rely on any other WPAN physical and link layer mechanism like BLE. Note that Thread is a full-stack architecture that exploits and extends some IETF mechanisms. As such, Thread is comparable to other proprietary stacks like ZigBee, BLE, and Z-Wave [14]. Obviously, Thread supports end-to-end IPv6 connectivity and consequently it is fully IoT compliant.

Thread is a lot more than a protocol stack; by being both an architecture and a framework, it provides a recipe to deploy IoT WPANs. When compared to other IoT architectures, Thread has been designed for highly reliable wireless device-to-device communication because it does not have a single point of failure. Moreover, Thread specifies the interaction between the layers and the supported mechanisms needed to deliver an end-to-end solution. Although Thread is typically intended for home automation, it can be used in the context of other industries ranging from asset monitoring to IIoT. Home automation is associated with the concept of connected home where the devices consist of AC powered lights, door and window sensors, motion detectors, door locks, appliances, as well as *heating, ventilation, and air conditioning* (HVAC) equipment and battery-operated thermostats, smoke, CO, and CO2 detectors. Thread device reliability is very important, and it implies that, as simple as the devices can be, they must still be able to store certain persistent information like PAN identifiers and other security parameters.

Thread attempts to provide simple and low-cost network installation, startup, as well as operation with full IPv6 support for integration with other Internet applications. Additionally, Thread supports several topologies that increase flexibility in most connected home deployments of hundreds of devices. By relying on end-to-end IP connectivity, device installation is performed using a smartphone, a computer, or a tablet through an application or through the web. Product installation codes enable only authorized devices to join a Thread network. Additionally, all traffic is authenticated and encrypted with AES. Although device connectivity is typically through IEEE 802.15.4, users rely on Wi-Fi to communicate with the network directly through a HAN or through a cloud-based application [12].

The architecture relies on a series of simple mechanisms that complement routing and allow the network to self-configure and repair itself. Thread inherits most of the characteristics from the IETF and IEEE protocols that form its stack; they include long battery life, security, and mesh routing support among other features. The coverage is given by the physical layer protocol, IEEE 802.15.4, and as such it is good enough to cover a large home. Any equipment used to extend coverage of IEEE 802.15.4 networks can be used also in Thread scenarios. Battery life is extended by means of power duty cycles that enable a single device to operate over several years on a few AA batteries.

R. Herrero, *Fundamentals of IoT Communication Technologies*, Textbooks in Telecommunication Engineering, https://doi.org/10.1007/978-3-030-70080-5_9

Fig. 9.1 IETF stack

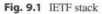

DEVICE DATA		application layer
CoAP		
DTLS		transport layer
UDP		
IPv6	RPL	network layer
6LoWPAN		
IEEE 802.15.4		link layer
IEEE 802.15.4		physical layer

Fig. 9.2 Thread stack

DEVICE DATA		application layer
CoAP		
DTLS		transport layer
UDP		
IPv6	DV	network layer
6LoWPAN		
IEEE 802.15.4		link layer
IEEE 802.15.4		physical layer

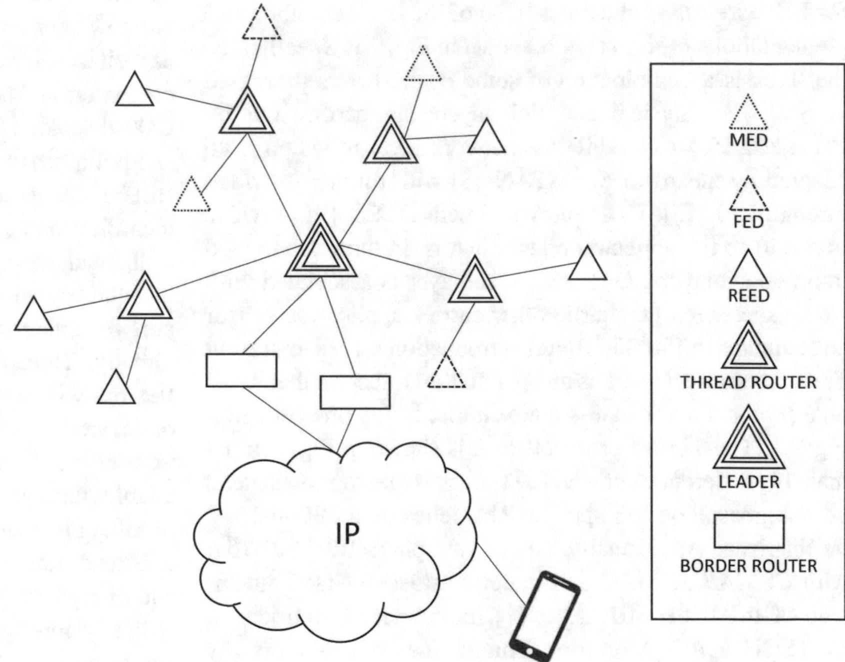

Fig. 9.3 Thread network

9.2 Topology

Figure 9.3 shows the standard Thread topology including all key devices: (1) border routers, (2) Leaders, (3) *Router Eligible End Devices* (REEDs), (4) *Thread routers*, (5) *Full End Devices* (FEDs), and (6) *Minimal End Devices* (MEDs). Border routers play the role of IoT gateways by providing connectivity between IEEE 802.15.4-based access and the IP core. In general end devices, of any time, do not talk to other devices, and connectivity is with other routers and the network core. The border routers are, therefore, responsible for providing the routing capabilities between device traffic and public Internet cloud applications as illustrated in Fig. 9.4. As such, they usually include both an IEEE 802.15.4 link layer interface and one or more supplemental IPv6 link layer interfaces for access to an external network. Border routers have at least one assigned GUA. Note that the rest of the devices in a Thread network typically support ULA type unicast addresses. Outgoing datagrams from the Thread network interface are forwarded to the exterior interfaces, while datagrams from the exterior interfaces are forwarded to the Thread interface and then routed over the Thread network to the destination. The border router may also perform filtering and NATting based on system settings. Moreover, the border router may participate in an exterior routing protocol where

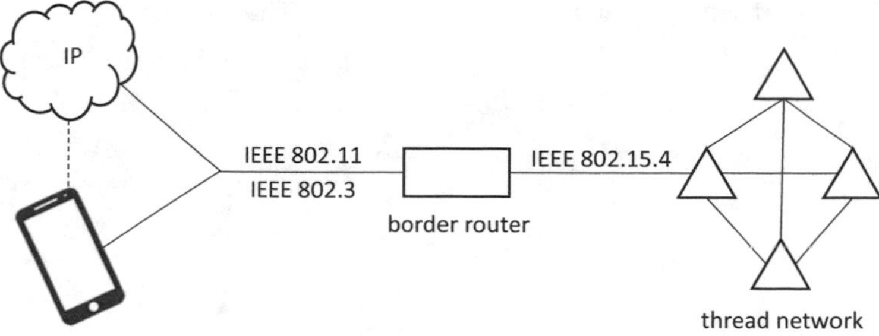

Fig. 9.4 Border router operation

it advertises global IPv6 prefixes and handles global unicast address allocation of devices in the Thread network. Essentially, users typically connect from adjacent IP networks by means of IEEE 802.11 (Wi-Fi) or IEEE 802.3 (Ethernet), and they access devices through the core interface of the border router. In order to support redundancy and improve reliability, a Thread network typically has multiple border routers. The *Thread Management Framework* (TMF) protocol, which uses CoAP for transport, enables the notification of network data from border routers and DHCPv6 servers to Thread leaders [13].

Thread routers provide routing capabilities to devices and coordination for devices to securely join the network. Because of their critical operation, Thread routers do not typically run on constrained hardware and therefore do not perform power duty cycles that affect performance. Thread routers can become REEDs by downgrading their functionality. REEDs are end devices, and as such they do not provide any routing or any services to other devices. Depending on topology changes and other network conditions, REEDs can become Thread routers or leaders without user interaction. REEDs cannot become border routers since they do not usually include an interface to the core side that is needed to provide gateway capabilities. Leaders are routers that keep track, through a registry, of other routers and their assignment. Consequently, leaders respond to requests from REEDs to become routers. For the most part, leaders rely on CoAP to assign and manage router addresses. The information stored in the leader registry is backed up in other Thread routers such that if a leader becomes out of service, any Thread router can automatically become the new leader. End devices that are not REEDs can be either FEDs or MEDs. Both FEDs and MEDs only communicate to Thread routers and cannot serve as routers or become routers. MEDs do even need to explicitly synchronize with their Thread routers to transmit.

The Thread topology overcomes single points of failure by relying on an automated and controlled mechanism in which devices can take over functionality provided by failing devices, or they can change their operation to become independent of the failing devices. If a FED is communicating with a Thread router and the Thread router fails, then the FED can select a new Thread router. If a Thread router fails, a close-by REED can become Thread router so that all MEDs and FEDs associated with the failing Thread router become children of the transitioning REED. In most cases, all these topology changes are transparent to the end user. There are certain conditions, however, under which a single point of failure cannot be prevented. Border routers are limited by hardware characteristics that force them to have at least two interfaces: one to the access WPAN side and another to the IP core side. This implies that if a border router fails, communication to the IP core side will not be possible unless another redundant border router exists in the topology.

A Thread network is typically composed of a single partition, but in certain cases, connectivity loss can lead to disconnected sections of the network that end up grouped into separate partitions. The idea is that each partition is a Thread network with its own leader and network data. When connectivity between partitions is back, they merge back, and the original Thread network topology is restored. Figure 9.5 illustrates how a single partition with one leader becomes two partitions with two leaders when connectivity is lost between one side and the other of the Thread network.

One important concept in the context of Thread networks is commissioning. Commissioning is the process by which a user deploys a new device in a Thread network. Border routers are essential in relaying messages between a management application, known as commissioner, and network devices. Some of these network devices can be leader routers that typically arbitrate the petitioning of multiple commissioners [11].

9.3 Routing

Thread supports mesh forwarding that enables devices to talk to other devices by indirectly forwarding frames through intermediate nodes. All routing entities (i.e. routers, border routers and leaders) keep track and maintain routes to guarantee that mesh forwarding is always available and fully functional. Although up to 64 router addresses are supported in a single Thread network, not all these addresses can be used simultaneously in order to delay the reuse of addresses

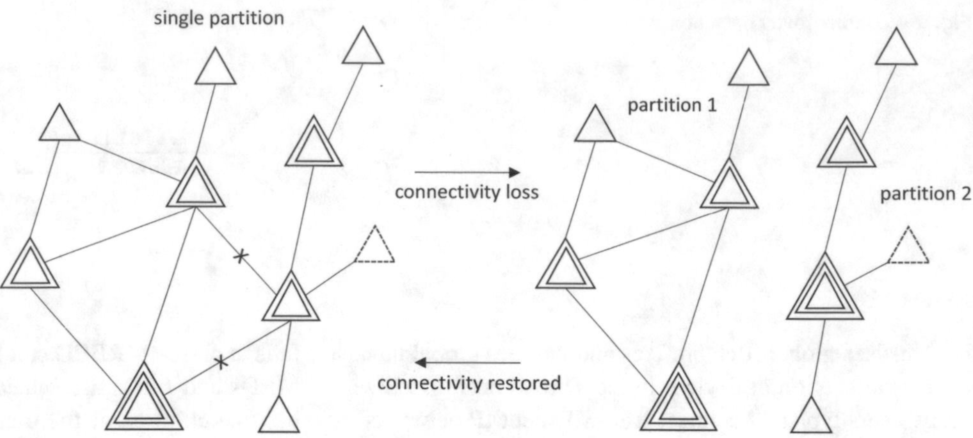

Fig. 9.5 Multiple partitions

of deleted devices. Since REEDs, FEDs, and MEDs do not support routing, all traffic they generate is handled by their parent Thread router.

In a single Thread network, there are up to 32 active routers that rely on DV to compute routing tables. Specifically, Thread routing relies on standard *RIP Next Generation* (RIPng) standardized under IETF RFC 2080. Note that RIPng supports a maximum of 15 hops per end-to-end path and uses hop count as routing metric. As with any DV mechanism, each router is aware of the next hop routers and the cost of transmitting traffic to each of them. This enables routers to independently build routing tables that guarantee full network connectivity. Note that as opposed to RPL, where routing is hierarchical, routing under Thread is mostly flat. Moreover, by virtue of relying on DV routing, routing repairs tend to act fast, although temporary loops are likely to occur. Routing information is exchanged by means of *mesh link establishment* (MLE) messages. MLE messages are compressed for efficiency, and they are used not only to propagate link costs between routers but also to establish, identify, and maintain secure links and to detect neighbors. MLE messages also enable the propagation of network configuration parameters like the PAN identifier. Note that link costs are typically asymmetrical as in most IoT networks, because the traffic from a device to a router and from the same router to the same device is affected by different impairments and channel capacities. Since MLE messages operate at link level, they are sent as link-local unicast and multicast frames. Multicasting of MLE messages is carried out by simple flooding in accordance with the *multicast protocol for low-power and lossy Networks* (MPL). MPL enables multicast forwarding in constrained networks while avoiding the need of maintaining any multicast forwarding topology. MPL introduces two modes of execution. Under proactive mode, MPL data messages are transmitted and triggered by a trickle timer without any prior indication that nodes are intended to receive the messages. Similarly, under reactive mod MPL data messages are multicast transmitted when it is discovered that nodes have not received the messages. In the context of Thread, MPL is mostly supported in reactive mode. The idea behind a trickle timer is to transmit network parameters and attributes while minimizing, at the same time, the load of the network.

DV routing is different from other routing mechanisms like RPL typically associated with on-demand routing. Although under IoT RPL is more stable and exhibits better performance than DV mechanisms, DV routing has a much lower overhead. Specifically, RPL DIO, DAO, and DIS messages are propagated throughout the network, while DV link costs are only propagated between neighboring routers. The link costs represent all the information routers need to build comprehensive routing tables. Due to the dynamic nature of the link costs, routing tables are adjusted on-the-fly to compensate for connectivity issues in a comparatively faster way than RPL. Also, when compared to RPL, DV routing fast reactions can lead to loops and other routing instabilities that can affect short term performance.

Routing relies on a bitmap lookup where the address of a REED, FED, or MED parent is determined by looking at the *most significant bits* (MSB) of the device address. In general, a sender determines the router associated with the receiver and forwards datagrams to the neighbor that provides the lowest-cost path. Again, the cost information between devices and routers is propagated by means of single-hop MLE messages. The cost associated with a specific link between two entities is based on the power of the signals originated at the neighbors. Specifically, the incoming RSSI is mapped to a link quality index between 0 and 3 where 0 indicates that the cost is unknown.

If a device or a router receives an MLE message from a neighbor, it recomputes its routing table by updating or creating a new entry based on the information carried by the message. The MLE message contains the cost associated not only to the close neighbor but also to all other routers that are reachable from the neighbor. Unfortunately, the IEEE 802.15.4 frame payload size limits to 32 the number of routers for which link cost information can be included. Routing tables are traditional IP routing tables with entries that specify network addresses and the corresponding next hop information.

Example 9.1 What is a valid addressing scheme for the end devices of the following Thread network?

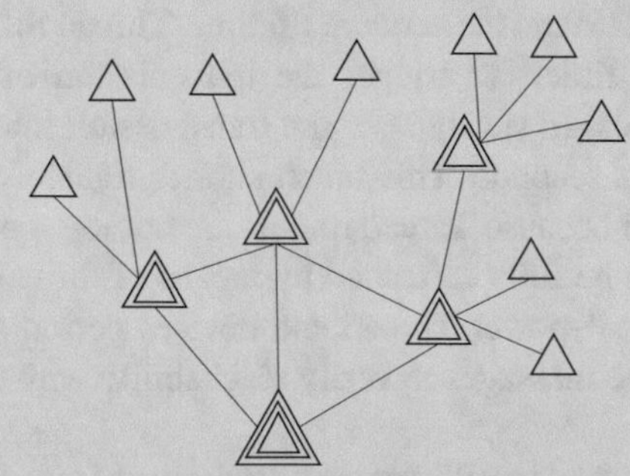

Solution Each device 16-bit address is given by a 6-bit router identifier and a 10-bit device identifier:

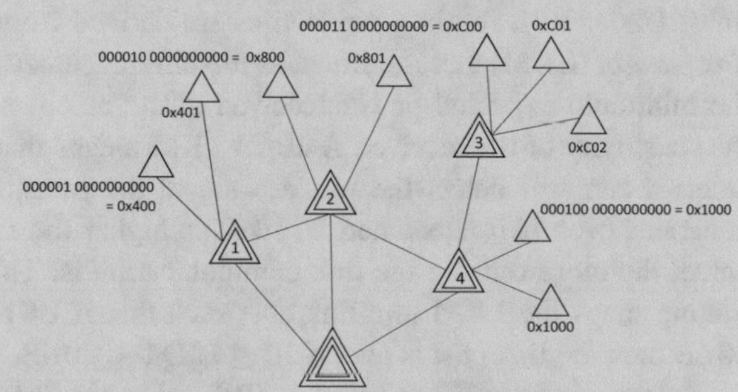

Note that 16-bit addresses are shown as both binary and hexadecimal numbers. Other schemes are also possible.

As mentioned before, the six MSBs of the 16-bit address indicate one of the possible 64 destination routers. Essentially, these six bits are used for the datagram to traverse the network from the source to the destination router. Once it arrives at the router, the remaining ten *least significant bits* (LSBs) of the address are used to reach the destination device.

Since the border router interacts with both access and core networks, it typically forwards to the leaders the list of prefixes the router can provide external connectivity to. The prefix, the 6LoWPAN context, the border router address, as well as the DHCPv6 server associated with the prefix if it is available. A device can configure a global unicast address relying on the DHCPv6 server or, if supported, can rely on SAA to do so. DHCPv6 is used by devices to retrieve network and multicast addresses needed for them to work. Since DHCPv6 is centralized with a server assigning addresses, there is no chance for overlapping and therefore no need for any conflict resolution. SAA, on the other hand and as indicated in Sect. 4.2, is fully distributed, and it enables devices to independently configure their own address without any human or nonhuman intervention based on their own link layer addresses and the network prefix provided by the router. By virtue of being distributed, SAA requires DAD for contention and conflict resolution.

Leaders keep track of the 16-bit addresses of the border routes, and they identify the REEDs that are upgraded to become Thread routers and the Thread routers that are downgraded to become REEDs. Additionally, leaders rely on CoAP to assign and manage router addresses. As already indicated, however, all information stored at the leader is propagated to other routers so they can automatically become new leaders if problems prevent the original leader from working. Note that because devices rely on 6LoWPAN for IPv6 adaptation, stateful header compression and decompression are widespread throughout the network.

9.4 Why Not RPL?

The two design assumptions of RPL, indicated in Sect. 7.3, are upward-focused routing and data-reactive route update. Upward-focused routing implies that most traffic is asymmetric and flows from the devices to the border router. Sometimes, depending on the transport and session protocols under consideration, traffic can be symmetric. For example, confirmable CoAP relies on the transmission of acknowledgments whenever a request is received. Other traffic types, like actuation, are asymmetric but mostly downward from the border router to the devices. Moreover, upward reliability of RPL is at the expense of downward reliability. Specifically, upon changes of the physical connectivity, downward routes are comparatively updated much more slowly than upward routes. Additionally, RPL networks configured with non-storing nodes are likely to be affected by overhead due to source routing that affects the scaling of downward traffic [6].

Data-reactive route update implies that, in certain cases, data traffic is used to detect connectivity issues. This therefore minimizes the need of transmitting routing control messages. One issue with this approach, however, is that the detection takes time as multiple missing datagrams are typically required to declare lack of connectivity. Of course, these missing datagrams are sacrificed as instruments of routing signaling.

The Thread approach to routing attempts to avoid these problems by relying on two other design goals: energy efficiency and scalability. Regarding energy efficiency, Thread specifies that routers must neither sleep nor exhibit power cycles. To this end, it defines sleepy end devices that can sleep but are not allowed to forward messages from other devices. Regarding scalability, Thread supports up to 32 routers per network partition.

When compared to RPL, a Thread network reduces the number of devices from thousands to a few hundred with routers always powered on. Routers can therefore focus on routing without being affected by energy efficiencies. The assumption is a home automation scenario where there are enough power outlets to support several routers. More importantly, Thread is not just a routing mechanism. It implements a full stack that has under consideration information from multiple layers. Note that although IEEE 802.15.4, when enabled with the right power configuration, provides a long-range coverage, Thread coverage is limited to support connectivity within homes and buildings.

IEEE 802.15.4 and 6LoWPAN support mesh connectivity by means of mesh-under and route-over approaches. RPL, however, is a fully hierarchical routing mechanism where parent nodes talk to multiple children, thus, not really supporting mesh connectivity. Thread routing is based on a full mesh topology where some nodes have a path cost not just to the border router but to other nodes in the network. A node can select the next hop by itself by looking at its routing table. RPL routing paths are symmetric so datagrams from the clusterhead to a device traverse the same path but in opposite directions that traffic from that device to the clusterhead. This is because downward routes are derived from upward routes by reversing them. Moreover, this can be a problem since for a given link, the cost in one direction may be different than the cost in the opposite direction. Thread routing is asymmetric; it relies on traditional flat DV techniques, and, therefore, a route from the border router to a node may follow a path that is different from that of the route from the node to the border router. Thread also exhibits some hierarchical features, as end devices only interact with a single parent router and all traffic generated or terminated at the device goes through this parent. This model minimizes the computational complexity and the usage of memory resources improving the overall energy consumption. Therefore, under Thread there are two approaches to routing. Routers are complex devices that are always powered on and rely on flat routing to support mesh connectivity. End devices, on the other hand, are basic low-power devices that run on batteries and rely on hierarchical routing where any interaction is with a parent router.

Thread routers broadcast advertising messages that are used to update routing metrics in the context of the full mesh. The maximum number of 32 routers is intended to limit the amount of control traffic sent throughout the network. This contrasts with RPL routing where the DIO messages include the path cost to the clusterhead. Thread routing enables a router to determine the cost of paths and links to neighbors and other routers. Thus, it complements ND mechanisms. All 32 router entries fit in a single control message with each entry being carried by a single byte. An entry, that includes a 2-bit incoming link cost, a 2-bit outgoing link cost, and a 4-bit path cost, is positioned within the payload in a location that is associated with the address of the node in the network. End devices, by not supporting flat routing, do not broadcast advertising messages.

Both RPL and Thread routing rely on trickle timers. RPL uses the trickle timer for DIO transmissions that enable repairs and routing integrity. The period of this timer, in steady state, is in the order of minutes to minimize power consumption and extend the network lifetime. Thread routing relies on trickle timers to trigger the transmission of advertising messages. In steady state the transmission interval is in the order of seconds. This much shorter transmission period is possible because Thread routers are not low-power. The network can be fully refreshed by means of these control messages. Low-power Thread end devices periodically transmit keepalive messages to verify reachability with their parents.

The link costs under RPL are mostly derived from ETX estimations obtained by inverting the PDR calculated from the number of received datagrams. This is not always accurate, and it can lead to assigning low costs to unreliable links. Under RPL routing, the metrics are derived from the *link margin* (LKM) that is defined as the difference between the minimum expected power received at the receiver and the sensitivity of the receiver. A 20 dB LKM means that the receiver can still detect the weakest signal the transmitter generates even if it attenuated 20 dB. The higher the LKM value, the more reliable the link communication is. Thread routing maps the LKM to a link cost such that if LKM $\geq$ 20 dB then the link cost is one, if $10 \leq$ LKM $<$ 20 dB, then the link cost is two; if $2 \leq$ LKM $<$ 10 dB, then the link cost is four; and otherwise the link cost is infinity.

RPL supports two modes of operation: non-storing and storing that provide a trade-off between memory overhead and network overhead. Thread routing, by supporting an approach similar to that of traditional DV routing, uses tables for routers to find paths. This limits the memory use to 32 entries per table. In order to reach an end device, a sender must first reach the device parent router such that when the datagram arrives, it can be delivered to the device. Specifically, the IEEE 802.15.4 short 16-bit address of each end device, as described in Sect. 9.3, is assigned to have two components: a 6-bit parent router identifier and a 10-bit end device identifier. The parent router identifier is used to determine the path from the source to the router. The device identifier is used by the parent router to forward the datagram to the right child.

Since routers are always powered on, the only Thread entities that are subjected to duty cycles are end devices. As opposed to RPL, under Thread routing, there is no need for coordination between leaf nodes and parent routers. Specifically, an end device can sleep, but it is always guaranteed that its parent is on; therefore no complex coordination is needed. If the device needs to transmit datagrams, it just wakes up and sends them. If the parent router needs to send datagrams down to the device, it queues them up until the device wakes up. An end device can proactively ask the parent router to receive any queued datagrams. Specifically, it transmits a data request and then waits for a short interval of time for the parent to transmit any pending datagrams.

Although RPL supports multiple instances of the routing topology associated with different DODAGs, a device can only join one of this instances and rely on a single root clusterhead that interacts with the Internet. There is, therefore, a single point of failure in most RPL topologies. On the other hand, Thread routing by virtue of supporting full mesh topologies can rely on multiple simultaneous border routers that communicate with the Internet.

9.5 Thread Stack

One key factor in IoT networks is reliability. Reliability is critical because it is a response to the well-known limitations of LLNs that affect network loss and latency. Thread has three levels of reliability: (1) IEEE 802.15.4 natively provides, as indicated in Sect. 3.3.3.2, link layer retransmissions supported by acknowledged transactions; (2) CoAP provides, as indicated in Sect. 5.2.4, session layer retransmissions supported by the CoAP confirmable mode of operation; and (3) Thread itself relies on out-of-band communication that supports sharing information between devices and applications running on smartphones or on the web. This latter information known as commissioning information is used to provision devices so that they can securely join a specific network. CoAP, in the context of Thread, is used to provision addresses needed by the devices. Moreover, since CoAP is a REST protocol, it can be used to get and set data on Thread routers including diagnostic data.

Example 9.2 Consider the following Thread network where DV routing is affected by service disruption of one of routers as shown below:

(continued)

(continued)

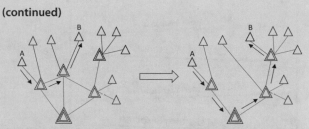

If device A sends a 100-byte packet to device B, what is the extra latency due to the new routing scenario?

Solution The per-hop delay is given by $D_{\text{h}} = \frac{8 \times L}{R} = 3.2\,\text{ms}$ because the frame length is $L = 100$ bytes long and the nominal IEEE 802.15.4 transmission rate is $R = 250{,}000$ bps. Assuming the best-case routing scenario, the topology change introduces two extra hops:

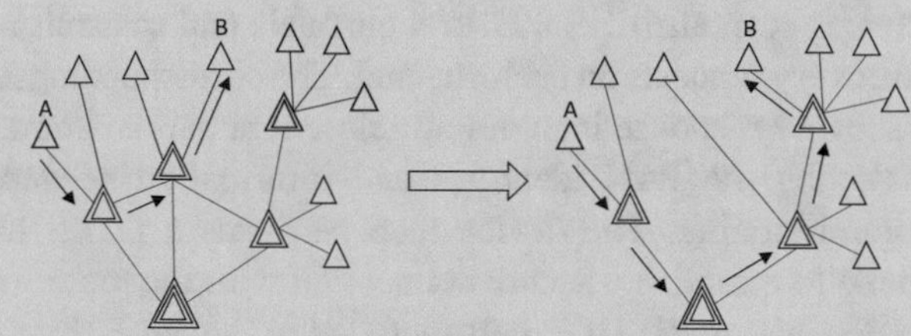

This implies an additional latency of 9.6 ms.

Under traditional IEEE 802.15.4, PAN coordinators are responsible for transmitting beacon frames that enable devices to join a network. This way of joining a network is device-driven. Essentially devices have a button that when it is pushed, it guarantees that the device securely joins the WPAN. In scenarios with multiple overlapping IEEE 802.15.4 PANs, however, this can lead to security breaches. The Thread architecture introduces instead a user-driven approach for devices to join networks. Specifically, commissioning information is used to enable a device to join and authenticate against the network. Devices, in turn, at joining time can be MEDs, FEDs, or REEDs with the latter able to become Thread routers once they learn the network configuration. Devices receive a 16-bit address with the MSBs identifying the parent Thread router and the remaining LSBs identifying the device itself. When a REED becomes a Thread router, the leader assigns it a router address. The leader includes a registry of all routers, and therefore it prevents overlapping router addresses. Thread routers, in turn, are responsible for preventing the address overlapping in children devices.

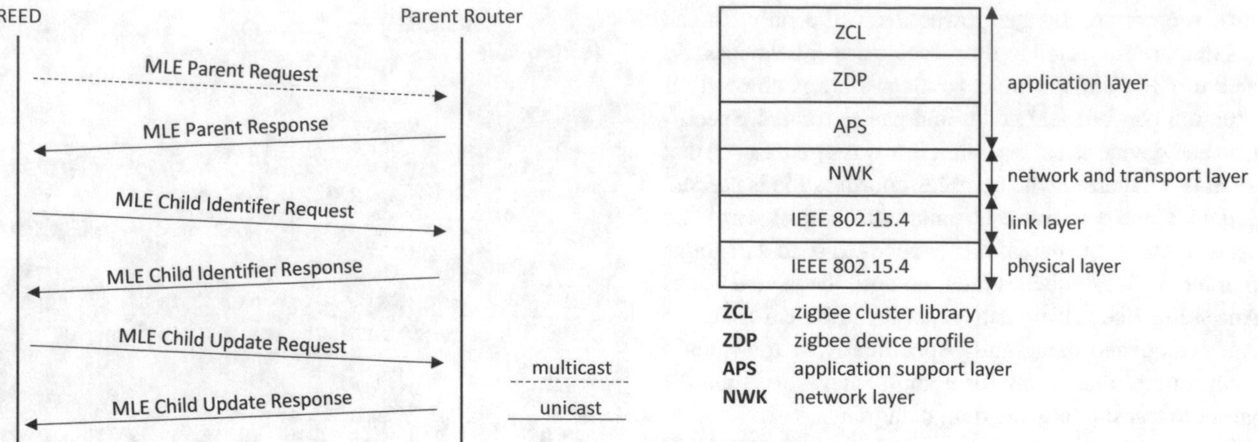

Fig. 9.6 REED attaching to a thread network

Fig. 9.7 ZigBee pro stack

Before users can select a device to join a network, the device itself must discover it. In order to accomplish this, devices scan all IEEE 802.15.4 channels and transmit MLE discovery requests on each channel. The corresponding MLE discovery response includes all relevant network parameters including the PAN identifier and information needed for commissioning. The device then performs a DTLS handshake to establish a secure connection with the router using as destination the UDP port included in the MLE discovery response. The router then relays the DTLS traffic from the device to the commissioner. Essentially, device and commissioner exchange tokens to establish trust. The commissioner inspects the device IID and other credentials, and if it is satisfied, it provisions the device with the appropriate services as well as with the *key encryption key* (KEK) that is used to encrypt other keys. Essentially, a network-wide key is secured through the KEK and delivered to the joiner. Once authenticated by the commissioner, the router gives the device network credentials that are used by the device to attach to the Thread network. If the device is preconfigured with the physical channel information and the PAN identifier, then discovery is not needed. Additional security considerations are included in Sect. 9.7.

Once the device has the network credentials, it can try to periodically attach to the network by multicasting MLE parent requests to routers and REEDs. REEDs may react by upgrading to a router role in order to provide routing capabilities to the attaching device. Figure 9.6 illustrates the MLE message exchange between a REED attaching itself to a router. In turn, the device and the router use MLE messages to configure a secure link and provision IPv6 addresses. All devices attach as end devices and can later upgrade by requesting a router identifier from the leader.

Regardless, after a device is attached to the network, it can start exchanging MLE messages with neighbors to update network parameters and link costs. Each MLE message includes several fields; (1) 16-bit and 64-bit IEEE 802.15.4 addresses of neighboring devices, (2) device capability information specifying the power duty cycle, (3) parent router link cost and link costs of all other Thread routers in the network, and (4) security and frame counter information. Note that all MLE messages, other than those used to retrieve network credentials, are encrypted during the joining process.

9.6 Application Layer

The highest layer prescribed by Thread is CoAP that mostly provides session layer functionality within the application layer itself. In the context of IoT IP protocols, application traffic originated and terminated at devices is not really standardized as it is transparent to the network and only visible by the applications. Depending on the industry under consideration, different proprietary mechanisms are typically used. This departs from the approach taken by full-stack technologies like BLE or ZigBee that rely on predefined profiles for devices and applications to exchange data.

Thread follows *dotdot* that is a universal and standard application language for IoT devices to communicate with other devices and entities in LLNs. Dotdot is an evolution, in turn, of the *ZigBee Cluster Library* (ZCL) which provides a standard mechanism for ZigBee devices to exchange data by means of clusters [4]. ZCL is just the highest layer of the ZigBee Pro stack as illustrated in Fig. 9.7. The stack also includes within the application layer domain the *ZigBee Device Profile* (ZDP) which provides device discovery services through specific commands and the *application support layer* (APS) that provides session layer functionality. Specifically, APS filters out duplicate frames and frames from non-registered devices or profiles that do not match. APS also supports reliability by means of retransmissions, and it carries several application layer tables that provide bindings between devices. Note that the use of profiles is

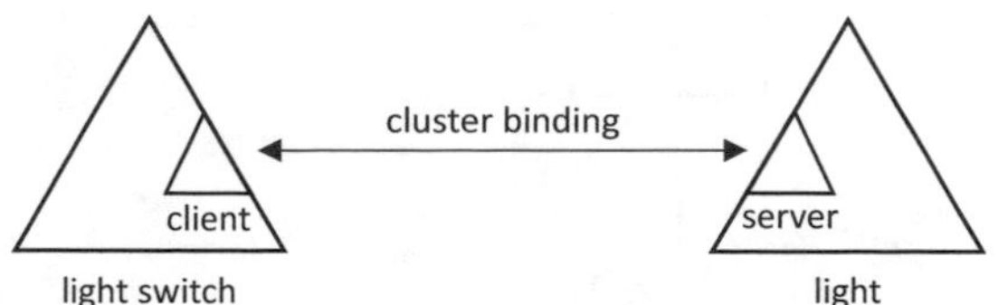

Fig. 9.8 On/off cluster

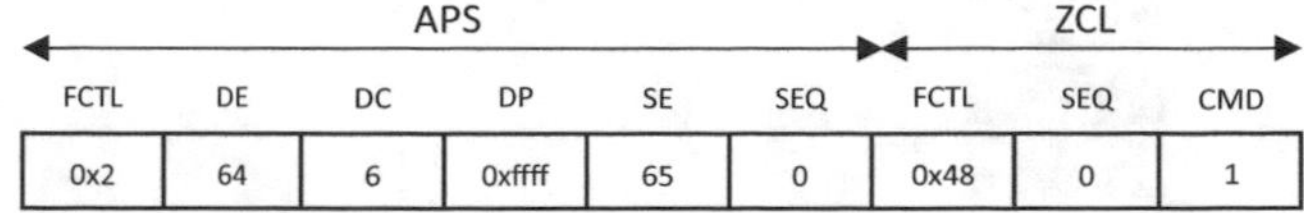

FCTL frame control
DE destination endpoint
DC destination cluster
DP destination profile
SE source endpoint
SEQ sequence
CMD command

Fig. 9.9 ZCL message

quite common with many full-stack M2M and hybrid IoT technologies like BLE and ANT+. But profiles force devices to only interact with those of their own type. Essentially, BLE devices and applications only interact with BLE devices and applications, ANT+ devices and applications only interact with ANT+ devices and applications, and, of course, ZigBee devices and applications only interact with ZigBee devices and applications.

Under ZCL, each cluster defines a communication interface that specifies a collection of commands and attributes. Devices implement client and server sides of each cluster. Cluster commands typically expose actions of devices, while cluster attributes usually expose the state of devices. Clusters are objects with attributes that follow flat structure that is indexed by identifiers. ZCL defines clusters intended for applications ranging from temperature sensing to dehumidification control. ZCL is not RESTful by design as it relies on numerous different general commands and manufacturer-specific commands per cluster. Because of the ZigBee LLN origin, ZCL messages are quite small and optimized to fit in a single IEEE 802.15.4 frame such that the unnecessary transmission of messages between devices is minimized.

Figure 9.8 shows an example of the ZCL *on/off cluster* with a light switch client transmitting an *on* command to a light server.

The evolution of ZCL from ZigBee to generic IoT IP network stacks defines the aforementioned dotdot application language for device communication. This is analogous to the situation of the link layer of ZigBee that was ported to other network stacks as IEEE 802.15.4.

With Thread, dotdot requires access control by means of *Authentication and Authorization for Constrained Environments* (ACE) that allows the provisioning application to specify the resources that are available each device.

Dotdot in the context of Thread is based on ZCL over IP that exploits CoAP for both session management, service discovery, and RESTful access. It encodes ZCL payloads with *Concise Binary Object Representation* (CBOR) standardized as IETF RFC 7049. CBOR introduces a data format that provides small message size and extensibility without the need to version negotiation.

The idea with ZCL over IP is to map ZCL binary messages as close as possible when encoded over CoAP. For example, Fig. 9.9 shows a ZCL message, encapsulated over APS, that is transmitted from the light switch to the light of Fig. 9.8. The message specifies that it is intended for destination endpoint, destination cluster, and command 64, 6, and 1 respectively. As per ZCL specifications, cluster identifier 6 is associated with the on/off cluster, while command 1 is associated with the *on* command. Note that the message has no payload. Under ZCL over IP this message is translated as a CoAP POST request to URI */zcl/e/64/s6/c/1* where */zcl* specifies ZCL protocol, */e/64* indicates the endpoint, *s6* indicates the cluster identifier, and *c/1* specifies the command. The POST request includes no body as the ZCL message has no payload. ZCL service discovery is carried out by ZDP, while ZCL over IP service discovery is carried out by IETF RFC 6690 *Core Link Format* described in Sect. 6.4.1.

In general, ZCL commands are accessible through REST CoAP requests: GET, PUT, POST, and DELETE. ZCL responses are returned as CoAP responses, typically as *2.05 Content* for success or as *4.05 Method Not Allowed* to indicate errors. URIs specify a combination of resources that are indicated by */e* for endpoints, */a* for attributes, */c* for commands, */b* for bindings, */r* for configurations, */n* for notifications, and */g* for group notifications.

Service discovery is invoked by means of the */.well-known/core* URI. Figure 9.10 illustrates an example, where an application attempts to discover endpoints implementing the on/off cluster by transmitting a multicast GET CoAP request to */.well-known/core?rt=urn:zcl:c:6.c* where the resource type attribute rt specifies ZCL for cluster identifier 6. The unicast response is a *2.05 Content* response that carries the link format associated with the endpoint *coap://[2001::21:5]/zcl/e/64/c6>;rt=urn: zcl:c:6.c*. Of course, Thread does not rely on standardized mDNS and DNS-SD for service discovery as CoAP is probably a better fit to map ZCL interactions.

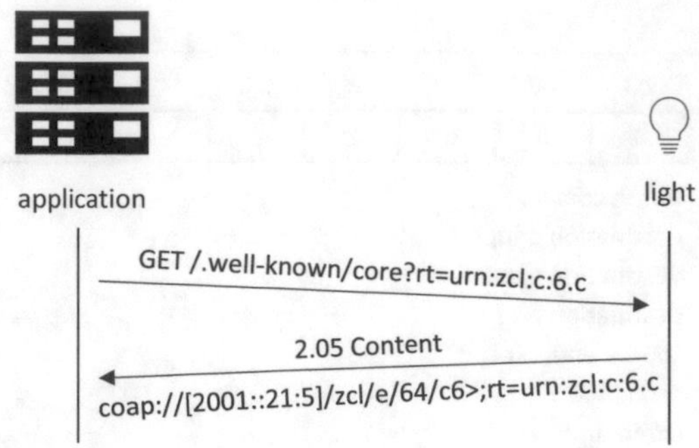

Fig. 9.10 Service discovery

9.7 Security Considerations

Security is a main concern with Thread. As most IoT architectures, Thread security is provided at many different layers. Specifically, as almost all link layer LLN technologies, IEEE 802.15.4 has built-in security mechanisms, described in Sect. 3.3.3.2, that enable hop-by-hop security. At the same time, application layer security by means of DTLS enforces security that complies with the end-end principle.

Thread IEEE 802.15.4 security provides encryption and 4-byte MIC authentication that are supported by a configuration that is based on a common key that is distributed to all devices in the network. One problem with this approach is that any security breach of a single device can lead to a network-wide security compromise. Additionally, and although IEEE 802.15.4 does provide some basic replay protection, Thread includes counters in its MLE messages in order to bring additional replay protection.

As indicated in Sect. 4.4, IEEE 802.15.4 can be the target of attacks due to the universal and static nature of the IIDs. Specifically, if the IID, that is vendor-dependent, becomes known, devices can be compromised. Moreover, in the context of IEEE 802.15.4 and 6LoWPAN, IIDs are used to derive IPv6 global addresses through SAA. Thread departs from this approach by allowing devices to use addresses that are either associated with the 16-bit node identifier or random.

DTLS is also used for security between commissioners and border routers over, for example, Wi-Fi. In this case, once a DTLS connection is established, a passphrase is used to validate the application against the border router. Once the application is validated, it can add new devices to the network. This mechanism is part of the *mesh commissioning protocol* (MeshCoP) that rules how devices securely authenticate and join the network. The process starts when a commissioner candidate, that is an application running on a smartphone or tablet, discovers a Thread network being advertised by its border router. The discovery mechanism enables the candidate to learn about the network name and the connectivity parameters needed to contact the router.

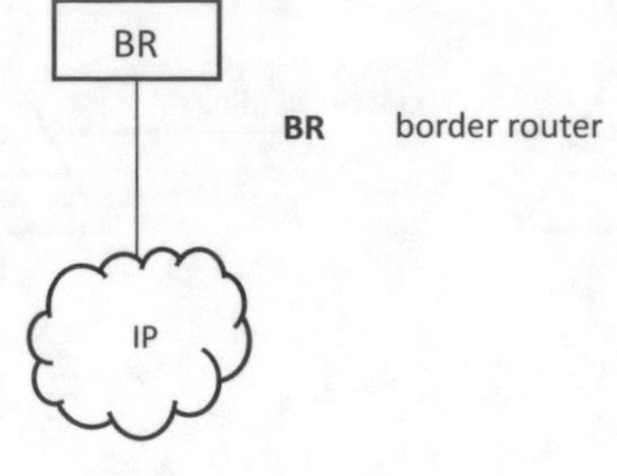

Fig. 9.11 Border router starting network

The candidate then connects by means of DTLS to the border router establishing a commissioning session that is authenticated by means of a *pre-shared key for the Commissioner* (PSKc). Next, the candidate registers with the border router that is simultaneously accepted by the leader as point of contact to the commissioner. The router then sends a confirmation message to the candidate that now becomes a commissioner.

The commissioner is responsible for joining devices to the Thread network. Devices that are no part of the network are called joiners. The joiner creates a DTLS connection to the commissioner in order to receive commissioning material that is used to join the network as described in Sect. 9.5.

9.8 Thread Network Formation

A Thread network can be started by a border router that initially performs many of the functions typically carried out by other devices in the network. Specifically, the border router connects to the public Internet over Wi-Fi (IEEE 802.11) or Ethernet (IEEE 802.3) and also takes over providing Thread leader functionality [10].

This scenario is representative of the topology shown in Fig. 9.11 where the network is made of a single device that connects to the Internet. The border router, **BR**, also acts as a server for the delegation to network devices of a global routable IPv6 prefix and as internal commissioner that can be used to enable other devices to attach to the network.

This is illustrated in Fig. 9.12 where the border router through a user interface starts a commissioning session with a joining router **R**. Router and border router now talk to each other through their corresponding WPAN IEEE 802.15.4 interfaces. After the router joins the Thread network, it starts receiving multicast MLEs from the border router that acts as a Thread leader. The border router provides DHCPv6 server functionality, allowing the router to negotiate a globally scoped address in order to interact with other nodes in the Internet.

Later, as shown in Fig. 9.13, a IEEE 802.11 (Wi-Fi)-based commissioning application, running on a smartphone, physically connects to the exterior interface of the border

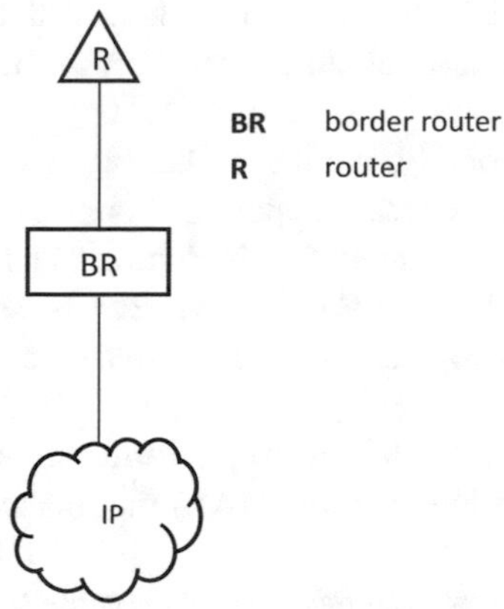

Fig. 9.12 Border router as internal commissioner

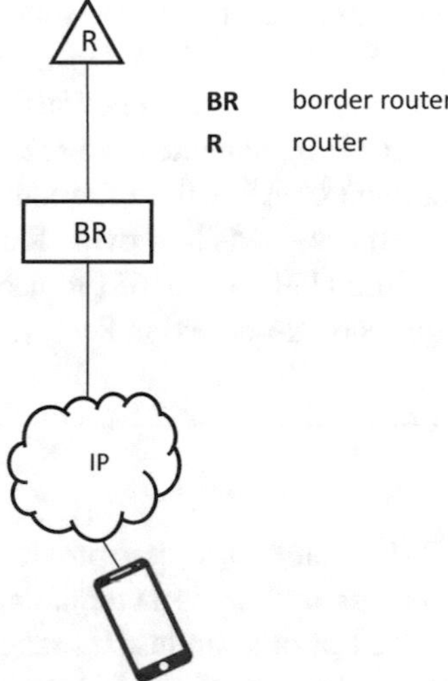

Fig. 9.13 External commissioner

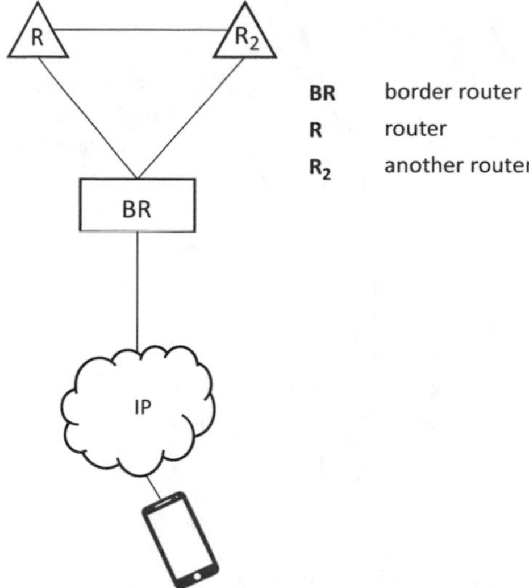

Fig. 9.14 New router joins network

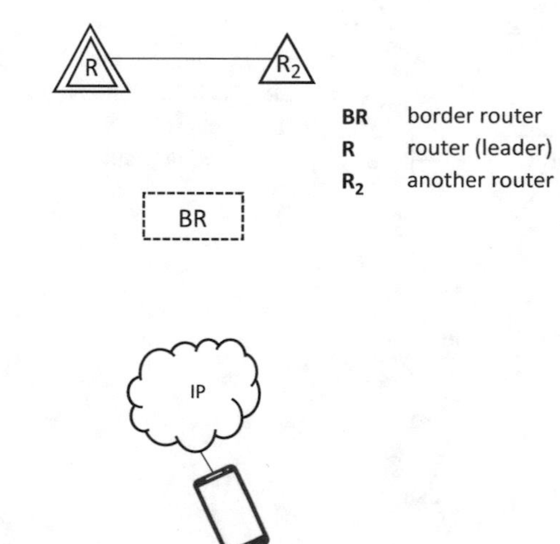

Fig. 9.15 Border router unavailable

router. Since the border router plays the role of Thread leader, a commissioning session is established between the commissioner and the border router.

The commissioner, then, enables another router $\mathbf{R}_2$ to join the Thread network. This is illustrated in Fig. 9.14. A DTLS connection is established between this new router and the border router that, in turns, relays the handshake to the commissioner. The border router then securely distributes network parameters to $\mathbf{R}_2$, which, in turns, uses DHCPv6 to get a globally scoped address.

If the border router is powered off, then another router in the Thread network typically takes over the role of leader. As shown in Fig. 9.15, router **R** becomes the new leader and it starts advertising network data that no longer includes external routing information.

Later, when the border router is back in service as shown in Fig. 9.16, it acknowledges router **R** as Thread leader. The border router uses TMF traffic to forward its external route information. Now, router **R** advertises network data that includes external routing information.

Finally, a new border router $\mathbf{BR}_2$ joins the Thread network as illustrated in Fig. 9.17. This new border router communicates with the leader by means of TMF traffic to inform it has taken over the role of DHCPv6 server.

9.9 OpenThread

As opposed to many other IoT architectures, Thread is available as a complete open-source implementation developed by Google/Nest called OpenThread initially released in May 2016 [5]. The goal of OpenThread is to enable faster deployments of Thread and to encourage chip manufacturers to standardize devices as a mechanism to bring better, cheaper, and more secure devices to end consumers [7].

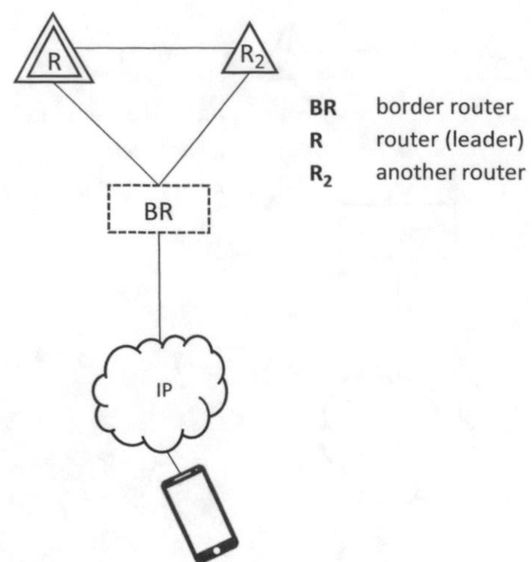

Fig. 9.16 Border router available again

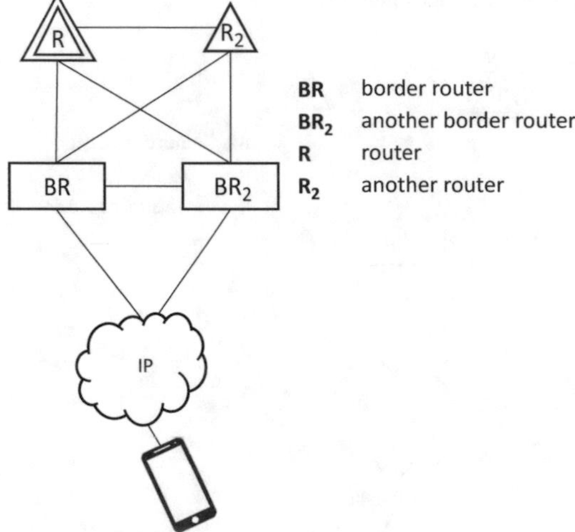

Fig. 9.17 New border router joins network

► **OpenThread RTOS Device Support** There are several implementations of OpenThread intended for RTOS-based devices. The list below details some of these implementations.

OT RTOS	Enabled constrained devices from integrating with Thread networks
	Integrates OpenThread over well-known FreeRTOS
	Relies on lwIP for networking and on MBed TLS for TLS
RIOT Open Thread	LGPLv2 licensed open-source C implementation
	Implemented as part of the RIOT RTOS
Zephyr Open Thread	Apache 2.0 licensed open-source C implementation
	Implemented as part of the Zephyr RTOS

OpenThread includes all stack layers and supports two modes of operation: as an event-driver kernel that can run in a bare-metal system or on top of an embedded operating system. Additionally, it runs on hardware with radios that enable Thread physical support and simultaneously implements all architecture layers including (1) IEEE 802.15.4 with security, (2) 6LoWPAN, (3) MLE and associated RIPng-based routing, (4) security key management, (5) specific device roles in Thread, (6) DTLS, and (7) CoAP. Physical differences between hardware platforms are supported by a *hardware abstraction layer* (HAL) that makes OpenThread highly portable.

The framework includes several separate functional units. Specifically, it provides a networking module that implements IPv6, MPL, and ICMPv6. It also includes a routing module that implements MLE to forward routing information and RIPng to maintain routing tables. OpenThread has additional modules that provide all the functions associated with DTLS as well as CoAP management. In regard to this latter functionality, although CoAP is used for control messaging, OpenThread also exposes APIs to provide REST functionality on Thread-specific UDP port 61631 to take advantage of 6LoWPAN compression described in Sect. 4.3.4.2.

Summary

Thread is a WPAN technology that provides a full stack primarily intended for IoT home automation that can also be used in the context of other industries ranging from asset monitoring to IIoT. The chapter started by introducing the main characteristics of Thread and how Thread compares to IETF IoT stacks. Details of the Thread topologies and DV routing were then explored. Specifically, DV routing was compared against RPL in order to explain why it is the best alternative for the Thread architecture. The chapter continued to explain the Thread stack, its application layer including resource discovery, and other security considerations. The chapter ended by describing the network formation and details of OpenThread.

Homework Problems and Questions

9.1 What is a problem of Thread is routers were constrained devices?

9.2 Consider Fig. 9.6 and assume an average transmission rate and packet size of 120 Kbps and 56 bytes, respectively. If no delays other than the transmission delay, how long does it take for a REED to attach to a Thread Network?

9.3 Why do you think ZCL messages are mapped as CoAP messages under Thread?

9.4 At least theoretically, how many end devices can a single Thread router support?

References

1. IEEE Standard for Local and Metropolitan Area Networks–Part 15.4: Low-Rate Wireless Personal Area Networks (lR-WPANS) amendment 1: Mac sublayer. IEEE Std 802.15.4e-2012 (Amendment to IEEE Std 802.15.4-2011) pp. 1–225 (2012)
2. IEEE Standard for Low-Rate Wireless Networks. IEEE Std 802.15.4-2020 (Revision of IEEE Std 802.15.4-2015) pp. 1–800 (2020)
3. Alexander, R., Brandt, A., Vasseur, J., Hui, J., Pister, K., Thubert, P., Levis, P., Struik, R., Kelsey, R., Winter, T.: RPL: IPv6 Routing Protocol for Low-Power and Lossy Networks. RFC 6550 (2012). https://doi.org/10.17487/RFC6550. https://rfc-editor.org/rfc/rfc6550.txt
4. Alliance, T.Z.: Zigbee cluster library specification (2019). https://zigbeealliance.org/wp-content/uploads/2019/12/07-5123-06-zigbee-cluster-library-specification.pdf
5. Google/Nest: Openthread (2020). https://openthread.io/
6. Kim, H.S., Kumar, S., Culler, D.: Thread/openthread: A compromise in low-power wireless multihop network architecture for the internet of things. IEEE Commun. Mag. **57**, 55–61 (2019)
7. Marksteiner, S., Exposito Jimenez, V.J., Valiant, H., Zeiner, H.: An overview of wireless IoT protocol security in the smart home domain. In: 2017 Internet of Things Business Models, Users, and Networks, pp. 1–8 (2017)
8. Montenegro, G., Schumacher, C., Kushalnagar, N.: IPv6 over Low-Power Wireless Personal Area Networks (6LoWPANs): Overview, Assumptions, Problem Statement, and Goals. RFC 4919 (2007). https://doi.org/10.17487/RFC4919. https://rfc-editor.org/rfc/rfc4919.txt
9. Rescorla, E., Modadugu, N.: Datagram Transport Layer Security Version 1.2. RFC 6347 (2012). https://doi.org/10.17487/RFC6347. https://rfc-editor.org/rfc/rfc6347.txt
10. ThreadGroup: Thread border routers (2015). https:https://openthread.io/
11. ThreadGroup: Thread commisioning (2015). https:https://openthread.io/
12. ThreadGroup: Thread 1.2 base features (2019). https://openthread.io/
13. ThreadGroup: Thread network fundamentals (2020). https://openthread.io/
14. Unwala, I., Taqvi, Z., Lu, J.: IoT protocols: Z-wave and thread. In: 2018 IEEE Green Technologies Conference (GreenTech). IEEE, Piscataway (2018)

Glossary

6Lo IPv6 over networks of resource-constrained nodes refers to the native IPv6 support of constrained IoT devices.

6LoWPAN IPv6 over low-power wireless personal area networks is a mechanism that enables IPv6 over several constrained IoT link layer technologies.

Access It is the portion of the network between the devices and the gateway.

Access point (AP) Under IEEE 802.11, it plays the role of IoT gateway.

ad hoc A mode of operation of a single BSS.

ACK Acknowledgment.

Active scanning Under IEEE 802.11, a device sends a probe frame indicating the SSID of the BSS it wants to be associated with.

Active sensor It is a sensor that emits sounds or generates electromagnetic waves that can be detected by means of external observation.

Actuator It is a logical device that perform some external change of an asset of the physical environment.

ADC Analog-to-digital converter.

Aggregation It implies that readouts from different sensors are collected by a sensor closer to the gateway or clusterhead.

ALOHA It is a MAC mechanism that enables devices to send datagrams without having to wait for the channel to be free.

AMQP The advanced message queuing protocol is a session layer protocol, similar to MQTT, that was initially designed for enterprise applications like financial businesses but due to its simplicity and small footprint has become an integral part of many IoT solutions.

Application layer Layer that involves application specific services as well as the conversion of information between the digital and analog domain.

Asset An asset is an element of the environment that an IoT device interacts with.

At least once One of the reliability modes supported by AMQP and MQTT.

At most once One of the reliability modes supported by AMQP and MQTT.

Authentication It dictates that only trusted sources can transmit data.

Autonomous cells Under 6TiSCH it provides proactive cell scheduling without any type of negotiation.

Backbone Same as core.

Beamforming It is a signal processing technique to support directional signal manipulation with the goal of minimizing interference.

BLE Bluetooth Low Energy is a physical and link layer technology for transmission of data over wireless channels. BLE is also known as Bluetooth Smart.

Block code It is a mechanism where a payload is partitioned into chunks of data called messages, and each message, when controlled redundancy is added, becomes a codeword that is used for error control.

Bundle Under TSCH, it is a scheme where multiple cells carry traffic between two devices.

Bus A type of network topology.

CBOR The Concise Binary Object Representation introduces a data format that provides small message size and extensibility.

Cell Under TSCH, it is the combination of the timeslot and channel.

Channel It is the medium that enables the propagation of signals between devices and applications.

Channel bonding It consists of transmitting frames of the same physical device over multiple nonoverlapping subchannels.

Channel capacity theorem It states that the maximum achievable transmission rate of a communication system is a direct function of SNR.

Channel coding It is an optional mechanism that embeds FEC information in headers.

Channel decoder A communication system component that removes controlled redundancy to improve reliability against channel impairments.

Channel encoder A communication system component that adds controlled redundancy to improve reliability against channel impairments.

Cloud computing Application processing performed at the network cloud.

R. Herrero, *Fundamentals of IoT Communication Technologies*, Textbooks in Telecommunication Engineering, https://doi.org/10.1007/978-3-030-70080-5

Cluster A group of devices that interact with a gateway for core side communication.

Clusterhead A gateway associated with a cluster.

CoAP The Constrained Application Protocol is a lightweight and highly efficient protocol that enables the management of IoT sessions.

CoAP resource discovery It is a distributed mechanism for CoAP service discovery.

CoAP resource directory It is a centralized mechanism for CoAP service discovery.

Code rate It is defined as the ratio k/n where n is the total number of transmitted packets (redundant and original) and k is the number of original packets.

Cognitive radio It is used to dynamically select the portions of the spectrum that minimize interference.

Complex device It is a usually a gateway.

Confidentiality It implies that information must be encrypted to make sure that it is inaccessible to unauthorized users.

Confirmable mode A CoAP transmission mode that compensates the inherent lack of reliability of UDP transport.

Continuous A type of DNS query that is made by fully compliant mDNS queriers and responders in order to support asynchronous operations like IoT load balancing.

Controller It is a logical device that performs some internal change in the physical device to assist sensing or actuation.

Convolutional code It is similar to a block code, but it operates on continuous streams of bits instead of on blocks of bits.

Cookie It is an add-on mechanism that can be used to introduce a state in the interaction between clients and servers.

Core It is the portion of the network between the gateway and the application.

CPS A cyber-physical system represents the interaction between a device and an asset.

CSMA/CA Carrier sense multiple access with collision avoidance is a MAC mechanism.

CSMA/CD Carrier sense multiple access with collision detection is a MAC mechanism.

D7AP The DASH7 Alliance Protocol is an open source protocol stack that enables LPWAN communications.

DAC Digital-to-analog converter.

Data-centric routing It involves clients sending queries to a specific network regions in order to retrieve a specific readouts and data associated with specific capabilities.

Data mining Knowledge extraction mechanism.

Data-information conversion Stage intended to lower throughput in order to optimize channel utilization and lower power consumption to extend battery life.

Datagram Name of a packet that is processed by a network layer.

DECT ULE Digital Enhanced Cordless Telecommunications Ultra Low Energy is a physical and link layer technology for transmission of data over wireless channels.

Device A device is a sensor, a controller, or an actuator running on small constrained embedded computer.

Demodulator A communication system component that reverses the signal transformation introduced by the modulator.

Detectability It indicates the probability that a given network section is able to detect a specific asset or physical phenomenon.

Differential signaling It consists of transmitting data by means of two complementary electrical signals that, when subtracted at the receiver, minimize noise.

Directed diffusion It is a flat data-centric routing mechanism that relies on response aggregation to accomplish power consumption efficiency.

Dispatch value Under 6LoWPAN, it is used as datagram type indicator.

Distribution services Under IEEE 802.11, they deal with station services that span beyond communication between devices in a given BSS.

Distribution system Under IEEE 802.11, it provides the backbone for connectivity to applications performing analytics.

DNS Domain name system is a well-established IP suite protocol that is mainly used for address resolution.

DODAG A destination-oriented directed acyclic graph is a graph used under RPL that enables devices to keep track of the routing topology.

DTLS Datagram Transport Layer Security is the preferred security mechanism in the context of 6LoWPAN and other IoT technologies.

DV Distance vector routing consists of devices sending their entire routing tables to their connected neighbors.

Downlink Direction from application to device.

Duty cycle It refers to devices that sleep in order to reduce power consumption at preprogrammed intervals.

Endpoint It is a network component also known as host that serves as the source or destination of messages.

End-to-end principle It states that, whenever possible, certain functions like security must be deployed on an end-to-end basis typically at the application layer.

Ethernet It is a physical and link layer technology for transmission of data over wireline channels.

Exactly once One of the reliability modes supported by AMQP and MQTT.

Exposed station It results from station C transmitting to station D and preventing station B to transmit to station A.

FEC Forward error correction is a mechanism that helps detecting and correcting errors due to channel distortion.

Flat architecture It is an architecture where all functions are performed by a single layer.

Flat routing It relies on devices interacting with each other without a single device acting as parent that concentrates and aggregates traffic of children devices.

Flooding It is a routing mechanism that is based on devices forwarding received datagrams through all possible neighbors.

Flow label IPv6 field that identifies the datagram as part of a flow of datagrams.

Fog computing Application processing performed at a gateway.

Forwarding It involves moving a datagram from an incoming to an outgoing link in a device or router.

Frame Name of a packet that is processed by a link layer.

Framing It is a generic technique that consists of adding fields and special synchronization markers to data propagated down from upper layers.

G3-PLC PLC standard.

Gateway It is a logical device that serves as an interface between access side IoT devices and core side applications.

Gossiping It is an alternative to flooding that relies on intermediate devices forwarding a received datagram to a single randomly selected neighbor.

Hidden station It results from stations A and C transmitting simultaneously to station B and not detecting each other since they are out of range.

Hierarchical routing It groups devices in clusters with clusterheads that acting as parent devices concentrate and aggregate traffic of children devices.

HIP Host Identity Protocol is an IP-based security mechanism.

HOL Head-of-line blocking occurs when the processing of a request prevents other responses from being transmitted.

Hop limit IPv6 field that provides a counter that is decremented as the datagram is forwarded throughout the network.

HTTP The HyperText Transfer Protocol is an application protocol that enables the management of IoT sessions. Specifically, it provides session layer management of web applications including client and server support.

Hyperspectral image Images made of data cubes that include a few dozens of spectral bands.

IEEE 1901.2 PLC standard.

IEEE 802.2 LLC protocol that is transported on top of standard IEEE 802.3.

IEEE 802.3 Ethernet standard.

IEEE 802.11 It is a physical and link layer technology for transmission of data over wireless channels.

IEEE 802.11ah An IEEE 802.11 developed for generic IoT support.

IEEE 802.15.4 It is a physical and link layer technology for transmission of data over wireless channels.

IEEE 802.15.4k It is a standard for low energy, critical infrastructure monitoring applications that relies on the 2.4 GHz, the 915/868 MHz, and the 433 MHz ISM bands for transmission.

IEEE 802.15.4g It is a new standard that targets smart metering applications like gas metering.

IIoT Industrial Internet of Things is a term associated with IoT in the context of industrial applications.

Industry 4.0 See IIoT.

Information-knowledge conversion Stage where information is processed by the application to generate knowledge.

Infrastructure A mode of operation of a single BSS.

Integrity It implies that the received messages are not altered in transit.

IoT Internet of Things is a term used to describe technologies, protocols, and design principles associated with Internet-connected things that are based on the physical environment.

IP Internet Protocol. It is a network layer protocol and fundamental building block for IoT communication.

IPSec/IKE IP security/Internet Key Exchange are IP-based security mechanisms.

IQRF It is an open LPWAN framework that includes devices, gateways, and applications addressing scenarios ranging from telemetry and industrial control to home and building automation.

ISM Instrument, scientific, and medical bands are unlicensed spectral bands used in many IoT wireless solutions.

ITU-T G.9903 PLC standard.

ITU-T G.9959 It is a physical and link layer technology for transmission of data over wireless channels.

ITU-T H.265 Video codec.

JTAG Joint Test Action Group standardized as IEEE 1149.1. It is used to perform a boundary-scan of an integrated circuit to enable circuit testing

Known answer suppression It is a mechanism by which known answers are suppressed by not being transmitted.

Layered architecture A communication architecture that segments functionality into different layers.

LDPC Low-density parity-check. Block code.

LEACH Low-energy adaptive clustering hierarchy is a hierarchical routing mechanism that by means of aggregation, it transmits device data to an application that acts as a sink.

Line code It is used to modulate digital bits of a link layer frame into an electrical signal that can be transmitted over the channel.

Linear regression Knowledge extraction mechanism.

Link Network component that connects endpoints and routers.

Link layer Layer that provides error control mechanisms for reliable transmission of information over the channel.

LLN Low-power and lossy network typically associated with IoT.

LOADng Lightweight On-demand Ad hoc Distance Vector Next Generation is a technology that provides reactive flat routing in LLNs.

LoRa Long range is the generic name for a full protocol stack that provides LPWAN capabilities that enables devices to run on a single battery for more than 10 years.

LoRaWAN It is the LoRa networking mechanism that enables devices to access the channel.

Location-based routing It consists of a client or device transmitting datagrams to another client or device by forwarding the traffic based on the geographical or physical location of the destination.

Logical devices Software-based devices than run on physical devices.

LPWAN A low-power wide-area network is an IoT network associated with very low transmission rates (up to 50 Kbps) over long distances (in the order of kilometers)

LS Link state routing consists of devices sending the state of their links to all the devices in the network.

LTE-M Standardized together with NB-IoT as part of the 3GPP Release 13, LTE-M, also known as LTE Cat M1 is a simplified version of 4G LTE that attempts to provide IoT support by reducing power consumption while extending signal coverage.

M2M Machine-to-machine communication refers to mechanisms that enable simple interaction between devices and applications. In most cases, these mechanisms are neither standardized nor Internet-based.

MAC Media access control is a set of rules that determine how frames are received and transmitted over the physical channel.

Machine learning Knowledge extraction mechanism.

MANET Mobile ad hoc network.

Master-slave A type of network topology.

mDNS Multicast DNS provides an alternative to traditional DNS by addressing some of the limitations of the latter in the context of IoT.

Mesh-under It is forwarding that occurs in the 6LoWPAN layer.

Media It refers to the transmission of the audio, speech, video, and images.

Message Name of a packet that is processed by an application layer.

MIMO multiple input-multiple output is a type of multi-antenna.

Mission planner Application that resides at a ground station and calculates flight paths of UAVs.

Mist computing Application processing performed at a device.

Modulator A communication system component that transforms a signal for efficient transmission over the channel.

Modulator A communication system component that transforms a signal for efficient transmission over the channel.

MQTT Message Queue Telemetry Transport is a protocol that enables the management of IoT sessions.

MS/TP Master-Slave/Token-Passing is a physical and link layer technology for transmission of data over wireline channels.

Multipath It is a scenario that usually results in signal fading that causes network packet loss.

Nagle's algorithm It is a mechanism by which TCP natively buffers packet payloads until it has enough data to make it efficient to send a datagram.

NB-Fi Narrowband fidelity is an open full-stack protocol that is the base of a commercial LPWAN turn-key solution that includes devices and networks.

NB-IoT Narrowband IoT is an LPWAN technology based on cellular communications and first introduced in the 3GPP Release 13.

ND Neighbor discovery provides communication with hosts and routers to enable connectivity beyond the local link.

Neighbor discovery attack It is an attack where neighbor discovery messages typically used in the context of WPANs are either dropped or corrupted in order to induce reachability issues.

Network layer Layer that ensures that information packets are delivered to the destination.

Next header IPv6 field that specifies the protocol under which the payload is encoded.

NFC Near-Field Communications is a physical and link layer technology for transmission of data over wireless channels.

Node coverage It presents the level of device redundancy available to capture data if a sensor failure occurs.

Non-beacon mode It is an IEEE 802.15.4 mode of transmission when they are infrequent enough that contention based access results in a lack of collisions.

Non-confirmable mode A CoAP transmission mode that is based on a fire-and-forget approach where messages are sent without expecting any acknowledgment.

Non persistent An HTTP session that relies on a single TCP connection for each transaction.

Non-storing node It is a node where a device does not store prefix information from all its children

Nwave It is a commercial LPWAN solution intended mobile devices associated with smart parking.

Observation It is a feature associated with REST that enables an observed device to transmit readouts whenever parameter changes are detected.

OFDM Orthogonal frequency division multiplexing is mechanism that enables the modulation of binary streams in communication channels.

On-shot A type of query that is typically associated with legacy DNS queriers and responders.

OPC UA The Open Platform Communications United Architecture provides a protocol stack that complies with the request/response paradigm.

Packet A basic unit of data transmitted in packet switched networks.

Packet bursting It consists in buffering a number frames before transmitting them all together in order to lower average channel contention delay.

Packetization To chunk a media stream into several packets.

PAN coordinator In the context of IEEE 802.15.4 and other technologies, it is the device that plays the role of gateway.

Passive scanning Under IEEE 802.11, it consists of devices sequentially listening for beacon frames transmitted by other devices over channels.

Passive sensor It is a sensor that is not active.

PEGASIS Power-Efficient Gathering in Sensor Information Systems is a hierarchical routing mechanism that relies on data aggregation.

Persistent An HTTP session that relies on a single TCP connection for all transactions.

Physical layer Layer that is dedicated to the transmission and reception of data over the channel including the conversion of information between the digital and the analog domains.

Physical devices Hardware-based devices.

Piconet A type of network topology.

PID A proportional integral derivative is a control loop mechanism that relies on feedback.

PLC Power line communication is a physical and link layer technology for transmission of data over wireless channels.

Presentation layer Layer that provides formatting of information for further processing including security extensions for encryption and decryption.

Proactive routing It consists of building routing tables that ultimately define forwarding behavior by means of the periodic transmission of routing information across all nodes in the network.

Proxy server It is a device that sits between client applications and sensors and devices acting as servers in order to respond to client requests on behalf of these servers.

Publish/subscribe A model that relies on a broker that queues and delivers messages between an application and a device.

Qowisio It is a UNB turn-key commercial LPWAN solution supporting a wide range of applications ranging from asset tracking and management to lighting and power monitoring.

Querier Also known as resolver or questioner transmits a query that is answered by the mDNS server.

Reactive routing It relies on dynamic on-demand building of routes from sources to specific destinations by relying on route discovery queries that flood the network.

Reliable delivery It is an optional mechanism that provides the infrastructure to signal and support retransmissions.

Request/response This model bases its interaction between application and device by means of requests and responses.

Responder Also known as DNS server or answerer, it responds to queries.

REST It is an architecture that formalizes a series of requirements and interfaces that are necessary for client/server interaction.

Route-over It is routing that occurs above the 6LoWPAN layer.

Router It is a network component that assists in the propagation of messages throughout the network.

Routing It involves determining a route or path that datagrams must follow from source to destination.

Routing information spoofing It is an attack where the intruder spoofs, alters, or replays routing information in order to create loops that lead to datagram loss.

RPL Routing for low power is a hierarchical and IPv6 address centric routing protocol.

RPMA is a media access scheme that serves as the base of a robust LPWAN technology.

RTCP Real-time control protocol is used to provide quality control over media streams.

RTP Real-time protocol is the preferred standard for the transmission of media. RTP sessions are typically established by means of SIP.

SCADA Supervisory control and data acquisition is an industrial control system architecture.

Scatternet A type of network topology.

Schedule Under TSCH, a schedule specifies what devices communicate with each other.

SD-DNS It is a standard that it is used with mDNS to provide service discovery.

Scrambler A mechanism that randomizes in a controlled fashion the sequence of transmitted bits.

Segment Name of a packet that is processed by a transport layer.

Sensor It is a logical device that interacts with the environment by sampling and generating readouts.

Selective forwarding It is an attack where an affected device only forwards certain datagrams in order to cause connectivity problems.

SigFox It is an LPWAN protocol stack that relies on an unique commercial network called SigFox.

Signaling It refers to the exchange of information between entities in order to establish a session.

Simple device It is a basic sensor, actuator, or controller.

Sinkhole attack It is an attack where an affected device attempts to become the destination of all traffic in a certain location.

SIP The Session Initiation Protocol provides a mechanism to create, manage, and finish sessions.

Slotframes Under TSCH, timeslots are grouped into slotframes that are periodically transmitted.

Source decoder A communication system component that performs, among other things, digital to analog conversions.

Source encoder A communication system component that performs, among other things, analog to digital conversions.

SPIN Sensor protocols for information via negotiation is a flat data-centric routing mechanism that through data negotiation and resource adaptation enables devices to forward datagrams from a source to a sink in a much more controlled and more energy efficient way than flooding and gossiping.

Spreading factor In SS, spreading factor or processing gain is the ratio between the symbol width and the chip width.

SS Spread spectrum is mechanism that enables the modulation of binary streams in communication channels.

Stateful compression It relies on associating redundant information in uncompressed IPv6 headers with a context identifier that is transmitted in the datagrams instead.

Stateless compression It is a compression mechanism that is simple, and, as opposed to stateful compression, it does not require to keep track of the state of inter-datagram redundancy.

Station Under IEEE 802.11, it plays the role of IoT access device.

Station services Under IEEE 802.11, they deal with authentication and privacy between devices.

Storing node It is node where a device stores prefix information from all its children.

SWD Serial Wire Debug is used to perform a boundary-scan of an integrated circuit to enable circuit testing

Sybil attack It is an attack where a single device assumes multiple identities in order to become a destination of many other nodes in the network.

Reed-Solomon Block code.

REST Representational state transfer is an architecture where transactions are destined to optimize the interaction of the entities involved in the communication.

Ring A type of network topology.

RS-232 Recommended Standard 232 defines signals connecting terminals and other equipment.

RS-485 Recommended Standard 232 defines signals connecting terminals and other equipment.

Session layer Layer that is in charge of managing multiple sessions between applications.

SNOW Sensor network over white spaces is an experimental LPWAN technology that relies on transmission over white space spectrum with modulation over unoccupied frequency guard bands between TV channels typically between 547 and 553 MHz.

Star A type of network topology.

Synchro It is a variable coupling transformer where the magnitude of the magnetic coupling between the primary and secondary varies in accordance to the position of a rotable element.

Telensa It is a fully proprietary LPWAN technology that focuses mainly on smart city applications with emphasis on smart lighting and smart parking as it does not support indoor communications.

Thread It is a protocol stack, an architecture and a framework that support IoT home automation.

Traffic class IPv6 field that provides QoS information.

Transmission rate Rate at which an application generates rate. Measured in units of bps.

Transport layer Layer that provides support of multiplexing of traffic from different applications.

Turbo code Block code.

Uplink Direction from device to application.

UPnP The Universal Plug and Play architecture provides a framework for the discovery of network elements ranging from computers and printers to gateways and access points.

Weightless It is an LPWAN protocol stack that relies on transmissions over the white space spectrum.

Wormhole attack It is an attack where the attacker captures datagrams in one location and retransmits them in another leading to routing and connectivity problems.

WPAN A wireless personal area network is an IoT network that comparatively provides higher transmission rates (in the order of Mbps) but supports shorter distances (in the order of hundreds of meters).

WSN Wireless sensor network a term that designates a system of very large number of wireless sensors that interact with an application that monitors events and other phenomena occurring in the physical environment.

WSN category 1 A type of network topology that includes a very large and highly dense deployment of devices. Transmission is multi-hop with communication from devices to gateways relying on intermediate sensors and actuators forwarding and aggregating packets.

WSN category 2 A type of network topology that is very simple and includes fewer devices that directly connect to a single gateway.

XMPP The eXtensible Messaging and Presence Protocol is an application layer protocol that loosely follows the REST paradigm.

Zero-configuration Mechanism where network deployment and provisioning are carried out without human intervention.

ZCL The ZigBee Cluster Library is a standard mechanism for ZigBee devices to exchange data.

ZDP The ZigBee Device Profile provides device discovery services through specific commands.

ZigBee Protocol stack with IEEE 802.15.4-based physical and link layers.

Index

R. Herrero, *Fundamentals of IoT Communication Technologies*, Textbooks in Telecommunication Engineering, https://doi.org/10.1007/978-3-030-70080-5

E

M

Printed in the United States
by Baker & Taylor Publisher Services